SOIL ENZYMES

SOIL
ENZYMES

Edited by

R. G. BURNS

*Biological Laboratory, University of Kent,
Canterbury, Kent*

1978

ACADEMIC PRESS

LONDON   NEW YORK   SAN FRANCISCO

A Subsidiary of Harcourt Brace Jovanovich, Publishers

ACADEMIC PRESS INC. (LONDON) LTD.
24/28 Oval Road,
London NW1

United States Edition published by
ACADEMIC PRESS INC.
111 Fifth Avenue
New York, New York 10003

Library of Congress Catalog Card Number: 77-93196
ISBN: 0-12-145850-4

Printed and bound in Great Britain by
Cox & Wyman Ltd, London, Fakenham and Reading

CONTRIBUTORS

BREMNER, J. M. Department of Agronomy, Iowa State University of Science and Technology, Ames, Iowa 50011, USA

BURNS, R. G. Biological Laboratory, University of Kent, Canterbury, Kent CT2 7NJ, England

CERVELLI, S. Laboratorio per la Chimica del Terreno del CNR, Via Corridoni, 78-56100 Pisa, Italy

DRAGĂN-BULARDA, M. Department of Plant Physiology, Babeş-Bolyai University, Cluj-Napoca, Romania

KISS, S. Department of Plant Physiology, Babeş-Bolyai University, Cluj-Napoca, Romania

LADD, J. N. Division of Soils, CSIRO, Private Bag No. 2, Post Office, Glen Osmond, South Australia 5064, Australia

McLAREN, A. D. Department of Soils and Plant Nutrition, University of California, Berkeley, California 94720, USA

MULVANEY, R. L. Department of Agronomy, Iowa State University of Science and Technology, Ames, Iowa 50011, USA

NANNIPIERI, P. Laboratorio per la Chimica del Terreno del CNR, Via Corridoni, 78-56100 Pisa, Italy

RĂDULESCU, D. Department of Plant Physiology, Babeş-Bolyai, University, Cluj-Napoca, Romania

ROBERGE, M. R. Laurentian Forest Research Centre, 1080 Route du Vallon, P.O. Box 3800, Sainte-Foy, Québec G1V 4C7, Canada

ROSS, D. J. Soil Bureau, Department of Scientific and Industrial Research, Private Bag, Lower Hutt, New Zealand

SEQUI, P. Laboratorio per la Chimica del Terreno del CNR, Via Corridoni, 78-56100, Pisa, Italy

SKUJIŅŠ, J. Department of Biology, UMC 55, Utah State University, Logan, Utah 84322, USA

SPEIR, T. W. Soil Bureau, Department of Scientific and Industrial Research, Private Bag, Lower Hutt, New Zealand

PREFACE

This is a book about the activity of plant, animal and microbial enzymes in soil. In that enzymes are essential for the degradation of organic debris it is also, in the broadest sense, a book about the biology of terrestrial ecosystems. There are chapters dealing with fundamental and applied aspects of the subject; the former embraces amongst others, colloid science, enzymology, microbiology and biochemistry; the latter agronomy. In some instances these approaches overlap considerably.

Successive chapters detail the evolution of soil enzymology during the twentieth century; the types of enzymes found in soil and their origins; and the important developments in the kinetics of soil enzyme reactions. Professor Kiss and his colleagues contribute a chapter which puts polysaccharidases in an agricultural context whilst Bremner and Mulvaney discuss the vitally important enzyme, urease. Enzymes crucial to the phosphorus and sulphur cycle are reviewed at length in Chapter 6. Professor Sequi and his co-authors have described the interactions between pesticides and fertilizers on one hand and soil enzymes on the other whilst some varied aspects of enzyme location, function and measurement are recounted in Chapter 8. Finally Dr. Roberge has compiled an appendix chapter which presents some of the methods currently used for assaying and extracting soil enzymes. In editing these chapters I have not attempted to rationalize contrasting viewpoints and the reader will deduce that considerable controversy exists, especially in relation to interpretative and methodological issues.

There are two momentous suggestions, made some centuries apart, which connect soil to the origins of life (Genesis II, v.7; A.G. Cairns-Smith "The Life Puzzle", Oliver and Boyd, Edinburgh, 1971). Whilst this volume may not totally support such claims the hope is that, to the uninitiated, the vitality of the subject will emerge in all its varied ecological, biochemical, microbiological and agricultural guises. To the committed soil biologist it may act as a catalyst (what else for a book concerned with enzymes?) for future research.

My thanks are due to the contributors for delivering their manuscripts with respectable promptness and for their tolerance of my editorial demands. Professor Hackenbush has, as always, given a considerable amount of emotional support.

R.G. BURNS,
University of Kent

July 1977

CONTENTS

6. SOIL PHOSPHATASE AND SULPHATASE

T. W. Speir and D. J. Ross

7. INTERACTIONS BETWEEN AGROCHEMICALS AND SOIL ENZYMES

S. Cervelli, P. Nannipieri and P. Sequi

8. ENZYME ACTIVITY IN SOIL: SOME THEORETICAL AND
PRACTICAL CONSIDERATIONS

R. G. Burns

APPENDIX—METHODOLOGY OF SOIL ENZYME MEASUREMENT
AND EXTRACTION

M. R. Roberge

I

History of Abiontic Soil Enzyme Research

J. SKUJIŅŠ

Ecology Center, Utah State University, Logan, Utah, USA

I. Introduction

With the advent of plant and animal biochemistry it was recognized that many reactions involving soil organic matter transformations may be catalysed by enzymes existing outside the microorganisms and plant root systems. In the first decade of this century, soil scientists compared the soil solution to the blood of animals and it was noted by Quastel in 1946 that the soil may be looked upon as a biological entity, i.e. as a living tissue.

The first reports on extracellular soil enzyme activity (catalase) appeared before the turn of the century. For the subsequent fifty years reports on soil enzymatic activities sporadically appeared in the literature. These reports involved not only catalase but also urease, phosphatase and several other enzymes. The research work in this field was confounded by a lack of appropriate methodologies and no understanding of the true nature of enzymes; although a number of enzymatic activities have been demonstrated in extracted, enriched concentrates since the last decades of the nineteenth century, Sumner first isolated an enzyme (urease) in crystalline form from jack bean meal in 1926. Extracellular or, more specifically, *abiontic* (Skujiņš, 1976) enzymatic activities in the soil were and are of profound interest to soil biologists and biochemists who need to elucidate the biochemical activities in soils. Among the questions asked in the early years and to which complete answers have not been found yet, are the following: How are enzymes in soil distributed and localized? What are their stabilizing mechanisms? What are the origins of enzymes in soil? What is the significance of enzymes in plant nutrition and what are the extracellular enzyme-root interrelationships? What are the catalytic properties of the inorganic soil constituents? What is their contribution to the soil organic matter turnover and to the humic matter formation? Today's emphasis is on applying current soil enzymology knowledge to soil fertility considerations, pollution and use of modern agricultural chemicals.

The principal difficulty in studying soil enzymes has been the problem of separating microbial activities from the extracellular enzymatic activities. In approaching this problem experimentally, it is desirable not to disturb the chemical and physical soil properties. Commonly, antiseptic agents have been added to the soil. The idea was to inhibit microbial activities while retaining the abiontic enzymatic activities in presumably nondisturbed condition. Other later methods have included high-energy radiation sterilization and antibiotics. However, a perfect agent has not yet been found.

Soil enzymology still desperately needs new methodologies. When they become available many of the earlier reports should be re-evaluated since their abiontic enzyme activity determinations may not have successfully separated the microbial from abiontic enzyme activities. Such past measurements can be of value, however, if the procedures and results are properly evaluated in the light of present knowledge.

Present research in soil enzymology has several facets. One seeks basic understanding of the activity characteristics of free enzymes in the heterogenous soil matrix and their stabilization mechanisms associated with the soil particulate matter. Another is defining their involvement in soil organic matter turnover and in plant nutrition. These efforts may indicate whether the results of soil enzyme research could be used to judge soil fertility. Other pertinent questions concern the interactions of abiontic enzymes with various pesticides, their use in forensic science and the relevance of soil enzymology to planetary exploration for detection of extraterrestrial life.

The real advancement in soil enzymology started after 1950 with the increased understanding of enzymatic reactions in general and with the applicable methodologies that became available from plant, animal and microbial biochemistry. The increase in the number of papers published since 1950 is evident from Fig. 1.

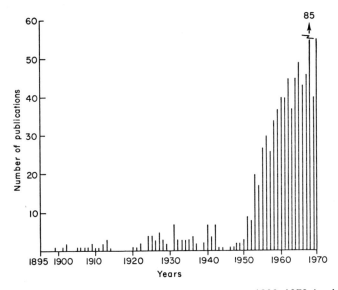

Fig. 1. Number of Reports Published on Soil Enzymes, 1899–1970 (revised from Kiss and Boaru, 1965).

The listing excludes most modern meeting abstracts. The 1968 number of 85 includes reports originating from the Symposium on Soil Enzymes, Minsk, BSSR, June 27–30, 1967, published in *Sbornik Dokladov Simpoziuma po Fermentam Pochvi*, Nauka i Tekhnika, Minsk, 1968.

This chapter mainly considers the earlier, historical aspects of the emergence of soil enzymology as a field of science. Recent advances and the state of knowledge are described elsewhere in this book and in recent reviews (Kiss *et al.*, 1975a; Skujiņš, 1976).

II. The Long Childhood

The first report on the presence of extracellular enzymes in soil appears to be that by A. F. Woods who, in 1899, wrote:

I have also determined by experiment that oxidizing enzymes, especially the peroxidase may occur in the soil, and as a rule, are not destroyed by the ordinary bacteria of decay. These enzymes enter the soil through the decay of roots and other parts of plants which contain them.

Wood's report was presented at the American Association for the Advancement of Science annual meeting in Columbus, Ohio in August 1899. Although evidence of enzymes in soils was practically nil at that time, their importance in biological processes in soils was correctly inferred and appreciated. Conn (1901) stated:

> Fermentations, whether caused by the metabolism of bacteria or yeasts, or by enzymes, secreted by these organisms or by higher plants, are of vital importance in agricultural processes. Without their agency in breaking up organic compounds the soil would rapidly become unfit for supporting life.

Most of the early attention was directed to catalase, apparently because of its easy detection and because of limited knowledge of other enzymes. The presence of oxidases (peroxidases) in soil was indicated by Cameron and Bell in 1905, who treated soil with a guaiac solution and suggested that a resultant blue colour demonstrated the presence of these enzymes. This method was amplified and improved by Schreiner and Sullivan (1910). König et al. (1906, 1907) used biological inhibitors, e.g. cyanide, to show the enzymatic nature of catalase activity in soils and concluded, as did May and Gile (1909), that activity was correlated with organic and inorganic fractions and with microbial activity, while the inorganic components contributed considerably to hydrogen peroxide decomposition. These results have been repeatedly verified by modern methods.

Further reports on catalase activity in soil were published by Osugi (1922) who emphasized the importance of inorganic catalysis; by Balks (1925), who indicated that the catalase activity was contributing to the humus turnover in soils and by Smolik (1925) and Barasio (1927), who studied catalase in Czechoslovakian and Italian soils, respectively. In 1924 Stoklasa described determination of catalase activity in Abderhalden's "Methods on Biochemical Examination of Soils". Catalase measurement as an index for soil fertility was also suggested by Waksman and Dubos (1926). The conclusion, however, was that the catalytic phenomena in soils is very complex and cannot be used as a simple method of determining soil productivity (Waksman 1927).

Early reports on catalase activity in soils include those by Wachtel (1912), Penkava (1913) and Kappen (1913). These investigators again demonstrated the considerable contribution of inorganic soil components to the soil catalytic activity. An iodometric method to demonstrate the oxidizing power of soils was introduced by Gerretsen (1915). The evidence that much of the soil catalytic power is derived from the inorganic components and that the biological contribution cannot very well be correlated with soil fertility led the agronomists to believe that inorganic "catalytic fertilizers" might be the answer to increase soil fertility (Lyon et al., 1917). Such conclusions were also based on the extensive work on soil catalysis by Sullivan and Reid

(1912). Chouchack (1924), however, determined that the hydrogen peroxide decomposed by normal soil differed from that decomposed by sterilized soil and used the difference as an index for biological activity. A later work described soil fertility determinations using enzymatic activities, especially catalase (Rippel, 1931).

In the 1930s, numerous papers describing catalase activity and trying to link it with soil fertility appeared in various parts of the world, notably by Kurtyakov (1931) in the Soviet Union, Radu (1931) in Rumania, Rotini (1931) and Galetti (1932) in Italy; Scharrer in Germany published a series of papers (Scharrer, 1927, 1928a, 1928b, 1936), including a report (Scharrer, 1933) on catalase kinetics. Matsuno and Ichikava (1934) reported on hydrogen peroxide decomposition in Japanese soils.

Baeyens and Livens (1936) in Belgium and Valy (1937) in Hungary, concluded again that soils rich in organic matter owe their high catalase activity to fungi and bacteria, which may be inhibited by antiseptics, and verified that much of the activity in soils low in organic matter is due to ferric and manganese hydroxides. The influence of manganese and iron compounds in the soil on catalase activity was also studied by Pfeil (1932) and by Novak and Crha (1940). The effect of fertilization with natural manure on catalytic activity in soil was described by Kobayashi (1940) in Japan and in 1949 Velasco and Levy described the activation energies and the average life of activated enzyme-substrate complex in soils at various temperatures. They reiterated that the effect of a soil on hydrogen peroxide decomposition is a rough measure of its fertility.

It has been evident, since the early years of soil enzymology, that:

Some decomposition of protein and of phenol and hydrolysis of cane sugar also occurs under conditions where living organisms seem excluded (from soil)—(Russell, 1921).

The lack of understanding of enzymatic characteristics precluded further interpretations of the observed phenomena at that time. Fermi (1910) reported that a proteolytically active fraction which hydrolysed gelatin was extracted from soil by phenol. The proteolytic activity was shown in fresh water sediments by Messineva in 1940. She demonstrated that proteolytic activity exists to about 2 m depth in fresh water deposits and that of catalase to about 7·5 m. She also measured lipase activity (Messineva, 1941). The next significant work on proteinases in the soil appeared in the 1950s. The presence of a deaminase in waterlogged soils was reported by Subrahman-yan (1927). Later work, however, has discounted the presence of extra-cellular deaminases in soils.

In 1933, Rotini reported the presence of pyrophosphatase in soils. However, this work later underwent a considerable re-evaluation. Rotini worked not only with catalase and phosphatases but also showed the presence of urease in soils (Rotini, 1935a). A review on ureolytic action in soils was published by Rotini and Sessa in 1943. Rotini (1935b) demonstrated the enzymatic decomposition of cyanamide into urease. The presence of urease and its enzymatic characteristics in soils were reported in a series of papers by Conrad (1940a, b; 1941, 1942a, b; 1944) and Conrad and Adams (1940). The dephosphorylation of organic phosphorus compounds in soil by extracellular enzymes was examined by Rogers and his co-workers (Rogers 1942a, b; Rogers *et al.* 1941, 1942).

Toluene has been used as an antiseptic to inhibit microbial activity and facilitate the study of extracellular enzymes since the early days of soil enzymology. As early as 1914, Buddin, however, had reported the incomplete effect of toluene for this purpose. The still controversial use of toluene in soil enzyme work was thoroughly re-examined in the 1950s.

Scharrer (1928a) irradiated soils with ultraviolet light but failed to achieve much inactivation of catalase.

Sterilization of soil by high energy electron beams or gamma radiation was introduced in the 1950s.

From the early days it has been of interest to determine the sources of "free" enzymes in soils. The first report on such enzymes in soil (Woods, 1899) points to plant roots as a source of catalase. Knudson and Smith (1919) reported the excretion of amylase by plant roots, and Knudson (1920) indicated similar release of invertase. Plant roots were cited as a source of a number of enzymes such as catalase, tyrosinase, asparaginase, urease, amylase, invertase, protease and lipase by Kuprevich (1949). Rogers, *et al.* (1942) showed that phosphatase was excreted from corn and tomato roots. In the ensuing years the release of enzymes by plant roots has been well substantiated by work in plant physiology. It is recognized now, however, that microbes constitute a major contribution to abiontic enzyme activities in soil (as reviewed by Ladd in this volume, chapter 2).

III. The Learning Years

Since 1950 interest in soil enzyme research has been continually expanding, imaginative theoretical approaches and methods have been introduced and a wealth of information regarding various enzymatic reactions in soil has been collected. Soil enzymology has emerged as a field of vigorous scientific inquiry. Two laboratories should be noted for giving the impetus to soil enzymology research, that of V. F. Kuprevich of the Soviet Union and especially that of E. Hofmann, G. Hoffmann and their colleagues in Germany.

The work by V. F. Kuprevich and his colleagues has been documented in a monograph (Kuprevich and Shcherbakova, 1966). This publication contains numerous references to early Soviet sources which are otherwise unavailable to Western readers; many of the references have short abstracts. A more concise version describing advances by V. F. Kuprevich's group is also available to the Western reader (Kuprevich and Schcherbakova, 1971). E. Hofmann and G. Hoffmann, with their coworkers in Germany, have been among the most fruitful in creating interest in soil enzymology among Western scientists and in developing new approaches and methodologies, many of which are still in use (Hofmann and Hoffmann, 1966). Another important contributor is A. SH. Galstyan, who started to publish in 1957 in Armenia. His work has been recently summarized (Galstyan, 1974). The laboratory of I. S. Kiss and his coworkers in Rumania has been most productive; they have made valuable contributions to the understanding of the significance of enzymatic reactions in soil. Most of their work has been summarized in recent reviews (Kiss *et al.*, 1975a; Kiss *et al.*, 1975b).

Generally, most contributions to the soil enzyme research in the 1950s and 1960s came from Western and Eastern Europe, especially Belgium, Germany, France, Rumania and the USSR. During those years research in soil enzymology was considered too "exotic" in the US and mostly ignored. There are no references to enzymatic activities in soils in such classics as Waksman's "Soil Microbiology" (1952) and Alexander's "Introduction to Soil Microbiology" (1961). In 1965, however, a method for enzymatic examination of soils (urease) was included in the "Methods of Soil Analysis" published by the American Society of Agronomy (Porter, 1965). A number of reviews regarding advancements in soil enzyme research were published notably those by Kiss (1958a), Durand (1965), Voets and Dedeken (1966), and Skujiņš (1967). The current status of soil enzymology around the world has been reviewed by Sequi (1974), Kiss *et al.* (1975a), and Skujiņš (1976). The European work in the 1950s was directed to developing methodologies for identifying extracellular enzymes. Most attention was directed to the various carbohydrases, proteinases, phosphatase and urease. Methods for catalase were re-evaluated and determination of dehydrogenase activity in soils was introduced. Concurrently, enzymatic activity distribution in the soil profiles of various ecosystems, especially agricultural, was examined and various attempts were made again to correlate the enzymatic activities with soil fertility and with the soil "biological index".

1. Imperfect methods

A prime problem in soil enzymology has been finding an ideal sterilizing agent to facilitate the separation of extracellular enzyme activities from those of microorganisms. The perfect agent, which would completely inhibit all microbial activities but not affect the extracellular abiontic enzymes in any way, remains unavailable.

The antiseptic and bacteriostatic agents tried and examined include phenol, acetone, thymol, chloroform and ether (Rotini, 1935a). The most acceptable, however, appears to be toluene, the most widely used microbial inhibitor in soil enzymology studies. Toluene is decomposed by a number of soil microorganisms; however, this happens only at concentrations below 0·1%. For enzymatic studies in soils, toluene is added in much higher amounts, usually 10–25% by volume. Toluene is believed to prevent the synthesis of enzymes, the proliferation of microorganisms and the assimilation of substrate and reaction products, thus excluding contributions by active organisms from the enzymatic activity measurements. The use of toluene in soil enzyme studies was controversial during the 1950s and 1960s, producing debates in the literature, for example, between Claus and Mechsner (1960) and Hofmann and Hoffmann (1961). Critical evaluations have been made by several authors, notably Beck and Poschenrieder (1963) and Kiss and Boaru (1965). Kiss and Boaru (1965) demonstrated that toluene can completely prevent the assimilation of substrate or its reaction products by the microorganisms. They concluded that toluene is a highly effective agent for inhibition of microbes and useful for enzymatic determinations in soil. They showed that under the accepted conditions of use (10–25% in the sample), toluene also prevents the synthesis of new enzymes. Toluene does not destroy microorganisms in soils. At these concentrations, however, it does prevent microbial proliferation. Even at lower concentrations, when the multiplication of some microorganisms might take place, the enzymatic activity of the microflora present is often negligible in comparison with the activity of the accumulated abiontic enzymes. Kiss and Boaru (1965) concluded that the lysis of microorganisms by toluene leads only to negligible or no increased activity in the enzyme systems tested, as exemplified by invertase, suggesting that further studies are necessary to clarify other enzymatic systems. Kiss et al. (1962a) successfully used toluene as a microbial inhibitor for incubations of up to 14 days at 35°C. Beck and Poschenrieder (1963) showed that the inhibitory effect and the concentration needed depend on pre-treatment procedures and the moisture content of any particular soil. In air-dried or naturally moist samples, at least 20% toluene present is necessary. In soil suspensions, however, 5–10% toluene concentration is sufficient. In 1914, Buddin had reported that toluene had an incomplete effect as an inhibitor, but apparently a low toluene concentration was used in his experiments. Claus and Mechsner (1960) based their criticism of toluene on a growth of microorganisms observed in toluene-treated soils.

Drobník (1955) noted variations in the inhibitory effectiveness of toluene on soil microorganisms and stated that toluene prevented the assimilation of metabolic products by microflora. He suggested (Drobník, 1961) that further studies were necessary to evaluate the worth of toluene. Galstyan (1965) also examined toluene and found no assimilation of reaction products. He concluded that any possible autolysis of microbial cells during treatment did not affect the measured enzymatic activities. The effect of toluene on enzymatic determinations in soil has been examined by a number of other authors. Nowak (1964), for example, stated that treatment with toluene decreased but not totally suppressed soil bacteria; Katsnelson and Ershov (1958) used toluene without any negative effects on the enzymatic reactions in soil. Toluene is presently widely used as a microbial inhibitory agent for soil enzymatic activity measurements and the only enzymatic activity which is reported to be seriously affected by toluene is dehydrogenase.

A useful approach to determining abiontic enzyme activities in soil should be through antibiotic inhibition of microbes, especially by the antibiotics that inhibit protein synthesis. This method has been used occasionally and is successful if proper antibiotics are chosen. The first use of an antibiotic preparation (BIN 7) was reported by Kuprevich (1951), who reported no sign of influence on invertase activity in soil. Kiss (1958c) used benzylpenicillin, streptomycin sulphate, and sulphanilamide in determining soil invertase and observed strong inhibition of invertase synthesis by microorganisms. Other studies (Kiss 1961, Kiss and Drăgan-Bularda 1968b) showed that levan was formed from sucrose in soil in the presence of chloromycetin, indicating that levan sucrase pre-existed in the soil and did not form during its incubation with chloromycetin. They concluded that enzymatic activities observed in soils under such experimental conditions indicate the pre-existence of extracellular abiontic enzymes. Such experiments, however, do not resolve the problem of enzymatic activity attributable to metabolizing cells and to enzymes released from cells with destroyed membranes.

Benefield (1971) used penicillin G (benzylpenicillin) to inhibit microorganisms in cellulase activity assays in soils, and Kunze (1970) used streptomycin during determinations of catalase activity. Antibiotics have also been successfully used by Kiss, et al. (1972) in studies of several extracellular soil carbohydrases. They established that chloromycetin diminished invertase activity but increased transferase activity. Chloromycetin had no effect on maltase and cellobiase, but diminished the hydrolytic action of lactase in the presence of toluene. Observed enzymatic activities in the presence of chloromycetin were considered due to enzymes pre-existing in the soils rather than any synthesized by proliferating bacteria during the incubation of the soil samples (Kiss et al., 1972).

The first attempt to observe radiation effects on soil enzymatic activities was performed by Scharrer (1928a) with ultraviolet radiation. The UV irradiation had minimal effects on the catalase activity in soil. After a seven-day continuous infra-red irradiation, Dommergues (1960) noted that the microbial count in one soil was reduced by 50%, while invertase activity remained unchanged. In several other soils, however, invertase activity dropped to 29%. The effects of 40 and 90 MHz microwave irradiation on soils was tested by Voets and Dedeken (1965). The irradiation resulted in a considerable decrease in microbial counts as well as in invertase and protease activities.

The high energy ionizing radiation appears to be the most nearly ideal agent for soil sterilization in enzymatic research. Dunn *et al.* (1948) were first to use ionizing radiation for sterilization of small quantities of soil. In 1957 McLaren *et al.* showed that soil may be sterilized by an electron beam of sufficient energy and intensity. By using a 5–10 MeV electron beam and also hard x-rays and gamma-rays from a ^{60}Co source, a 2 Mrep dose was necessary to sterilize 1 g soil samples (McLaren *et al.*, 1962). The sterilized soil retained its enzymatic activity. Radiation is a differential sterilization method. Generally, multicellular organisms are more sensitive to irradiation than unicellular organisms and, in turn, extracellular enzymes are less sensitive than microbial cells. Thus, microbial proliferation can be inactivated while soil enzymatic activities remain intact. This has been demonstrated with urease (McLaren *et al.*, 1957; Skujiņš and McLaren, 1969), phosphatase (Skujiņš *et al.*, 1962) and other enzymes as reviewed by McLaren (1969). The usual criterion for sterility is a lack of microbial multiplication. In radiation sterilized soils, however, microbes might be biochemically active for several days (Peterson, 1962) as the irradiation may affect their genetic mechanism but not certain of their enzymatic activities.

Enzymes differ widely in their sensitivity to irradiation. The dosage necessary for microbial inactivation also eliminates some enzymatic activities in soils, for example, urease (McLaren *et al.*, 1957; Skujiņš and McLaren, 1969; Vela and Wyss, 1962). Such increases may be due to an enhanced permeability of dead cell membranes. Generally, 1 g soil samples can be sterilized with about 2 Mrad dosage, large quantities require larger doses. The number of live cells exposed to irradiation diminishes as a logarithmic function with respect to radiation dosage. For the naturally mixed soil microbe population each species has a unique constant for its inactivation. Survival curves are therefore difficult to predict, as are the doses required for complete sterilization of particular soils. Thus, a considerable variability in minimum necessary dosages has been demonstrated.

Sterilizing doses of ionizing radiation also induce some chemical and physical changes in a soil, manifested in decreased soil aggregate stability or an increased release of inorganic nutrients. It has been demonstrated, however, that irradiation of soils produces soil samples that are quite useful for studies in soil enzymology, as it disturbs soil less than other sterilizing agents. The irradiation technique and the characteristics of radiation-sterilized soils have been reviewed by McLaren (1969) and Skujiņš (1976).

2. The search for specific activities

2.1. Carbohydrases

Carbohydrases have been a favourite hunting ground for soil enzymologists. The enzymatic reaction products (most often glucose) are easily discernible in reaction mixtures where the substrate usually is insoluble and neither the products nor substrates react with soil particles.

The extensive work done on invertase activity in soils has been promoted, without doubt, by the ease of its detection. A method for assaying invertase in soil was published by Hofmann and Seegerer (1950), another method was introduced by Kuprevich (1951). They demonstrated that hydrolysis of sucrose takes place in the presence of toluene. Their method was adopted for photometric determination of glucose by Gettkandt (1956). In the 1950s much effort was directed toward examining the relationship of invertase activity to soil fertility and determining soil biological activity (e.g. Kroll, 1953; Hofmann and Bräunlich, 1955; Drobník and Seifert, 1955). Invertase activity was shown to be related to microbial numbers and metabolic activity, decreasing with the depth of profile, in common with other enzymes. The highest invertase activity was found in neutral calcareous soils. Cultivated soils had above average activity levels, while sandy soils had low activities. Although invertase activity was generally correlated with microbial numbers and activities, Nowak (1964) indicated that toluene did not inhibit the microorganisms sufficiently to separate their activities from that of invertase.

Kiss (1958a) critically examined invertase activity in soil by comparing the hydrolysis of sucrose in the presence and the absence of toluene. His work and that by Peterson and Astafyeva (1962) and Galstyan (1965), showed the rate of invertase activity to be the same in the presence and absence of toluene. The later work by Kiss et al. (1971), Kiss et al. (1972), Voets and Dedeken (1965) and Voets et al. (1965) demonstrated that invertase activity accumulated in soils as an extracellular enzyme and that proliferation of microorganisms during experimental determinations contributed very little to the total invertase activity measured. Invertase investigations expanded considerably in the 1960s. Among the numerous authors the contributions by Ross (1965, 1966, 1968a) and Ross and Roberts (1968) deserve special mention. These scientists presented a detailed examination of invertase activity in grassland soils. They also studied amylases and verified that enzymatic activities were considerably influenced by vegetative cover, period of sampling and depth of soil, while the clay content had less effect. Variations in annual rainfall and soil organic carbon when considered together but not separately, explained much of the variability in sucrose hydrolysing activities, which seem to be related more to soil composition than to its organic matter content.

Similar variations in activities of other soil enzymes have been shown, and the lack of correlation among the enzymes and specific soil properties has become more evident. Research on soil invertase activity showed some incongruities that were duplicated later in work on other soil enzymes, for example, Galstyan (1958b) showed that high invertase activity was associated with low catalase activity. On the other hand, Verona (1960) demonstrated that germinating seeds in the soil increased both invertase and catalase activity. Irrespective of microbial contributions to invertase activity in the soil, a considerable contribution is made also by plant roots. The general interest in invertase activity in soils has somewhat diminished in the 1970s compared to research on other soil enzymes.

In a series of papers, Hofmann and Hoffmann introduced methodologies and demonstrated the presence of other carbohydrases in soils. α-Glucosidase (maltase) was demonstrated with α-phenylglucoside (Hofmann and Hoffmann, 1953, 1954) or maltose as substrate (Hofmann and Hoffmann, 1955b). Maltose was also used as a substrate by Drobnik (1955).

β-Glucosidase (emulsin, cellobiase, gentiobiase) was detected in toluene-treated soils by using salicin, arbutin, and β-phenylglucoside (Hofmann and Hoffmann, 1953) or cellobiose (Hoffmann 1959b). α-Galactosidase (melibiase) was studied by using α-phenylgalactoside or melibiose by Hofmann and Hoffmann (1954) and Hoffmann (1959b). For β-galactosidase (lactase) determination, β-phenylgalactoside and lactose were used (Hofmann and Hoffmann, 1954; Hoffmann, 1959b). Kiss re-examined the presence of α-glucosidase in soils (Kiss, 1958b; Kiss and Péterfi, 1960a, b). Although some β-glucosidase is accumulated in soil, Kiss et al. (1972) demonstrated that, following prolonged incubation in non-sterile conditions, cellobiose degradation was largely due to cellobiase produced by proliferating organisms rather than by previously accumulated enzymes.

The presence of β-galactosidase in soil was re-examined by Ryšavý and Macura (1972a, b). The carbohydrase activity rates in soils are dependent on the use of specific substrates (Kiss and Péterfi, 1960a, b; Hofmann and Hoffmann, 1953, 1954). After re-examining carbohydrase activity in soils Kiss et al. (1975a) emphasized the need to compare enzymatic activities in soils incubated with and without toluene to distinguish the contribution of the microbial activity versus the activity of accumulated enzymes.

Amylase activity in soil was first noted by Hofmann and Seegerer (1951b). Methods for detecting amylases were developed by Hofmann and Hoffmann (1955) and Drobník (1955). It was shown in the early work that soils contained considerably higher concentration of β-amylase than α-amylase (Hofmann and Hoffmann 1955). Galstyan (1965) concluded that amylase activity in soils was due to extracellular, accumulated enzymes. Starch decomposition in the soil has been studied quite extensively and has been reviewed by Reese (1968), and Kiss and Drăgan-Bularda (1972). It is evident that invertase and amylase are produced adaptively in soil as shown by Drobník (1955). Amylase has been extracted from soil (Shcherbakova et al., 1970, 1971).

Although it has been assumed for a long time a priori that cellulase exists in soil extracellularly, the first attempt to show its extracellular existence in the soil was reported by Márkus (1955), who used cellophane pieces in toluene-treated soil. In the 1960s and 1970s a number of investigators have used various substrates including cellophane, cellulose powder and carboxymethylcellulose to demonstrate extracellular cellulase activity in soils. Kong et al. (1971) and Kong and Dommergues (1972) used cellulose powder and carboxymethylcellulose to distinguish C_1 and C_x components of the cellulolytic activity. They correlated the ratio of C_1 to C_x activities with the reaction of the soil tested. C_x activity predominated in acidic and C_1 in calcareous soils. Specific ratios may depend on nutrient availability. C_x and C_1 activities in the soil have been studied also by Drozdowicz (1971) and by Tomescu (1970). Cellulolytic activity appears to be inducible in soils (Ambrož 1973), and many soils might not have measurable amounts of extracellular cellulases. Kiss et al. (1962b) showed the presence of cellulase activity in only 1 out of 10 soils. Similarly, when 14 soils were tested for cellulase activity by Seetharaman et al. (1968), 11 showed cellulase. The experimental results, relative to cellulase activity in soils generally, are conflicting. Kiss et al. (1975a) concluded from the literature that cellulase accumulation in soil is not a general phenomenon and recommended further studies for a better understanding of cellulase accumulation in soils.

The extracellular presence of xylanase activity in soils has been demonstrated by Sørensen (1955, 1957). In γ-irradiated soils, xylanase activity was about 25 % lower than in toluene treated samples. Xylanase apparently is a radiation-sensitive enzyme in soils.

Lichenase activity was reported by Kiss et al. (1962b) in 9 out of 10 soils tested.

Inulase was first detected in toluene-treated soil samples by Hoffmann (1959b) and later studied by Kiss and Péterfi (1961) who concluded that inulase is an accumulated enzyme in soil.

Upon incubating soil with toluene and dextran, dextranase activity was shown to be present by Drăgan-Bularda and Kiss (1972). Although it was evident that dextranase exists in soil extracellularly, its appearance in soil is inducible. The inducibility of extracellular enzymes in soil has been demonstrated for xylanase by Sørensen (1955), for levan sucrase by Kiss (1961), and for amylase and invertase by Drobník (1955). Extracellular levanase was studied by Kiss (1961) and Kiss *et al.* (1965). It was concluded, however, that most of the substrate was hydrolysed not by accumulated enzymes but rather by enzymes produced by proliferating microorganisms.

The presence of polygalacturonase (pectinase) in soils was demonstrated by Hoffman (1959b) and Kaiser and Monzon de Asconegui (1971). The polygalacturonase activity was correlated with *Azotobacter chroococcum* proliferation (Monzon de Asconegui and Kaiser, 1972).

2.2. *Esterases*

Phosphatase (phosphomonesterase) in soil has been studied by numerous investigators. In 1933, Rotini suggested that transformations of organic phosphates in soil were caused by enzymatic activities. Rogers (1942a, b) and Rogers *et al.* (1941, 1942) showed that dephosphorylation of calcium glycerophosphate occurred rapidly under conditions presumably excluding activities of proliferating microbes and suggested that the source of phosphatase was from plant roots. Phosphatase activity in soil was originally assayed by determining the release of inorganic phosphate from the substrates (Rogers, 1942a, b; Rogers *et al.*, 1941, 1942; Jackman, 1951; Jackman and Black, 1951, 1952a, b; Mortland and Gieseking, 1952). It was soon recognized, however, that such assays did not account sufficiently for soil fixation of released phosphate. An analytical method involving the determination of released phenol from phenylphosphate was introduced by Kroll *et al.* (1955) and used by Kramer and Erdei (1958, 1959). In numerous subsequent methods the non-phosphate moiety is determined upon hydrolysis of phosphate. For example, β-naphthylphosphate was used by Ramírez-Martínez and McLaren (1966a), and Skujiņš *et al.* (1962) determined the non-reacted glycerophosphate upon incubation. Another variation of the phenylphosphate method has been described by Khaziev (1972) and the use of nitrophenylphosphate has been introduced by Tabatabai and Bremner (1969). This last method appears to be one of the most precise. The main difficulty in obtaining valid data regarding phosphatase activity, as with other soil enzymes, has been in the identification of the proper method.

With the introduction of radiation sterilization of soils, it was demonstrated that phosphatase is resistent to irradiation and may thus be measured in the absence of microbial activity after irradiation (Skujiņš *et al.* 1962; Voets *et al.*, 1965; Voets and Dedeken, 1965).

The phosphatase activity in soils has been critically re-evaluated by Ramírez-Martínez and McLaren (1966b). They concluded that the analytical method used, determined the results, with considerable variations being introduced by handling and drying procedures.

Organic phosphate compounds in soil usually constitute 30–70% of its total phosphate. The assimilation of organically bound phosphate by plants and microorganisms is preceded by their hydrolysis by extracellular phosphatases. Although phosphatases accumulate in soils, the intensity of the excretion of phosphatases by plant roots and microorganisms is apparently dependent upon the organisms' needs for phosphates. A number of reports in the literature, as reviewed by Kiss *et al.* (1975) and by Skujinš (1976), indicate conflicting results regarding responses by microorganisms to phosphate relationships in soil following diverse treatments. It has been suggested that the release of phosphatases in soil may be induced where the mineral phosphate concentration is low as compared to that of the organically bound phosphorus. Phosphatase activity in soils often greatly exceeds the activity expected to be contributed by the microorganisms present (Ramírez-Martínez and McLaren, 1966b).

Although alkaline and acid orthophosphoric monoesterphosphohydrolases have been recognized, it has been difficult to distinguish their activities in the soil. Hoffmann (1968) suggested three types of soil phosphatases; acid, neutral and alkaline. The released phosphatases apparently are from widely diverse sources, have different pH optima and also reflect the pH values of soils. The resultant research problems include the need for assays that would reflect a range of pH values for each individual soil. Phosphatase activity has responded differentially to inorganic (NPK) and organic soil amendments, usually by increasing its activity (Rankov and Dimitrov, 1971). Muresanu and Goian (1969) showed that phosphatase activity increased with moderate fertilization but decreased upon higher dosages. Halstead (1964) indicated that its activity was higher with increasing amounts of organic matter. Addition of inorganic phosphate diminished the activity and in organic soils it was associated with phosphate availability (Geller and Dobrotvorskaya, 1961). The current information, however, indicates that phosphatase activity is directly related to the level of organic phosphorus in soil (Gavrilova *et al.*, 1974). Information about phosphatase activity in soils has been reviewed by Ramírez-Martínez (1968). Kiss *et al.* (1975a) and Skujinš (1976).

The hydrolyses of other types of inorganic phosphates added to soil have been examined by a number of investigators. Hydrolysis of metaphosphates was studied by Rotini (1951) and Rotini and Carloni (1953). They demonstrated that some hydrolysis of metaphosphate might be due to extracellular enzymes; however, a considerable contribution by inorganic catalysis was evident. Rotini (1933) reported evidence for hydrolysis of pyrophosphate to orthophosphate by soil extracts in the presence of toluene and the work of Sutton *et al.* (1966) also indicated that hydrolysis of pyrophosphate in soils may be due to enzymatic activity. Recently a number of investigators have reported considerable abiotic contribution to the hydrolysis, for example, Gilliam and Sample (1968) and Juo and Maduakor (1973). On the other hand, Khaziev (1972) concluded that non-enzymatic, abiotic factors might have a limited role in pyrophosphate hydrolysis.

Phytase activity in soils was examined by Jackman (1951) and Jackman and Black (1951; 1952a, b). The activity followed microbial activity in soils. The amount of toluene present was apparently insufficient to inhibit the microorganisms.

Phytates are relatively insoluble substrates. They are slowly decomposed in soil and tend to accumulate in large amounts. Also, the phytates that are commercially available for experimental work might contain penta- and hexa-phosphophytates which might be extremely difficult to decompose. The published results of experiments on phytase activity, which involved commercial preparations, should be evaluated accordingly.

It has been demonstrated that nucleotidases (nucleases) also accumulate in soils. Burangulova and Khaziev (1965a) have suggested that nucleic acid/lignin complexes are formed which are resistant to enzymatic attack. Nilsson (1957) reported low rates of nucleotide decomposition in soils.

Haig (1955) found that most of the phenylacetate and phenylbutyrate hydrolysis associated with acetylesterase activity in soil was due to microbial, rather than accumulated enzyme activity.

Extracellular lipase activity in peat and mud was first indicated by Pokorná (1964). Pancholy and Lynd (1971, 1973) found lipase activity in extracts of loamy sand. Pokorná (1964) used tributyrin as the substrate, whereas Pancholy and Lynd (1971, 1973) determined lipase by its activity on 4-methylumbelliferone butyrate. Pokorná (1964) measured the lipase activity in the absence and in presence of toluene and showed only minimal differences. Generally, measurements taken in the absence and presence of toluene are considered primary indicators for the presence of accumulated enzymes in soil.

Extracellular arylsulphatase was first shown to be present in soils by Tabatabai and Bremner (1970a, b) with p-nitrophenylsulphate as the substrate. Their results indicated that much of the arylsulphatase was bound to cell constituents of nonproliferating organisms.

2.3. Proteases and Amidases

Another group of enzymes that have been widely studied are the various proteases. The degradation of proteins is an important part of the nitrogen cycle in soils and there have been numerous investigations of extracellular proteases in the last decade. The first report on proteolytic activity in soil extracts was presented by Fermi (1910). A fraction having a cathepsin-like activity was isolated by Antoniani, et al. (1954) by using ammonium sulphate and tungstate. McLaren et al. (1957) demonstrated a trypsin-like activity in soil by assaying it with a specific substrate, benzoylarginineamide.

Albumin, gelatin and casein were used by Ambrož (1963, 1965, 1966). The most commonly used method involves hydrolysis of gelatin and was introduced by Hofmann and Niggemann (1953). Hoffman and Teicher (1957) developed the method further by determining the released amino acids photometrically as a cupric ion complex. Voets and Dedeken (1964a) bioassayed the release of arginine from gelatin by using *Leuconostoc mesenteroides*. The more modern methods include those of Ladd and Butler (1972) and Ladd and Paul (1973), who used benzoylarginineamide, dipeptides, casein and haemoglobin for short incubation times. Proteases exist in soils as accumulated enzymes. Proteolytic activity in soils is rather sensitive to storage (Hoffmann and Teicher, 1957) and to various agents applied to soils. Trypsin-like activity was destroyed by irradiating soil with sterilizing dosages (McLaren *et al.*, 1957) that did not appreciably destroy urease or phosphatase.

The presence of extracellular asparaginase was demonstrated by Drobník (1956) in toluene-treated soils. Asparaginase was extensively studied by Mouraret (1965). His work indicated that asparaginase is accumulated in the soil; however, it most likely exists in soils as an enzyme bound to cell constituents.

Glutaminase in soils was examined by Galstyan and Saakyan (1970) and Galstyan (1973), who showed that in toluene-treated soils glutamine was hydrolysed to glutamic acid and ammonia.

Except for asparaginase and glutaminase, it appears that other abiontic deaminases do not exist in soils (Ambrož, 1966; Voets and Dedeken, 1964b). Although Subrahmanyan (1927) reported that soil extracts de-aminate amino acids, especially glycine, his report in modern context appears to be erroneous.

Because of the agricultural importance of urease as a decomposing agent for urea which is used as a fertilizer, urease has been the most widely studied soil enzyme. It was first examined in soil by Rotini (1935a). Extensive reports on urease activities in soils were published by Conrad (1940a, b; 1941; 1942a, b; 1944) and Conrad and Adams (1940). Later, in the 1950s, urease was studied by Kuprevich (1951), Scheffer and Twachtmann (1953), Hofmann and Schmidt (1953), Galstyan (1958a), and Galstyan and Tsyupa (1959). Because of the interest in urease, a number of methods for urease determination in soil have been developed. Most of these are based on measuring released ammonia from toluene-treated, urea-amended soils. These include the Kjeldahl methods, Conway's micro-diffusion dish method, and titrimetry. Due to the difficulties in determining small amounts of ammonia released relative to the high levels of ammonia present in soils, methods have also been developed for determining the residual urea by using xanthydrol or p-dimethylaminobenzaldehyde. Other methods have been based on the release of carbon dioxide; these include ^{14}C labelled urea for the detection of $^{14}CO_2$ (Skujiņš and McLaren, 1969). Urease has been measured with or without buffers and with or without toluene. Sterilization methods have included radiation, as it has been shown that urease withstands high dosages without being substantially denatured. The methodologies have been reviewed and listed by Skujiņš (1976). Since Rotini's 1935 report the application of toluene was found not to diminish but in many cases to increase urease activity in soils (Tabatabai and Bremner, 1972). In a non-treated, native soil urease activity is the sum of extracellular and intracellular sources. Irradiation disrupts cellular membrane integrity and makes intracellular urease available to the substrate. This phenomenon might be used to indicate the relative amounts of intracellular and extracellular enzymes present in any particular soil (Skujiņš and McLaren, 1969). Increased activities were demonstrated also by Thente (1970), who irradiated soil and compared its urease activity with that of soil treated with toluene. A number of reports in the literature, however, show that toluene decreases urease activity in soil (Anderson, 1962; McGarity and Myers, 1967; Roberge, 1968; Paulson and Kurtz, 1969). Considering the persistence and stability of urease in soils such measurements, however, clearly indicate the difference between the intracellular and extracellular enzymes present in particular soils. Depending on the origin of the soil, the accumulated urease might vary approximately between 50 and 90%. Urease has been detected in soil stored for over 80 years and in permafrost soils over 9000 years old (Skujiņš and McLaren, 1968, 1969). Persistence and accumulation of large amounts of urease in soils might have dire consequences on the loss of urea nitrogen from applied fertilizers. Considerable attention has been devoted to applications of urease inhibitors to soils to inactivate their native urease, thus diminishing fast release of amonium nitrogen from urea and urea-derivatives fertilizer (Kiss et al., 1975a; Bremner and Mulvaney, this volume).

Since the first decade of this century, numerous investigators of the degradation of calcium cyanamide have established that it is mostly due to abiotic catalysis, especially by manganese dioxide, although ferric and aluminium hydroxides participate in the process. However, Ernst and Glubrecht (1966) and Ernst (1967) re-examined the biological degradation of calcium cyanamide by using ^{15}N labelled substrate. They showed that inorganic catalysis was responsible for less than 8% of the degradation in an agricultural soil, while about 20–50% was due to extracellular enzymes. The rest of the hydrolysis they ascribed to degradation by proliferating microorganisms indicating that, in soils which are low in manganese dioxide and other possible inorganic catalysts, the biological degradation of calcium cyanamide might be significant.

Rotini (1956, 1963) has described a process in soil whereby urea might be isomerized to ammonium cyanate. The ammonium cyanate might be again hydrolysed abiotically by manganese dioxide but an enzymatic process in soil also may participate.

2.4. Oxidoreductases

Dehydrogenase activity in soils provides correlative information on the biological activity and microbial populations in soil. The biochemical properties of dehydrogenases are such that free, abiontic dehydrogenases are not expected to be present in soil. Measurements of dehydrogenase activities in soil, nevertheless, are important to soil biology. Experimental procedures do not involve bacteriostatic agents to inhibit microbial activities at the time of measurement; instead the reduction of 2, 3, 5-triphenyltetrazolium chloride (TTC) to triphenyl formazan is measured. The method was first introduced by Lenhard (1956, 1957) and several modifications have been described, notably by Stevenson (1959), Galstyan (1962) and Kozlov and Mikhailova (1965). Early dehydrogenase measurements were done anaerobically but Casida et al. (1964) indicated that atmospheric oxygen did not affect the procedure. The difference between the actual and potential dehydrogenase activity in soils was described by Kiss et al. (1966, 1969) et al. The actual value was assumed to be obtained when no hydrogen donor was added, whereas the potential dehydrogenase activity was determined after an addition of glucose.

Currently, the real significance of dehydrogenase measurements is not clear. The positive correlation between dehydrogenase activity and either the rate of carbon dioxide production or oxygen uptake has been reported by Stevenson (1959), Casida et al. (1964), Skujiņš (1973), Bauzon et al. (1968), Rawald et al. (1968) and others. However, Schaefer (1963), Ross and Roberts (1970), Gilot and Dommergues (1967), Bauzon et al. (1969) did not find such relationships. Similar problems have arisen in correlations of dehydrogenase activity with microbial numbers. Raguotis (1967), Rawald et al. (1968) and Roth (1965) indicated positive correlations, whereas Stevenson (1959), Hirte (1963), Skujiņš (1973), Ross and Roberts (1970) and Gilot and Dommergues (1967) reported none.

Dehydrogenase was negatively correlated with phosphatase activity and positively correlated with proteolysis and with the nitrification potential (Skujiņš, 1973). No correlation was found between invertase and dehydrogenase activities (Kiss et al., 1969). Dehydrogenase activity is usually well correlated with soil humus content (Laugesen and Mikkelsen, 1973; Russel and Kobus, 1974), which prompted Moore and Russell (1972) to re-examine the correlative factors affecting dehydrogenase activity. They concluded that there was a lack of useful correlation between dehydrogenase activity and soil properties known to influence plant growth and yields. However, the activity did not respond to an addition of nutrients and they concluded that their method was not precise. Dehydrogenase activity measurements obviously represent the immediate metabolic activities of soil microorganisms at the time of the test. The specificity and content of dehydrogenases in a mixed population of various strains of microbial species in soil differ and respond differently to available substrates. Clearly dehydrogenase activity needs further evaluation.

Catalase was the first enzyme studied in soils (Woods, 1899). Since then, there have been nearly one hundred reports. It was practically the sole enzymatic activity in soils studied during the first part of the century, due to the lack of knowledge about other enzymes and due to the ease of catalase detection by manometric methods. A modern manometric method was first described by Kuprevich (1951) and later improved by Kuprevich and Shcherbakova (1956). An improved volumetric method was described by Weetall et al. (1965) and various modifications of the original method have been introduced since. A titrimetric procedure was used by Kappen in 1913 and later modified by Osugi (1922), Rotini (1931) and others. This method has been used by Katznelson and Ershov (1958) and was later modified by Johnson and Temple (1964). Titration with $KI-Na_2S_2O_8$ has been performed by Verona (1960). Factors affecting hydrogen peroxide decomposition in soils have been re-examined by Johnson and Temple (1964) and Beck (1971). Streptomycin has been used successfully to inhibit microorganisms so that abiontic catalase can be distinguished from the catalase activity of proliferating microbes (Kunze, 1970). Large amounts of hydrogen peroxide added to soil may be decomposed by inorganic catalysts. Catalase is inhibited by a number of chemicals produced by plants and deteriorating plant residues, for example, vanillic, syringic and gallic acids and tannin as reported by Kunze (1970, 1971). Next to the presence of inorganic catalysts in soil, these compounds introduce another complicating factor for the evaluation of catalytic activity and its significance in soil.

Ross (1968b) reported that some of the glucose released during assays of carbohydrases in soil was lost as it was oxidized to gluconic acid and to 2-ketogluconic acid in the presence of toluene. Further work (Ross, 1974) showed that glucose oxidase was present in the top 8 cm layer of the soil. The conclusion, however, was that the loss of glucose due to action by glucose oxidase during the enzymatic assays was negligible. The results also suggested the presence of glucose dehydrogenase in the experimental system as an extracellular enzyme.

It has been evident that soil contains various phenol oxidases such as o-diphenol oxidase, p-diphenol oxidase and polyphenol oxidase. The group has also been shown to include peroxidase, experimentally, but in the soil it is difficult to distinguish among the activities of specific oxidases due to the crossover specificities of some of the enzymes with respect to the substrates used. It appears, however, that o-diphenol oxidase is the commonly studied enzyme in soils. Phenol oxidase (tyrosinase or catechol oxidase) was detected by Kuprevich (1951) and methods have been published (Galstyan, 1958b; Kozlov, 1964a; Kuprevich and Shcherbakova, 1966, 1971). The difficulties encountered in detecting of phenol oxidases have been similar to those with phosphatases where the substrate or the product might be strongly sorbed to soil particles. The polyphenol oxidases apparently are involved in humification processes. Mayaudon and Sarkar (1974a, b; 1975) and Mayaudon *et al.* (1973a) successfully isolated diphenol oxidases from soil which they purified and examined in detail.

The presence of uricase (urate oxidase) in soil was demonstrated by Durand (1961) and later extracted from urate-enriched soils by Martin-Smith (1963). Further examination of the urate oxidase system in soil (Durand, 1963) showed that the substrate, as well as the enzyme, may be sorbed on clays, and that degradation proceeds at a higher rate when the substrate and the enzymes are desorbed from the clay surfaces. Although urate oxidase apparently exists extracellularly in the soil, the further metabolism of the products of uric acid degradation is an intracellular function of proliferating microorganisms (Durand, 1961).

2.5. Lyases

Drobník (1956, 1961) demonstrated the presence of an extracellular decarboxylase in toluene treated soils. β-Alanine was formed from aspartate. In the absence of toluene, β-alanine is apparently further metabolized by microorganisms and becomes non-detectable.

Presence of glutamate decarboxylase was indicated by Kuprevich and Shcherbakova (1966) who showed that glutamic acid was decarboxylated to γ-aminobutyrate. A number of amino acids especially DOPA, tyrosine and tryptophan (but not phenylalanine), were decarboxylated by aseptic extracts of fresh soils (Mayaudon *et al.*, 1973b). Chalvignac (1968, 1971) and Chalvignac and Mayaudon (1971) reported the presence of an extracellular system in soil which decarboxylated and oxygenated tryptophan to β-indoleacetic acid. The enzyme system was thought to be bound to a clay-humic complex.

2.6. Transferases

Transaminase activity in soils was first indicated by Hoffman (1959, 1963) working with aspartic acid, leucine, valine, glutamic acid and pyruvate in toluene-treated soils. Synthesis of amino acids in soil has been reported by Kuprevich *et al.* (1964) and Kuprevich *et al.* (1968). Solutions containing keto analogues of alanine and glutamic acid were used in toluene-treated soils in the presence of ammonia and following incubation, chromatographic techniques detected increasing amounts of the corresponding amino acid. It was concluded that extracellular enzymes were present.

Levan sucrase (sucrose-6-fructosyltransferase) and dextran sucrase (sucrose-6-glucosyltransferase) have been studied by Kiss and Drăgan-Bularda (1968a, b). They reported that synthesis of levan in toluene- and sucrose-treated soils was due to extracellular levan sucrase in soil, however, proliferating microorganisms also might have participated. Drăgan-Bularda and Kiss (1972) reported that dextran sucrase, which converts sucrose to 1, 6-glucan and fructose, might also exist in the soil as an extracellular enzyme. It has been also hypothesized that phenyloxidases participate in synthetic reactions during the formation of humic acids; however, no direct evidence has been presented.

2.7. Negative results

A series of attempts by a number of investigators to detect extracellular occurrence of various enzymes in soil have given negative results. The studies included oxidation of sulphide to sulphate (Galstyan and Arutyun-yan, 1968), reduction of sulphates (Galstyan, 1966), enzymes participating in the nitrification process (Cawse, 1968), nitrate reductase (Galstyan and Markosyan, 1966; Cawse and Cornfield, 1969), nitrite reductase (Galstyan and Saakyan, 1970) and hydroxylamine reductase (Galstyan and Saakyan, 1971).

3. Sources: plants, animals or microbes?

Any study of enzymes in soil immediately invokes questions about their origins. In the first report of extracellular enzymes in soil (Woods, 1899) it was already suggested that the catalase observed was excreted by plant roots. Plant physiologists have accumulated voluminous data showing that plant roots do excrete certain enzymes into the rhizosphere soil (Collet, 1975) and numerous investigators have tried to correlate enzymatic activities in soil with the prevailing vegetation. Activities in the rhizosphere do not reflect only the plant root contribution but also that of the microbial flora and in most ecosystems it is doubtful whether a non-rhizosphere soil exists, the microbial population is regulated by its associated vegetation to a considerable degree.

It has been assumed, *a priori*, that enzymes excreted by plant roots would contribute to the make-up of the abiontic soil enzyme system. During the early years it was noted by Knudson and Smith (1919) that amylase is secreted by plant roots. Rogers (1942b) and Rogers *et al.* (1942)

indicated that phosphatase and nuclease activities in soils are contributed by excretions from corn and tomato roots. Kuprevich (1949) reported that plant roots excrete a number of enzymes. His sterile techniques, however, have been criticized by Estermann and McLaren (1961) who examined phosphatase, invertase and urease distribution between the roots and rhizoplane of barley plants and showed that the phosphatase and invertase could be attributed mainly to enzymes in the roots, while urease was a function of rhizoplane organisms. No protease, esterase, amylase and asparaginase activities were found in either roots or rhizoplane. A number of investigators in the USSR (Galstyan, 1958b; Davtyan, 1958; Latypova, 1961; Peterson, 1961) have reported elevated activities for many enzymes in the rhizosphere, these were not unexpected but the specific contributions of plant roots and microflora in the rhizosphere were not distinguished.

Vlasyuk et al. (1956, 1957, 1959) compared activities in the rhizosphere soil with those in the non-rhizosphere soil. They observed considerably increased urease activity in the rhizosphere of legumes but the differences in urease activity associated with other crops were less pronounced. Similarly, they noted that invertase, catalase, phosphatase and protease activities were much higher in the rhizosphere of legumes, particularly in clover and lupin, than in the rhizospheres of oats and wheat. It was noted also, that activities also depended on the preceding crop. Phosphatase activity values of different crops were similar but fertilizers quite markedly increased enzymatic activities in the rhizosphere. Nilsson (1957) however, reported that oats, barley and peas did not affect urease activity readings but phosphatase activity was greater in the oat rhizosphere than elsewhere and was slightly lower in pea and barley rhizospheres than in non-crop soils. Invertase and urease activities were higher in the rhizospheres of barley, cotton and lucerne than in non-rhizosphere soils (Davtyan 1958).

In the 1960s the examination of rhizosphere effects on soil enzymatic activities was initiated by a number of investigators, especially in Western Europe. Cellulase activity was lowest in non-rhizosphere soil and increased considerably in rhizospheres (Bernhard, 1965). The behaviour of microorganisms and enzymes in the rhizospheres of barley, rye and wheat was described by Voets and Dedeken (1966). They reported an appreciable increase of invertase, urease, β-glucosidase and phosphatase in the rhizosphere. After four weeks of growth, however, proteolytic activity was lower than in the non-rhizosphere soil although the proteolytic microflora were proliferating at a high rate. Invertase activity in soil has been repeatedly correlated with plant density and composition (Ross 1965, 1966, 1968a).

Much of the work in soil enzymology is based on the assumption that the source of most abiontic enzymes are the soil microflora. This notion may have been supported by the voluminous work on extracellular enzyme release which has appeared in medical, general and industrial microbiology. It may also be assumed, *a priori*, that enzymes participating in the degradation of insoluble substrates in soil, for example, cellulases, chitinase and proteases, are released by soil microorganisms extracellularly. Evidence also has been accumulated that a number of enzymes might be bound to the cellular particles of dead microbial cells. The methodological difficulties encountered in distinguishing microbial origins of abiontic enzymatic activities in soils are evident. Some inferences might be drawn from tests suggested by Kiss and Boaru (1965) in which the soil is examined in the presence and in the absence of toluene. However, activities exhibited in the presence of toluene also include enzymes released by plants and other sources.

Hofmann (1963) and Hofmann and Hoffmann (1955a) opined that microorganisms were the sole sources of abiontic enzymes in soils. In certain cases, abiontic enzymatic activity has been correlated with microbial proliferation (Daragan-Sushcheva and Katsnelson, 1963; Geller and Dobrotvorskaya, 1961).

The contribution of soil fauna to its abiontic enzyme content was studied by Kiss (1957). He indicated that earthworms contributed to the invertase activity in soil, especially in the surface layer of soil but that ants contributed little. A study by Kozlov (1965) indicated that the soil fauna did contribute to the enzyme content, but in a limited manner.

4. Soil enzymology and "Fertility Index"—a fallacy?

Increasing research on soil enzymes in the 1950s produced numerous empirical observations of horizontal and vertical enzyme distribution in various soil types and ecosystems, with respect to amendments, cultivation practices and responses to environmental and climatic factors. It was hoped that correlated information on extracellular enzymes in soil would provide a tool for determining the total biological activity in soil and, consequently, a "fertility index" of soils usable for practical purposes in agriculture. Such expectations were expressed by a number of investigators in the 1950s. Unfortunately, the observations produced conflicting, confusing data and some methodologies were questionable.

Hofmann and Seegerer (1950) suggested that invertase activity would indicate the quality of soil and would be usable as a measure of its fertility. They stated (Hofmann and Seegerer, 1951a) that the determination of soil enzyme content was more important to soil fertility ratings than a microbiological measurement. In a review of his soil enzyme work, Hofmann (1955) provided examples showing that the fertility of soil can be estimated better from extracellular enzyme assays than from its Carbon dioxide production. Lajudie and Pochon (1956) suggested that soils may be classified according to their proteolytic activity and that such classification agrees with the biological activity and the apparent fertility. Similar suggestions were expressed by Moureaux (1957), who examined invertase activity and populations of nitrogen fixing and nitrifying organisms in a wide range of soils. The results showed significant variations in activities according to soil type, topography and plant cover. The Moureaux values generally reflected the productivity of a given soil and it was suggested that the use of biological tests to compliment physical and chemical analyses of soils was justified for fertility determinations. Baroccio (1957) showed that catalase activity was much more closely related to the physical and chemical factors associated with soil fertility than to those of nitrifying capacity. Kuprevich (1958) re-emphasized the evidence for direct correlation between enzymatic activity and soil fertility.

Balicka and Sochacka (1959) showed that invertase activity in sandy soils depended on the type of crops being grown but asparaginase and urease activities were regulated more by the soil organic matter levels which thus provided a useful index of the productivity, whereas the total numbers of bacteria were poor parameters for this purpose. Invertase activity was also suggested by Bose et al. (1959) as a way to assess microbial activity or soil fertility.

A correlation between plant yield and biochemical properties of soil, including enzymatic activities, respiration and the rate of cellulose decomposition, was shown by Chang and Cheng (1962). In 1969, Maliszewska compared biological activities of different soils and suggested that soil respiration, proteolytic activity and cellulolytic activity are the most appropriate parameters for correlating soil fertility.

Enzymatic activities have been examined in soils of most geographic regions of the world and certain similarities have emerged. However, the general decrease of enzymatic activities with the depth of soil may be well correlated but a horizontal distribution correlation is seldom evident. Each soil type presents its own enzymatic pattern and most often, the enzymatic and other biochemical and soil microbiological activities cannot be correlated. Typically, enzymatic activities do not correlate with the soil respiration values, nor with the microbial numbers, although there are always exceptions.

Cultivation most often increases various activities, and as a rule, so does the application of fertilizers and organic amendments. Of the many authors examining these relationships, Koepf (1954a, b, c, 1956), Latypova (1960), Kudryachev (1968), Kvasnikov and Perov (1965), Kozlov (1964b), Novogrudskaya (1963) may be mentioned. Stenina (1968) studied the enzymatic activities in different cultivated soil types of the central taiga. Galstyan with his colleagues contributed considerable volume of data to the knowledge about the effects of cultivation on enzymatic activities in various types of soils, for example, Galstyan *et al.* (1962) reported on activity changes upon the reclamation of arid soils. Khan (1970) re-examined enzymatic activities in soil as influenced by various crops and fertilizers and Lupinovich and Yakushava (1965) noted that invertase, phosphatase and catalase activities in the surface horizon increased when virgin soils were cultivated. Applications of fertilizers caused an increase in catalase activity, whereas lucerne meal had little affect (Baroccio, 1958). A negative effect of reclamation ploughing and subsoiling on enzymatic activity was noted by Habán (1966, 1968). However, tillage increased microbial population and enzymatic activities below the ploughed layer.

In semi-arid soils biological activities were very low as compared to cultivated, high-rainfall agricultural, or irrigated soils (Galstyan and Astvatsatryan, 1958; Galstyan and Avundzhyan, 1966; Skujiņš, 1973). Especially low were the dehydrogenase, catalase and invertase activities. In salt-affected soils (solonchaks), invertase, amylase and β-glucosidase were not detected, urease activity was low but catalase activity was comparatively high (Galstyan, 1960). Catalase activity was higher in the sub-arable horizon and was also higher in the rhizosphere of halophytes than in the respective surface and non-rhizosphere soils. Overgrazing and erosion considerably decreased catalase, invertase and phosphatase activities in semi-arid soils (Sarkisyan and Shur-Bagdasaryan, 1967).

Phosphatase activity and its relationship to soil fertility has been exceptionally controversial. It was examined by Keilling *et al.* (1960) and Keilling (1964) in French and African soils, with about half of the French soils exhibiting low activity. The lowest activities occurred in degraded vineyard soils, the highest in soils with elevated organic matter content as in grasslands and forests. The work by Vlasyuk and Lisoval (1964) showed that cultivation with additional mineral and organic fertilizers increased phosphatase activity especially during the vegetative period. Generally, the activity rating was closely related to the conditions of plant nutrition. Burangulova and Khaziev (1965b) concluded that phosphatase activity depended on the genetic, physical and chemical properties of specific soils. Phosphatase activity of genetically identical or similar soils reflected the fertility, type of fertilizer used and the biological properties of the plants grown. Fertilizers can positively or negatively affect phosphatase activity: it is reduced by phosphate and potassium fertilizers but increased by nitrogen.

As more information on abiontic enzyme activities in soils became available, correlations with soil fertility became more difficult to support and generalizations more difficult to make. As early as in 1954, Koepf (1954a, b, c) noted that no close relationships exist between nutrient levels, enzyme activities and carbon dioxide evolution. That the information on soil enzyme activities cannot provide a complete picture of biological status of soils and cannot serve as a criterion for their fertility level was emphasized by Drobník (1957) and similarly by Galstyan (1959). Turkova and Srogl (1960) showed that the correlation between carbon dioxide production and amylase and invertase activities varied under different plant associations in the same habitat on identical soil types. They concluded that enzymatic activity measurements cannot provide constant values that could characterize the total biological activity of a soil. Vasyuk (1961, 1964) closely examined microfloral and enzymatic activities in a number of cultivated and non-cultivated soils. He concluded that the biological activity of such soils was represented by neither microbial nor enzymatic correlations. Similar conclusions were reached by Daragan-Sushcheva and Katznelson (1963).

Yaroshevich (1966) examined soils subjected to fifty years rotated fertilizer application. He showed that continued use of manure increased soil respiration and enzyme activity, whereas inorganic fertilizers of equivalent nutrient value had the opposite affect. In a number of test plots where soils had the same fertility value for plants, enzyme activities varied according to the fertilizer regime and had no obvious correlation with the yield of sugar beet, the test plant. Similarly, Habán (1967) concluded that there was no direct relationship between microfloral populations, enzyme activities and crop yields.

In the 1970s, obtaining a "fertility index" by the use of abiontic soil enzyme activity values seems unlikely. It is evident that enzymes are substrate specific and individual enzymatic measurements cannot reflect the total nutrient status of the soil. Individual soil enzyme measurements, however, might answer questions regarding specific decomposition processes in the soil or questions about specific nutrient cycles. For example, it is of value to know the proteolytic activity characteristics in soil if one is concerned with the nitrogen cycle, or to examine phosphatase activity and to correlate it with phosphate availability from organic soil phosphates. Similarly, urease is important agriculturally as an enzyme which might limit the nitrogen available to plants from fertilizers or natural sources. Cellulase and other carbohydrases might indicate decomposition rates of litters. Several measurements have emerged as important parameters of the general biological activity in soils, especially dehydrogenase activity, carbon dioxide release (or respiration rates) and ATP measurements. It should not be concluded, however, that a "fertility index" for soils based on soil biochemical properties, might not eventually become available. The necessary parameters and the correlative relationships are just not yet apparent. Future research in soil enzymology may provide such needed specific or generalized correlative factors (Howard, 1972).

IV. Approaching Maturity

Because much of the recent work and status of soil enzymology is described in other chapters of this volume and in recent reviews (Kiss *et al.*, 1975a; Skujiņš, 1976); this section will emphasize only current directions.

Presently, soil enzyme research seems to have two general objectives:

1. To understand the state and behaviour of abiontic enzymes in soil in modern terms. This involves developing modern methodologies, the extraction and characterization of abiontic enzymes and examining their behaviour in ecosystems.

2. To develop and apply knowledge in soil enzymology to current environmental and agricultural problems and other practical purposes. This area of research involves the examination of abiontic enzyme participation in the degradation of man-made substances applied to the soil, especially various pesticides and heavy metals pollutants. Similarly, knowledge of soil enzymology is being applied to forensic science and it has implications in extraterrestrial life exploration.

1. Kinetics in a heterogenous system

The abiontic enzymes in soil exist in a heterogenous system in which the enzymatic reactions take place at the solid/liquid interfaces. To characterize enzymatic activities in soil, several investigators have tried to determine the kinetics based on the Michaelis-Menten equation, expressed in K_m values, similar to those determined in homogenous experimental systems. However, the kinetics used for soluble enzymes in solution are not always applicable to heterogenous systems. The problems of enzyme kinetics in heterogenous molecular environments, as in soil, have been pointed out and described (McLaren, 1960; McLaren and Babcock, 1959; McLaren and Packer, 1970).

Attempts to determine the Michaelis constant for abiontic enzymes in soils have been made by several investigators. Tabatabai and Bremner (1971) determined the K_m for arylsulphatase and Tabatabai (1973) for urease. In both cases it was evident that the experimental conditions and the soil behaviour introduced parameters that were not measurable and were not included for the determination of K_m values and the maximum velocities. Another application of the Michaelis-Menten approach to determine the kinetics of urease activity showed that it was possible to distinguish the K_m values for the urease activities caused by the proliferating microbial and abiontic soil ureases (Paulson and Kurtz, 1970). A novel method to determine K_m values for phosphatase was introduced by Cervelli *et al.* (1973) who used the Freundlich absorption isotherm in the Michaelis-Menten velocity equation. The Michaelis-Menten type equations were also determined for phosphatase activity in dripping columns as compared to batch-type reactions in soil by Brams and McLaren (1974). The K_m values for the columns were about 2·5 times greater than those for the soil in the batch-type systems. It is obvious that the determination of K_m values for enzymes in soil is a complex problem and further theoretical considerations and methodologies are necessary.

2. The enzyme-humus complexes

In the 1950s and 1960s considerable attention was devoted to examinations *in vitro* of enzyme and clay interactions. The enzyme proteins were strongly sorbed by clays, which influenced enzymatic reaction rates and enzyme stability, as reviewed by McLaren and Packer (1970) among others. Several investigators were prompted to fractionate soil to obtain information on the localization of enzymatic activities according to the textures of soils. A study of esterase activity showed that the clay fraction had the strongest phenylacetate effect, considerable activity was also present in the silt fraction but very little in the sand (Haig, 1955). Fractionation experiments were performed also by Hoffmann (1959a), who demonstrated that carbohydrase activity peaked in the silt fraction and urease activity reached in the clay fraction. Because he demonstrated that the clay fraction contained a minimal number of microbes, it was concluded that extracellular urease had been adsorbed and remained active on the clay.

Further insights in the enzymatic activity location within the soil matrix came after the first few successful enzyme extractions. Because of the very strong sorptive and other chemical interactions between enzyme protein and organic, as well as inorganic, soil constituents, it has been extremely difficult to extract active enzymes until methods could be adopted from biochemistry. A number of unsuccessful attempts were reported in earlier years, specifically those by Conrad (1940a) on urease, Haig (1955) on urease and phosphatase and Hübner (1956) on cellulase and pectinase. A number of early reports on soil enzyme extractions do not describe the methodology in enough detail to indicate whether the extraction procedures are acceptable in modern terms especially regarding the maintenance of sterility. Fermi (1910) reported extracting a proteolytically active fraction, Subrahmanyan (1927) extracted glycine deaminase and Ukhtomskaya (1952) extracted several enzymes from soil with a phosphate solution. The first verified attempt to extract proteinases appears to be that of Antoniani *et al.* (1954) who reported extracting a cathepsin-like activity from soil with ammonium sulphate and sodium tungstate. Briggs and Segal (1963) reported isolating urease, although with a rather low specific activity.

Two enzymes decomposing uric acid were isolated from uric acid-enriched soil by Martin-Smith (1963) and Getzin and Rosefield (1968) reported extracting a malathion-decomposing enzyme from malathion-enriched soil. An esterase that hydrolysed malathion was also extracted by Satyanarayana and Getzin (1973), who concluded that the isolated enzyme was a carbohydrate-protein complex, contributing to the stability and persistence of the extracellular enzyme entity in soil.

Peroxidase extraction was reported by Bartha and Bordeleau (1969). Mayaudon and Sarkar (1974a, b) extracted diphenol oxidases which were separated on a DEAE cellulose column in three fractions. Their study indicated that the activities of diphenol oxidases were associated with a nucleo-protein and also that they were in a form of humic acid-enzyme complexes (Mayaudon and Sarkar, 1975). Dialysed humic acid fractions having tryptophan decarboxylase activity were extracted by Chalvignac and Mayaudon (1971). The humic fraction retained half of its activity after being subjected to the conditions of classic humic acid preparations, i.e., after an exposure to sodium hydroxide followed by precipitation with hydrochloric acid. Ladd (1972) demonstrated that proteases may be extracted from soils with a simple (Tris) buffer. Extracted proteases were studied also by Mayaudon *et al.* (1975) who concluded that the polysaccharides present in the extracts contributed to the stabilization of the humic-enzyme systems in soils. Two methods for extraction of urease have been developed by McLaren and his colleagues. One method (Burns *et al.*, 1972a) involved dispersion of soil organic matter with urea, whereas the other (Burns *et al.*, 1972b) employed sonification. In the first method the extractant urea was later decomposed by the extracted urease and upon examination of the extracted urease system, the authors concluded that soil urease is present in the organic colloidal particles when the pores are large enough for water, urea and ammonia to pass freely, but sufficiently small to exclude a proteolytic enzyme. In a further report, McLaren *et al.* (1975) concluded that both methods showed urease activity in soil constituents that may be regarded as native humus. A report by Ladd and Butler (1975) has further contributed to the understanding of the enzyme-humic systems. It appears that the enzyme proteins of the humic-protein complex could be bound to humus by ionic, hydrogen or covalent binding and that most of the abiontic enzymatic activity in the soil is stabilized as humus-protein complexes (McLaren, 1975).

3. In the modern environment

With the increased extensive use of agricultural chemicals and the appearance of industrial pollutants, including pesticides, considerable interest has been raised about research on the interactions of these chemicals in soil and with the soil enzymes. One of the first such attempts appears to be by Rotini and Galoppini (1967), who studied the decomposition of a synthetic detergent, sodium dioctylsulphosuccinate and by Gregers-Hansen (1964), who examined the decomposition of diethylstilboestrol in soils. In both cases, it appeared that extracellular enzyme moieties exist which decompose these substrates. Libbert and Paetow (1962) demonstrated that indole-3-acetonitrile and indole-3-aldehyde were hydrolysed to indole-3-acetic acid and oxidized to indole-3-carboxylic acid in soils incubated with toluene, suggesting that enzymes catalysing these reactions were existing abiontically in the soil. Formation of indole-3-acetic acid from trytophan in toluene-treated soils was first reported by Chalvignac (1968). It was also formed from trytophan by the enzyme system extracted from soil by Chalvignac and Mayaudon (1971). Similarly, peroxidase extracts from several soils (Bartha and Bordeleau, 1969) converted chloroanilines to chloroazobenzenes (i.e., degradation products of phenylamide herbicides). The numerous investigators who have reported on the influence of pesticides on extracellular enzyme activities in soils are listed by Kiss *et al.* (1975a). The effect of heavy metals on soil enzymatic activity has been described by several investigators, for example, Tyler (1974, 1976a, b).

Because of the potential for soil urease to volatilize ammonia from applied urea fertilizer and to diminish the nitrogen availability to plants, numerous investigators in the last decade have investigated various urease inhibitors for agricultural application. A detailed discussion on soil urease inhibitors and enzyme interactions with various agricultural and industrial chemicals, including pesticides, is presented by Kiss *et al.* (1975a).

Examination of extracellular enzyme activities in ecosystems has been emphasized and continued in the 1970s. Of the numerous investigators in this field, recent work by Pancholy and Rice (1973a, b), Neal (1973), Kunze *et al.* (1975) and Voets *et al.* (1975) may be mentioned. During the 1970s most of the studies associated with the International Biological Programme have used soil enzymology methods to model the nutrient turnover systems in soils. Soil enzymology methods can also be used in criminology to characterize the evidence (Thornton and McLaren, 1975) and for the examination of extracellular enzymes in aquatic situations, either free or in the sediments (Verstraete *et al.*, 1976).

V. Conclusions

It has become evident that so-called extracellular enzymatic activities in the soil include not only free extracellular enzymes and enzymes bound to inert soil components but also active enzymes within dead cells and others

TABLE 1

Conceptual scheme of the components of soil enzyme activities

	Enzymatic activity in soil							
	Abiontic enzymes							IN ORGANISMS
	Accumulated enzymes					Continuously released extracellular enzymes		Endocellular enzymes of proliferating micro-organisms, plant roots, soil fauna
	Bound to microbial cellular components		Not associated with cellular components			From micro-organisms	From plant roots	
	In intact dead cells	In cellular fragments	Originating from microorganisms and soil fauna		Originating from plant roots			
			Endocellular enzymes from disrupted cells	Extra-cellular enzymes				
Origin								
Location in soil	IN NON-PROLIFER-ATING CELLS		BOUND TO SOIL COMPONENTS		In liquid phase			IN ORGANISMS

Major components found experimentally in soils are capitalized.

associated with the dead cell fragments, collectively called abiontic soil enzymes. Soil enzymology studies also involve intracellular enzymes, for example, dehydrogenase and the study of co-factors such as ATP. The examination of adenylate energy change should become an important component in soil biochemical studies. The various apparent states of enzymes in soils are depicted in Table 1. It is evident that for a better understanding of the total enzymatic picture in the soil, each component should be studied individually, its source determined, its state specified and the importance in terms of the total soil environmental system described. Much of the overall biological activity in soils and its contribution to fertility and other applied aspects remains undefined and valid interpretations are still difficult to make. There is a dire need for more advanced methodologies, interpretations and theoretical background if we are to understand fully the total soil enzymatic complex. It is evident, however, that in many cases the interpretations of physiological activities in soil that have ostensibly been based solely on microbiological activities might include components of the accumulated abiontic enzyme contributions (e.g. phosphatase and urease). The current state of studies on extracellular, abiontic soil enzymes shows that this field of inquiry is on a sure scientific footing. As it approaches maturity, it will certainly provide us with a more thorough understanding of the total biological and biochemical processes of soils.

References

ALEXANDER M. (1961). "Introduction to Soil Microbiology". John Wiley and Sons, New York.

AMBROŽ Z. (1963). Several factors influencing protein decomposition in soil. *Rostl. Výroba* 36, 891–902.

AMBROŽ. V. (1965). The proteolytic complex decomposing proteins in soil. *Rostl. Výroba* 38, 161–170.

AMBROŽ Z. (1966). Some notes on the determination of activities of certain proteases in the soil. *Sb. vys. Šk. zeměd. Brně* 14, 57–62.

AMBROŽ Z. (1973). Study of the cellulase complex in soils. *Rostl. Výroba* 19, 207–212.

ANDERSON J. R. (1962). Urease activity, ammonia volatilization and related microbiological aspects in some South African soils. *Proc. 36th Congr. S. Afr. Sug. Tech. Ass.* 97–105.

ANTONIANI C., MONTANARI T. and CAMORIANO A. (1954). Soil enzymology I. Cathepsin-like activity. A preliminary note. *Ann. Fac. Agr. Univ. Milano* 3, 99–101.

BAEYENS J. and LIVENS J. (1936). Catalytic power of a soil and fertility. *Agricoltura* 30, 145–155.

BALICKA N. and SOCHACKA Z. (1959). Biological activity in light soils. *Zesz. Problem. Postep. Nauk rol.* 21, 257–265.

BALKS R. (1925). Research on the formation and degradation of humus in soil. *Landw. Versuchs-Stationen* **103**, 221–258.

BARASIO L. (1927). The catalase of Vercellese soils. *Giorn. Risicoltura* **17**, 12–15.

BAROCCIO A. (1957). Catalase activity as a soil-biological index of fertility. *Ann. Staz. chim.-agr. Roma. Ser.* **3** (134), 14.

BAROCCIO A. (1958). The catalase activity of soils as "bio-pedological index" of fertility. *Agrochimica* **2**, 243–257.

BARTHA R. and BORDELEAU L. (1969). Cell-free peroxidases in soil. *Soil Biol. Biochem.* **1**, 139–143.

BAUZON D., VAN DEN DRIESSCHE R. and DOMMERGUES Y. (1968). Respirometric and enzymatic characterization of surface horizons of forest soils. *Sci. Sol.* **2**, 55–78.

BAUZON D., VAN DEN DRIESSCHE R. and DOMMERGUES Y. (1969). Litter effect. I. The *in situ* influence of forest litters on several biological characteristics of soils. *Oecol. Plant.* **4**, 99–122.

BECK T. (1971). The determination of catalase activity in soils. *Z. Pflanzenernähr. Düng. Bodenkd.* **130**, 68–81.

BECK T. and POSCHENRIEDER H. (1963). Experiments on the effect of toluene on the soil microflora. *Pl. Soil* **18**, 346–357.

BENEFIELD C. B. (1971). A rapid method for measuring cellulase activity in soils. *Soil Biol. Biochem.* **3**, 325–329.

BERNHARD K. (1965). The dependence of microbial cellulase activity in the near rhizosphere and rhizoplane on species of cultivated and wild plants. *Zbl. Bakteriol. Parasitenkd., Abt. II*, **119**, 566–569.

BOSE B., STEVENSON I. L. and KATZNELSON H. (1959). Further observations on the metabolic activity of the soil microflora. *Proc. Nat. Inst. Sci. India, Biol. Sci.* **25**, 223–229.

BRAMS W. H. and MCLAREN A. D. (1974). Phosphatase reactions in columns of soil. *Soil Biol. Biochem.* **6**, 183–189.

BRIGGS M. H. and SEGAL L. (1963). Preparation and properties of a free soil enzyme. *Life Sci.* **2**, 69–72.

BUDDIN W. (1914). Partial sterilization of soil by volatile and non-volatile antiseptics. *J. Agr. Sci.* **6**, 417–451.

BURANGULOVA M. N. and KHAZIEV F. KH. (1965a). The nuclease activity of soils. *Nauch. Dokl. Vyssh. Shk. Biol. Nauki* (3), 198–201.

BURANGULOVA M. N. and KHAZIEV F. KH. (1965b). The phosphatase activity of the soil as affected by chemical fertilizers. *Agrokém. Talajtan* **14**, 101–110.

BURNS R. G., EL-SAYED M. H. and MCLAREN A. D. (1972a). Extraction of an urease-active organo-complex from soil. *Soil Biol. Biochem.* **4**, 107–108.

BURNS R. G., PUḴÏTE A. H. and MCLAREN A. D. (1972b). Concerning the location and persistence of soil urease. *Soil Sci. Soc. Am. Proc.* **36**, 308–311.

CAMERON F. K. and BELL J. M. (1905). The mineral constituents of the soil solution. *USDA, Bureau of Soils, Bulletin 30*, Washington, D.C.

CASIDA L. E. Jr., KLEIN D. A. and SANTORO T. (1964). Soil dehydrogenase activity. *Soil Sci.* **98**, 371–376.

CAWSE P. A. (1968). Effects of gamma radiation on accumulation of mineral nitrogen in fresh soils. *J. Sci. Fd Agric.* **19**, 395–398.

CAWSE P. A. and CORNFIELD A. H. (1969). The reduction of ^{15}N-labelled nitrate to nitrite by fresh soils following treatment with gamma radiation. *Soil Biol. Biochem.* **1**, 267–274.

CERVELLI S., NANNIPIERI P., CECCANTI B. and SEQUI P. (1973). Michaelis constant of soil acid phosphatase. *Soil Biol. Biochem.* **5**, 841–845.

CHALVIGNAC M. A. (1968). Evidence and preliminary study of a soil enzyme system degrading tryptophan in the absence of proliferating bacteria. *C. R. Acad. Sci. Ser. D* **266**, 637–639.

CHALVIGNAC M. A. (1971). Stability and activity of an enzyme system degrading tryptophan in diverse soil types. *Soil. Biol. Biochem.* **3**, 1–7.

CHALVIGNAC M. A. and MAYAUDON J. (1971). Extraction and study of soil enzymes metabolizing tryptophan. *Pl. Soil* **34**, 25–31.

CHANG H. W. and CHENG H. Y. (1962). Biochemical activity of soils giving high soybean yield. *Acta Pedol. Sin.* **10**, 1–12.

CHOUCHACK D. (1924). L'analyse du sol par les bactéries. *C.R.Acad Sci.* **178**, 2001–2002.

CLAUS D. and MECHSNER K. (1960). The usefulness of E. Hofmann's methods for the determination of enzymes in soil. *Pl. Soil* **12**, 195–198.

COLLET G. F. (1975). Exsudations racinaires d'enzymes. *Bull. Soc. Bot. Fr.* **122**, 61–75.

CONN H. W. (1901). "Agricultural Bacteriology". Blakiston's Son and Co. Philadelphia, Pennsylvania.

CONRAD J. P. (1940a). Hydrolysis of urea in soils by thermolabile catalysis. *Soil Sci.* **49**, 253–263.

CONRAD J. P. (1940b). The nature of the catalyst causing the hydrolysis of urea in soils. *Soil Sci.* **50**, 119–134.

CONRAD J. P. (1941). The catalytic activity causing the hydrolysis of urea in soils as influenced by several agronomic factors. *Soil Sci. Soc. Am. Proc.* **5**, 238–241.

CONRAD J. P. (1942a). The occurrence and origin of urease-like activities in soils. *Soil Sci.* **54**, 367–380.

CONRAD J. P. (1942b). Enzymatic versus microbial concepts of urea hydrolysis in soils. *J. Am. Soc. Agron.* **34**, 1102–1113.

CONRAD J. P. (1944). Some effects of developing alkalinities and other factors upon urease-like activities in soils. *Soil Sci. Soc. Am. Proc.* **8**, 171–174.

CONRAD J. P. and ADAMS C. N. (1940). Retention by soils of the nitrogen of urea and some related phenomena. *J. Am. Soc. Agron.* **32**, 48–54.

DARAGAN-SUSHCHEVA A. YU. and KATSNELSON R. S. (1963). Effect of meadow grasses on enzyme activity of soils. *Trudy Bot. Inst. Akad. Nauk SSSR. Ser.* **3** (14), 160–171.

DAVTYAN G. S. (1958). The principles of the agrochemical characterization of soils. *Pochvovedenie* **5**, 83–87.

DOMMERGUES Y. (1960). Effect of infra-red and solar radiations on the inorganic nitrogen content and on some biological characteristics of soils. *Agron. Trop.* (Nogent-sur-Marne) **15**, 381–389.

DRĂGAN-BULARDA M. and KISS S. (1972). Dextranase activity in soil. *Soil Biol. Biochem.* **4**, 413–416.

DROBNÍK J. (1955). The hydrolysis of starch by the enzymatic complex of soils. *Folia Biol.* **1**, 29–40.

DROBNÍK J. (1956). Degradation of asparagine by the soil enzyme complex. *Cesk. Mikrobiol.* **1**, 47.

DROBNÍK J. (1957). Biological transformations of organic substances in the soil. *Pochvovedenie* (12), 62–71.

DROBNÍK J. (1961). On the role of toluene in the measurement of the activity of soil enzymes. *Pl. Soil* **14**, 94–95.

DROBNÍK J. and SEIFERT J. (1955). The relationship of enzymatic inversion in soil to some soil-microbiological tests. *Folia Biol.* **1**, 41–47.

DROZDOWICZ A. (1971). The behaviour of cellulase in soil. *Rev. Microbiol.* **2**, 17–23.

DUNN C. G., CAMPBELL W. L., FRAM H. and HUTCHINS A. (1948). Biological and photo-chemical effects of high energy, electrostatically produced Roentgen rays and cathode rays. *J. Appl. Phys.* **19**, 605–616.

DURAND G. (1961). Sur la Dégradation Aérobic de l'Acide Urique par le Sol. Thèse doct., Université de Toulouse.

DURAND G. (1963). Degradation of the purine and pyrimidine bases in the soil: sorption on clays. A study in the function of pH and concentration. *C. R. Acad. Sci.* **256**, 4126–4129.

DURAND G. (1965). Les enzymes dans le sol. *Rev. Écol. Biol. Sol.* **2**, 141–205.

ERNST D. (1967). Transformation of cyanamide in arable soils. *Z. Pflanzenernähr. Düng. Bodenkd.* **116**, 34–44.

ERNST D. and GLUBRECHT H. (1966). Investigations on sterilization of cultivable soils. In *Biophysikalische Probleme der Strahlenwirkung* (E. Muth Ed.) ff. 163–167, Thieme, Stuttgart.

ESTERMANN E. F. and MCLAREN A. D. (1961). Contribution of rhizoplane organisms to the total capacity of plants to utilize organic nutrients. *Pl. Soil* **15**, 243–260.

FERMI C. (1910). Sur la presence des enzymes dans le sol, dans les eaux et dans les poussieres. *Zentralbl. Bakteriol. Parasitenkd., Abt.* II, **26**, 330–335.

GALETTI A. C. (1932). A rapid and practical test for the oxidizing power of soil. *Ann. Chim. Appl.* **22**, 81–83.

GALSTYAN A. SH. (1958a). Fermentative activity of some Armenian soils. IV. Urease activity in soil. *Dokl. Akad. Nauk Arm. SSR.* **26**, 29–32.

GALSTYAN A. SH. (1958b). Determination of the comparative activity of peroxidase and polyphenoloxidase in soil. *Dokl. Akad. Nauk Arm. SSR* **26**, 285–288.

GALSTYAN A. SH. (1959). Some questions of the study of soil enzymes. *Soobshch. Lab. Agrokhim. Akad. Nauk Arm. SSR* (2), 19–25.

GALSTYAN A. SH. (1960). Enzyme activities in solonchaks. *Dokl. Akad. Nauk Arm. SSR* **30**, 61–64.

GALSTYAN A. SH. (1962). Method of determination of dehydrogenase activity in soil. *Dokl. Akad. Nauk Arm. SSR* **35**, 181–183.

GALSTYAN A. SH. (1965). Method of determining activity of hydrolytic enzymes of soil. *Pochvovedenie* (2), 68–74.

GALSTYAN A. SH. (1966). Determination of the activity of sulphate reductase of soil. *Dokl. Akad. Nauk SSSR* **166**, 959–960.

GALSTYAN A. SH. (1973). Formation of easily hydrolysable nitrogen in the soil. *Biol. Zh. Arm.* **26**, 15–19.

GALSTYAN A. SH. (1974). "Enzymatic Activity of Armenian Soils". Ayastan, Erevan.

GALSTYAN A. SH. and ARUTYUNYAN E. A. (1968). A method for the determination of sulfide oxidase in soils. *Dokl. Akad. Nauk Arm. SSR* **47**, 287–289.

GALSTYAN A. SH. and ASTVATSATRYAN B. N. (1958). Biological activity of the stony, semi-arid soils of the foothill zone in Armenia. *Izv. Akad. Nauk Arm. SSR, Ser. Biol. Sel'skkoho, Nauki*, **11**, 89–98.

GALSTYAN A. SH. and AVUNDZHYAN Z. S. (1966). Enzyme activity of soils of the Ararat plain. *Biol. Zh. Arm.* **19** (10), 14–22.

GALSTYAN A. SH. and MARKOSYAN L. V. (1966). Determination of nitrate-reductase activity of soil. *Dokl. Akad. Nauk Arm. SSR* **43**, 147–150.

GALSTYAN A. SH. and SAAKYAN E. G. (1970). Determining the activity of soil nitrite reductase. *Dokl. Akad. Nauk Arm. SSR* **50**, 108–110.

GALSTYAN A. SH. and SAAKYAN E. G. (1971). Determination of the activity of soil hydroxylamine reductase. *Dokl. Akad. Nauk Arm. SSR* **53**, 184–186.

GALSTYAN A. SH., TSYUPA G. P. (1959). Some questions of the study of amidase activity in soil. *Izv. Akad. Nauk Arm. SSR. Biol. Nauki* **12**, No. 10, 83–87.

GALSTYAN A. SH., SARKISYAN, S. A. and BAKHALBASHYAN D. A. (1962). Alterations in biological activity of semi-arid stony soils during their reclamation. *Izv. Akad. Nauk Arm. SSR. Biol. Nauki* **15**, 29–37.

GAVRILOVA A. N., SAVCHENKO N. I. and SHYMKO N. A. (1974). Forms of phosphorus and the phosphatase activity of the principal soil types in the BSSR, *Trans. 10th Int. Congr. Soil Sci., Moscow* **4**, 281–289.

GELLER I. A. and DOBROTVORSKAYA K. M. (1961). Phosphatase activity of soils in sugar beet regions. *Akad. Nauk SSSR, Inst. Mikrobiol. Trudy* **11**, 215–221.

GERRETSEN F. C. (1915). Het oxydeerend vermogen van dem bodem in verband met het uitzuren. *Meddl. Proefsta. Java Suikerind.* **5**, 317–331.

GETTKANDT G. (1956). Colorimetric determination of glucose in soil solutions and its application to Hofmann's enzyme method. *Landw. Forsch.* **9**, 155–158.

GETZIN L. W. and ROSEFIELD I. (1968). Organophosphorus insecticide degradation by heat-labile substances in soil. *J. Agric. Fd Chem.* **16**, 598–601.

GILLIAM J. W. and SAMPLE E. (1968). Hydrolysis of pyrophosphate in soils: pH and biological effects. *Soil Sci.* **106**, 352–357.

GILOT J. C. and DOMMERGUES Y. (1967). Note on the calcareous stony mor of the subalpine station of R.C.P. 40. *Rev. Écol. Biol. Sol.* **4**, 357–383.

GREGERS-HANSEN B. (1964). Decomposition of diethylstilboestrol in soil. *Pl. Soil* **20**, 50–58.

HABÁN L. (1966). The effect of subsoiling on the activity of microbial processes in the soil. *Věd. Pr. Výzk. Úst. Rastl. Výroby Pieštanoch* **4**, 141–152.

HABÁN L. (1967). Effect of ploughing depth and cultivated crops on the soil micro-flora and enzyme activity of the soil. *Věd. Pr. výsk. Úst. Rastl. Výroby Pieštanoch* **5**, 159–169.

HABÁN L. (1968). Effect of soil deepening on some biological properties of degraded chernozem. *Sb. Ref. mezin. Věd. Symp. Běrn* **1966**, 393–396.

HAIG A. D. (1955). *Some Characteristics of Esterase- and Urease-like Activity in the Soil*. Ph. D. dissertation, University of California, Davis.

HALSTEAD R. L. (1964). Phosphatase activity of soils as influenced by lime and other treatments. *Can. J. Soil Sci.* **44**, 137–144.

HIRTE W. (1963). Notes on the determination of dehydrogenase activity in soil. *Zbl. Bakt. Parasitenkd. Abt. II*, **116**, 478–484.

HOFFMANN G. (1959a). Distribution and origin of some enzymes in soil. *Z. Pflanzenernähr. Düng. Bodenkd.* **85**, 97–104.

HOFFMANN G. (1959b). Investigations on the synthetic effects of enzymes in soil. *Z. Pflanzenernähr. Düng. Bodenkd.* **85**, 193–201.

HOFFMANN G. (1963). Synthetic effects of soil enzymes. *Rec. Progr. Microbiol.* **8**, 230–234.

HOFFMANN G. (1968). A photometric method for the determination of the phosphatase activity in soils. *Z. Pflanzenernähr. Düng. Bodenkd.* **118**, 161–172.

HOFFMANN G. and TEICHER K. (1957). The enzyme systems of cultivated soils. VII Proteases II. *Z. Pflanzenernähr. Düng. Bodenkd.* **77**, 243–251.

HOFMANN E. (1955). The enzymes in soil and their role in the biology and fertility of soil. *Z. Acker-u. Pflanzenbau* **100**, 31–35.

HOFMANN E. (1963). The origin and importance of enzymes in soil. *Rec. Progr. Microbiol.* **8**, 216–220.

HOFMANN E. and BRÄUNLICH K. (1955). The saccharase content of soils as affected by various factors of soil fertility. *Z. Pflanzenernähr. Düng. Bodenkd.* **70**, 114–123.

HOFMANN E. and HOFFMANN G. (1953). α- and β-glucosidases in the soil. *Naturwissenschaften* **40**, 511.

HOFMANN E. and HOFFMANN G. (1954). Enzyme system of cultivated soils. V. α- and β-galactosidase and α-glucosidase. *Biochem. Z.* **325**, 329–332.

HOFMANN E. and HOFFMANN G. (1955a). The origin, determination and significance of enzymes in soil. *Z. Pflanzenernähr. Düng. Bodenkd.* **70**, 9–16.

HOFMANN E. and HOFFMANN G. (1955b). The enzyme system of cultivated soils. VI. Amylase. *Z. Pflanzenernähr. Düng. Bodenkd.* **70**, 97–104.

HOFMANN E. and HOFFMANN G. (1961). The reliability of E. Hofmann's methods for determining enzyme activity in soil. *Pl. Soil* **14**, 96–99.

HOFMANN E. and HOFFMANN G. (1966). Determination of, soil biological activity with enzymatic methods. *Adv. Enzymol.* **28**, 365–390.

HOFMANN E. and NIGGEMANN J. (1953). The enzyme system of cultivated soils. III. Proteinase. *Biochem. Z.* **324**, 308–310.

HOFMANN E. and SCHMIDT W. (1953). Enzyme system of cultivated soils. II. Urease. *Biochem. Z.* **324**, 125–127.

HOFMANN E. and SEEGERER A. (1950). Soil enzyme as measure of biological activity. *Biochem. Z.* **321**, 97.

HOFMANN E. and SEEGERER A. (1951a). Soil enzymes as factors of fertility. *Naturwissenschaften* **38**, 141–142.

HOFMANN E. and SEEGERER A. (1951b). Enzyme systems of our cultivated soil I. Saccharase. *Biochem. Z.* **322**, 174–179.

HOWARD P. J. A. (1972). Problems in the estimation of biological activity in soil. *Oikos* **23**, 235–240.

HÜBNER G. (1956). Studies on enzymes in soils. *Wiss. Z. K.-Marx Univ. Leipzig* **6**, 425–427.

JACKMAN R. H. (1951). Mineralization of inositol-bound phosphorus in soil. *Iowa St. Coll. J. Sci.* **25**, 260–264.

JACKMAN R. H. and BLACK C. A. (1951). Hydrolysis of iron, aluminium, calcium, and magnesium inositol phosphates by phytase at different pH values. *Soil Sci.* **72**, 261–266.

JACKMAN R. H. and BLACK C. A. (1952a). Phytase activity in soils. *Soil Sci.* **73,** 117–125.

JACKMAN R. H. and BLACK C. A. (1952b). Hydrolysis of phytate phosphorus in soils. *Soil Sci.* **73,** 167–171.

JOHNSON J. L. and TEMPLE K. L. (1964). Some variables affecting the measurement of "catalase activity" in soil. *Soil Sci. Soc. Am. Proc.* **28,** 207–209.

JUO A. S. R. and MADUAKOR H. O. (1973). Hydrolysis and availability of pyrophosphate in tropical soils. *Soil Sci. Soc. Am. Proc.* **37,** 240–242.

KAISER P. and MONZON DE ASCONEGUI M. A. (1971). Measurement of the activity of pectinolytic enzymes in soil. *Biol. Sol* **14,** 16–19.

KAPPEN H. (1913). Die katalytische Kraft des Ackerbodens. *Fühlings Landw. Ztg.* **62,** 377–392.

KATSNELSON R. S. and ERSHOV V. V. (1958). Study of microflora in virgin and cultivated soils of the Karelian A.S.S.R. II. Biological activity. *Mikrobiologiya* **27,** 82–88.

KEILLING M. (1964). Soil biological data. *C. R. Acad. Agric. Fr.* **50,** 1131–1138.

KEILLING J., CAMUS A., SAVIGNAC G., DANCHEZ PH., BOITEL M. and PLANET (1960). Contribution to the study of the biology of soils. *C. R. Acad. Agric. Fr.* **46,** 647–652.

KHAN S. U. (1970). Enzymatic activity in a grey wooded soil as influenced by cropping systems and fertilizers. *Soil Biol. Biochem.* **2,** 137–139.

KHAZIEV F. KH. (1972). Determination of phosphatase activity in soils. *Biol. Sol* **16,** 22–23.

KISS I. (1957). The invertase activity of earthworm casts and soils from ant-hills. *Agrokém. Talajt.* **6,** 65–68.

KISS I. (1958a). Soil enzymes. *In* "Talajtan" (M. J. Csapo, Ed.) pp. 495–622, Agro-Silvica, Bucharest.

KISS S. (1958b). New data regarding the identity of soil saccharase and soil α-glucosidase (maltase). *Stud. Univ. Babeş-Bolyai Sér. Biol.* **3,** 51–55.

KISS S. (1958c). Experiments on the production of saccharase in soil. *Z. Pflanzenernähr. Düng. Bodenkd.* **81,** 117–125.

KISS S. (1961). Presence of levan sucrase in soils. *Naturwissenschaften* **48,** 700.

KISS S. and BOARU M. (1965). Some methodological problems of soil enzymology. *Symp. on Methods in Soil Biology, Bucharest,* 115–127.

KISS S. and DRĂGAN-BULARDA M. (1968a). Studies on soil levansucrase activity. *Ştiinţa Sol.* **6,** 54–59.

KISS S. and DRĂGAN-BULARDA M. (1968b). Levan sucrase activity in soil under conditions unfavorable for the growth of microorganisms. *Rev. Roum. Biol. Sér. Bot.* **13,** 435–438.

KISS S. and DRĂGAN-BULARDA M. (1972). Polysaccharidases in soil. *Contrib. Bot. Univ. Babes-Bolyai (Cluj),* 377–384.

KISS S. and PÉTERFI S. (1960a). Importance of substrates in determining and comparing maltase (α-glucosidase) and lactase (β-glucosidase) activities in soil. *Stud. Univ. Babeş-Bolyai Ser. Biol.* **2,** 275–276.

KISS S. and PÉTERFI S. (1960b). Inhibition of the activity of soil maltase. *Pochvovedenie* (8), 84–86.

KISS S. and PÉTERFI S. (1961). Presence of carbohydrases in the peat of a Salicea community. *Stud. Cercet. Biol.* **12,** 209–216.

40 J. SKUJIŅŠ

KISS A., BOSICA I. and MÉLIUSZ P. (1962a). Effectiveness of toluene as an antiseptic agent in the determination of the activity of enzymes in the soil. *Stud. Univ. Babeş-Bolyai, Sér. Biol.* (2), 65–70.

KISS S., BOSICA I. and POP M. (1962b) Enzymic degradation of lichenin in soil. *Contrib. Bot., Cluj.*, pp. 335–340.

KISS S., BOARU M. and CONSTANTINESCU L. (1965). Levanase activity in soil. *Symp. on Methods in Soil Biology*, Rumanian Soil Science Society, Bucharest, 129–136.

KISS S., DRĂGAN-BULARDA M. and OPREANU I. (1966). Enzyme activity in hopyard soils. *Symp. on Soil Biology, Cluj.*, Rumanian National Society of Soil Scientists, Bucharest, **1966**, 105–112.

KISS S., BOARU M. and STOIŢA M. (1969). Enzyme activity in vineyard soils. *Rev. Roum. Biol. Sér. Bot.* **14**, 127–132.

KISS S., DRĂGAN-BULARDA M. and RĂDULESCU D. (1971). Biological significance of enzymes accumulated in soil. *Contrib. Bot., Cluj.*, 377–397.

KISS S., DRĂGAN-BULARDA M. and KHAZIEV F. H. (1972). The influence of chloromycetin on the activity of some oligases in soil. *Lucr. Conf. Nat. Ştiinţa Solului* **1970**, 451–462.

KISS S., ŞTEFANIC G. and DRĂGAN-BULARDA M. (1974). Soil enzymology in Romania I. *Contrib. Botan. Cluj.*, 207–219.

KISS S., DRĂGAN-BULARDA M. and RĂDULESCU D. (1975a). Biological significance of enzymes in soil. *Adv. Agron.* **27**, 25–87.

KISS S., ŞTEFANIC G. and DRĂGAN-BULARDA M. (1975b). Soil enzymology in Romania II. *Contrib. Bot. Cluj.*, 197–207.

KNUDSON L. (1920). The secretion of invertase by plant roots. *Am. J. Bot.* **7**, 371–379.

KNUDSON L. and SMITH R. S. (1919). Secretion of amylase by plant roots. *Bot. Gaz.* **68**, 460–466.

KOBAYASHI M. (1940). Effects exerted by the continuous use of manures upon the buffer capacity and catalytic action of soil. *J. Sci. Soil, Japan* **14**, 789–796.

KOEPF H. (1954a). Investigations on the biological activity in soil. I. Respiration curves of the soil and enzyme activity under the influence of fertilizing and plant growth. *Z. Acker-und Pflanzenbau* **98**, 289–312.

KOEPF H. (1954b). Experimental study of soil evaluation by biochemical reactions. I. Enzyme reactions and CO_2 evolution in different soils. *Z. Pflanzenernähr. Düng. Bodenkd.* **67**, 262–270.

KOEPF H. (1954c). Experimental study of soil evaluation by biochemical reactions. II. Enzyme reactions and CO_2 evolution in a static fertilizer trial and with three principal cultural systems. *Z. Pflanzenernähr. Düng. Bodenkd.* **67**, 271–277.

KOEPF H. (1956). The dynamics of organic-matter decomposition in different soils. *Z. Pflanzenernähr. Düng. Bodenkd.* **73**, 48–59.

KONG K. T. and DOMMERGUES Y. (1972). Limitation of the cellulolysis in organic soils. II. Study of the enzymes in soil. *Rev. Écol. Biol. Sol* **9**, 629–640.

KONG K. T., BALANDREAU J. and DOMMERGUES Y. (1971). Measurement of the activity of cellulases in organic soils. *Biol. Sol.* **13**, 26–27.

KÖNIG J., HASENBÄUMER J. and COPPENRATH E. (1906). Several new properties of cultivated soils. *Landw. Versuchs-Stationen* **63**, 471–478.

KÖNIG J., HASENBÄUMER J. and COPPENRATH E. (1907). Relationships between the properties of soil and the nutrient uptake by plants. *Landw. Versuchs-Stationen* **66**, 401–461.

KOZLOV K. A. (1964a). Enzymatic activity of the rhizosphere and soils in the East Siberia area. *Folia. Microbiol.* **9**, 145–149.

KOZLOV K. A. (1964b). The fermentative activity of soils as an index of their biological activity. *Contributions of Siberian soil scientists to the 8th Int. Soil Sci. Congress Sib. Otd. Akad. Nauk SSSR*, Novosibirsk, pp. 96–106.

KOZLOV K. A. (1965). The role of soil fauna in the enrichment of soil with enzymes. *Pedobiologia* **5**, 140–145.

KOZLOV K. A. and MIKHAILOVA E. N. (1965). Dehydrogenase activity of some soils of eastern Siberia. *Pochvovedenie* (2), 58–63.

KRÁMER M. and ERDEI S. (1958). Investigation of the phosphatase activity of soils by means of disodium phenylphosphate. I. Method. *Agrokém. Talajt.* **7**, 361–366.

KRÁMER M. and ERDEI S. (1959). The application of the method of phosphatase activity determination in agricultural chemistry. *Pochvovedenie* (9), 99–102.

KROLL L. (1953). A biochemical method for determining the biological activity of soils. *Agrokém. Talajt.* **2**, 301–306.

KROLL L., KRÁMER M. and LÖRINCZ E. (1955). The application of enzyme analysis with phenylphosphate to soils and fertilizers. *Agrokém. Talajt.* **4**, 173–182.

KUDRYACHEV A. I. (1968). Effect of peat on the enzymatic activity of light soils. *Vestsi Akad. Navuk BSSR, Ser. Biyal. Navuk* No. 1, 49–52.

KUNZE C. (1970). The effect of streptomycin and aromatic carboxylic acids on the catalase activity in soil samples. *Zentralbl. Bakteriol. Parasitenkd. Abt. II*, **124**, 658–661.

KUNZE C. (1971). Catalase activity in soil samples as influenced by tannin, gallic acid and p-hydroxybenzoic acid. *Oecol. Plant.* **6**, 197–202.

KUNZE C., DÜRRSCHMIDT M. and LOTZ G. (1975). Seasonal changes in soil enzyme activity in two different forest associations. *Angew. Bot.* **49**, 229–235.

KUPREVICH V. F. (1949). Extracellular enzymes of roots of autotrophic higher plants. *Dokl. Akad. Nauk SSSR* **78**, 953–956.

KUPREVICH V. F. (1951). The biological activity of soil and methods for its determination. *Dokl. Akad. Nauk SSSR* **79**, 863–866.

KUPREVICH V. F. (1958). Problems of soil enzymology. *Vestn. Akad. Nauk. SSSR* (4), 52–57.

KUPREVICH V. F. and SHCHERBAKOVA T. A. (1956). Determination of the invertase and catalase activity of soils. *Vestsi Akad. Navuk Belarusk SSR, Ser. Biyal.* **2**, 115–116.

KUPREVICH V. F. and SHCHERBAKOVA T. A. (1966). *Pochvennaya Enzymologiya*. Nauk Tekh., Minsk (transl. from Russian: "Soil Enzymes", US Department of Commerce, National Technical Information Service, Springfield, Virginia).

KUPREVICH V. F. and SHCHERBAKOVA T. A. (1971). Comparative enzymatic activity in diverse types of soil. *In* "Soil Biochemistry", Vol. 2 (A. D. McLaren and J. Skujiņš, Eds.) pp. 167–201, Marcel Dekker, New York.

KUPREVICH V. F., TSYUPA G. P. and SHCHERBAKOVA T. A. (1964). Synthesis of glutamic acid in soil. *Dokl. Akad. Nauk. BSSR* **8**, 745–746.

KUPREVICH V. F., SHCHERBAKOVA T. A. and TSYUPA G. P. (1968). Synthesis of some amino acids in soil. *In* "Isotopes and Radiation in Soil Organic Matter Studies" pp. 227–230. International Atomic Energy Agency, Vienna.

42 J. SKUJIŅŠ

KURTYAKOV N. I. (1931). Characterization of the catalytic power of soil. *Pochvovedenie* (3), 34–48.

KVASNIKOV V. V. and PEROV N. N. (1965). Biochemical processes in leached chernozem in relation to tillage. *Dokl. vses. Akad. sel'.-khoz. Nauk* (10), 1–4.

LADD J. N. (1972). Properties of proteolytic enzymes extracted from soil. *Soil Biol. Biochem.* **4**, 227–237.

LADD J. N. and BUTLER J. H. A. (1972). Short-term assays of soil proteolytic enzyme activities using proteins and dipeptide derivatives as substrates. *Soil Biol. Biochem.* **4**, 19–30.

LADD J. N. and BUTLER J. H. A. (1975). Humus–enzyme systems and synthetic organic polymer–enzyme analogs. *In* "Soil Biochemistry", Vol. 5, (A. D. McLaren and E. A. Paul, Eds), pp. 143–194, Marcel Dekker Inc., New York.

LADD J. N. and PAUL E. A. (1973). Changes in enzymatic activity and distribution of acid-soluble, amino acid-nitrogen in soil during nitrogen immobilization and mineralization. *Soil Biol. Biochem.* **5**, 825–840.

LAJUDIE J. and POCHON J. (1956). Studies on the proteolytic activity of soils. *Trans. VI Int. Soil Sci. Congr.*, C, 271–273.

LATYPOVA R. M. (1960). Comparative activity of catalase and saccharase in peat and some mineral soils. *Dokl. Akad. Nauk BSSR* **4**, 357–359.

LATYPOVA R. M. (1961). Protease activity of derno-podzolic soils and effect on it of peat organic matter. *Dokl. Akad. Nauk BSSR* **5**, 582–584.

LAUGESEN K. and MIKKELSEN J. P. (1973). Dehydrogenase activity in Danish soils. *Tidsskr. Pl.* **77**, 516–520.

LENHARD G. (1956). The dehydrogenase activity in soil as a measure of the activity of soil microorganisms. *Z. Pflanzenernähr. Düng. Bodenkd.* **73**, 1–11.

LENHARD G. (1957). Dehydrogenase activity of soil as a measure of the quantity of microbially decomposable humic matter. *Z. Pflanzenernähr. Düng. Bodenkd.* **77**, 193–198.

LIBBERT E. and PAETOW W. (1962). Investigations of the enzymatic hydrolysis of indole-3-acetonitoile and of the oxidation of indole-3-aldehyde in the soil and in unboiled milk. *Flora, Jena* **152**, 540-544.

LUPINOVICH I. S. and YAKUSHAVA V. I. (1965). Change of the fermentative activity of dernogley soils in connection with their cultivation. *Vestsi Akad. Navuk BSSR, Ser. s.-gasp. Navuk* **3**, 23–25.

LYON T. L., FIPPIN E. O. and BUCKMAN H. O. (1917). "Soils—Their Properties and Management". MacMillan Co., New York.

MALISZEWSKA W. (1969). Comparison of the biological activity of different soil types. *Agrokém. Talajt.* **18**, 76–81.

MÁRKUS L. (1955). Determination of carbohydrates from plant materials with anthrone reagent. II. Assay of activity in soil and farmyard manure. *Agrokém. Talajt.* **4**, 207–216.

MARTIN-SMITH M. (1963). Uricolytic enzymes in soil. *Nature* **197**, 361–362.

MATSUNO T. and ICHIKAVA C. (1934). Catalytic action of Japanese soils. *Res. Bul. Gifu Imp. Coll. Agr.* No. 37, p. 22.

MAY D. W. and GILE P. L. (1909). The catalase of soils. *Puerto Rico Agr. Exp. Sta. Circular* No. 9, 3–13.

MAYAUDON J. and SARKAR J. M. (1974a). Study of diphenol oxidases extracted from a forest litter. *Soil Biol. Biochem.* **6**, 269–274.

MAYAUDON J. and SARKAR J. M. (1974b). Chromatography and purification of the diphenol oxydases of soil. *Soil Biol. Biochem.* **6**, 275–285.

MAYAUDON J. and SARKAR J. M. (1975). *Polyporus versicolor* laccases in the soil and the litter. *Soil Biol. Biochem.* **7**, 31–34.

MAYAUDON J., EL HALFAWI M. and BELLINCK C. (1973a). Decarboxylation of aromatic 1-^{14}C amino acids by soil extracts. *Soil Biol. Biochem.* **5**, 355–367.

MAYAUDON J., EL HALFAWI M. and CHALVIGNAC M. (1973b). Properties of diphenol oxidases extracted from soils. *Soil Biol. Biochem.* **5**, 369–383.

MAYAUDON J., BATISTIC L. and SARKAR J. M. (1975). Properties of proteolytically active extracts from fresh soils. *Soil Biol. Biochem.* **7**, 281–286.

MCGARITY J. W. and MYERS G. (1967). A survey of urease activity in soils of northern New South Wales. *Pl. Soil* **27**, 217–338.

MCLAREN A. D. (1960). Enzyme action in structurally restricted systems. *Enzymologia* **21**, 356–364.

MCLAREN A. D. (1969). Radiation as a technique in soil biology and biochemistry. *Soil Biol. Biochem.* **1**, 63–73.

MCLAREN A. D. (1975). Soil as a system of humus and clay immobilized enzymes. *Chemica Scripta* **8**, 97–99.

MCLAREN A. D. and BABCOCK K. L. (1959). Some characteristics of enzyme reactions at subsurfaces. *In* "Subcellular Particles" (T. Hayashi, Ed.) Ronald Press, New York.

MCLAREN A. D. and PACKER L. (1970). Some aspects of enzyme reactions in heterogeneous systems. *Adv. Enzymol.* **33**, 245–308.

MCLAREN A. D., RESHETKO L. and HUBER W. (1957). Sterilization of soil by irradiation with an electron beam, and some observations on soil enzyme activity. *Soil Sci.* **83**, 497–502.

MCLAREN A. D., LUSE R. A. and SKUJIŅŠ J. J. (1962). Sterilization of soil by irradiation and some further observations on soil enzyme activity. *Soil Sci. Soc. Am. Proc.* **26**, 371–377.

MCLAREN A. D., PUĶĪTE A. H. and BARSHAD I. (1975). Isolation of humus with enzymatic activity from soil. *Soil Sci.* **119**, 178–180.

MESSINEVA M. A. (1940). Enzymes in fresh water deposits. *Byul. Mosk. O-va Ispytatelyei prirodi* **49**, issue No. 5–6.

MESSINEVA M. A. (1941). Enzymatic activity in the deposits of Lake Zaluchye. *Tr. Labor. Genesisa Sapropelya* (2), 61–71.

MONZON DE ASCONEGUI M. A. and KAISER P. (1972). The utilization in soil of pectin breakdown products by *Azotobacter chroococcum*. *Ann. Inst. Pasteur, Paris* **122**, 1009–1028.

MOORE A. W. and RUSSELL J. S. (1972). Factors affecting dehydrogenase activity as an index of soil fertility. *Pl. Soil* **37**, 675–682.

MORTLAND M. M. and GIESEKING J. E. (1952). The influence of clay minerals on the enzymatic hydrolysis of organic phosphorus compounds. *Soil Sci. Soc. Am. Proc.* **16**, 10–13.

MOURARET M. (1965). "Contribution al'Etude de l'Activité des Enzymes du Sol: l'Asparaginase". ORSTOM, Paris.

MOUREAUX C. (1957). Biochemical tests on some Madagascaran soils. *Mem. Inst. Sci. Madagascar* **8**, 225–241.

MURESANU P. L. and GOIAN M. (1969). Investigations on phosphomonoesterase activity in soils treated with fertilizers and lime. *Agrokém. Talajt.* **18,** 102–106.

NEAL J. L., Jr. (1973). Influence of selected grasses and forbs on soil phosphatase activity. *Can. J. Soil Sci.* **53,** 119–121.

NILSSON P. E. (1957). Influence of crop on biological activities in soil. *K. Lantbrukshögsk. Annl.* **23,** 175–218.

NOVAK V. and CRHA B. (1940). The catalytic power of the iron-manganese concretions of the soil. *Sbornik ČAZ* **14,** 310–314.

NOVOGRUDSKAYA E. D. (1963). Accumulation of enzymes during ripening of soil-manure composts. *Agrobiologiya* (6), 880–885.

NOWAK W. (1964). Comparative studies on the bacterial population and saccharase activity of the soil. *Pl. Soil* **20,** 302–318.

OSUGI S. (1922). The catalytic action of soil. *Ber. Ohara Inst. Lhndw. Forsch. Japan* **2,** 197–210.

PANCHOLY S. K. and LYND J. Q. (1971). Microbial esterase detection with ultraviolet fluorescence. *Appl. Microbiol.* **22,** 939–941.

PANCHOLY S. K. and LYND J. Q. (1973). Interactions with soil lipase activation and inhibition. *Soil Sci. Soc. Am. Proc.* **37,** 51–52.

PANCHOLY S. K. and RICE E. L. (1973a). Soil enzymes in relation to old field succession: amylase, cellulase, invertase, dehydrogenase, and urease. *Soil Sci. Soc. Am. Proc.* **37,** 47–50.

PANCHOLY S. K. and RICE E. L. (1973b). Carbohydrases in soil as affected by successional stages of revegetation. *Soil Sci. Soc. Am. Proc.* **37,** 227–229.

PAULSON K. N. and KURTZ L. T. (1969). Locus of urease activity in soil. *Soil Sci. Soc. Am. Proc.* **33,** 897–901.

PAULSON K. N. and KURTZ L. T. (1970). Michaelis constant of soil urease. *Soil Sci. Soc. Am. Proc.* **34,** 70–72.

PENKAVA J. (1913). *Zemedelsky Arch.* **4,** 1–12. Cited in Kiss, 1958a.

PENKAVA J. (1913). *Zemedelsky Arch.* **4,** 99–106. Cited in Kiss, 1958a.

PETERSON G. H. (1962) Respiration of soil sterilized by ionizing radiations. *Soil Sci.* **94,** 71–74.

PETERSON N. V. (1961), Sources of enrichment of soil with enzymes. *Mikrobiol. Zh.* **23,** 5–11.

PETERSON N. V. and ASTAFYEVA E. V. (1962). Determination of saccharase activity in soil. *Mikrobiologiya* **31,** 918–922.

PFEIL E. (1932). The influence of the forms of manganese and iron on the catalytic power of healthy and "acid-sick" soils. *Z. Pflanzenernähr. Düng. Bodenkd. Teil A.* **23,** 129–139.

POKORNÁ V. (1964). Method for determining lipolytic activity upland and lowland peats and mucks. *Pochvovedenie* (1), 106–109.

PORTER L. K. (1965). Enzymes. *In* "Methods of Soil Analysis". Part 2 (C. A. Black, D. D. Evans, J. L. White, L. E. Ensminger and F. E. Clark, Eds), pp. 1536–1549. American Society of Agronomy, Madison, Wisconsin.

QUASTEL J. H. (1946). "Soil Metabolism". The Royal Institute of Chemistry of Great Britain and Ireland, London.

RADU I. F. (1931). The catalytic power of soils. *Landw. Versuchs Stat.* **112,** 45–54.

RAGUOTIS A. D. (1967). Biological activity of sod-podzolic forest soils of the Lithuanian SSR. *Pochvovedenie* (6), 51–56.

RAMÍREZ-MARTÍNEZ J. R. (1968). Organic phosphorus mineralization and phosphatase activity in soils. *Folia Microbiol.* **13**, 161–174.

RAMÍREZ-MARTÍNEZ J. R. and MCLAREN A. D. (1966a). Determination of soil phosphatase activity by a fluorimetric technique. *Enzymologia* **30**, 243–253.

RAMÍREZ-MARTÍNEZ J. R. and MCLAREN A. D. (1966b). Some factors influencing the determination of phosphatase activity in native soils and in soils sterilized by irradiation. *Enzymologia* **31**, 23–38.

RANKOV V. and DIMITROV G. (1971). Soil phosphatase activity as influenced by mineral and organic fertilization of the tomato and cabbage. *Pochvozn. Agrokhim.* **6**, 93–98.

RAWALD W., DOMKE K. and STOHR G. (1968). Studies on the relations between humus quality and microflora of the soil. *Pedobiologia* **7**, 375–380.

REESE E. T. (1968). Microbial transformation of soil polysaccharides. *In* "Study Week on Organic Matter and Soil Fertility". Wiley and Sons, Ltd., Interscience, New York, pp. 535–582.

RIPPEL A. (1931). Bakteriologisch-chemische Methoden der Fruchtbarkeitsbestimmung. *In* "Handbuch der Bodenlehre" (E. Blank, Ed.) pp. 670–671. Verlag Springer, Berlin.

ROBERGE M. R. (1968). Effects of toluene on microflora and hydrolysis of urea in a black spruce humus. *Can. J. Microbiol.* **14**, 999–1003.

ROGERS H. T. (1942a). Dephosphorylation of organic phosphorus compounds by soil catalysts. *Soil Sci.* **54**, 439–446.

ROGERS H. T. (1942b). The availability of certain forms of organic phosphorus to plants and their dephosphorylation by exoenzyme systems of growing roots and by soil catalysts. *Iowa State Coll. J. Sci.* **17**, 108–110.

ROGERS H. T., PEARSON R. W. and PIERRE W. H. (1941). Absorption of organic phosphorus by corn and tomato plants and the mineralizing action of exoenzyme systems of growing roots. *Soil Sci. Soc. Am. Proc.* **5**, 285–291.

ROGERS H. T., PEARSON R. W. and PIERRE W. H. (1942). The source and phosphatase activity of exoenzyme systems of corn and tomato roots. *Soil Sci.* **54**, 353–366.

ROSS D. J. (1965). A seasonal study of oxygen uptake of some pasture soils and activities of enzymes hydrolysing sucrose and starch. *J. Soil Sci.* **16**, 73–85.

ROSS D. J. (1966). A survey of activities of enzymes hydrolysing sucrose and starch in soils under pastures. *J. Soil Sci.* **17**, 1–15.

ROSS D. J. (1968a). Activities of enzymes hydrolysing sucrose and starch in some grassland soils. *Trans. 9th Int. Congr. Soil Sci., Adelaide* **3**, 299–308.

ROSS D. J. (1968b). Some observations on the oxidation of glucose by enzymes in soil in the presence of toluene. *Pl. Soil* **28**, 1–11.

ROSS D. J. (1974). Glucose oxidase activity in soil and its possible interference in assays of cellulase activity. *Soil Biol. Biochem.* **6**, 303–306.

ROSS D. J. and ROBERTS H. S. (1968). A study of activities of enzyme hydrolysing sucrose and starch and of oxygen uptake in a sequence of soils under tussock grassland. *J. Soil Sci.* **19**, 186–196.

ROSS D. J. and ROBERTS H. S. (1970). Enzyme activities and oxygen uptakes of soils under pasture in temperature and rainfall sequences. *J. Soil Sci.* **21**, 368–381.

ROTH G. (1965). The biochemical activity of soils; a new method of assessment. *In* *Proc. 39th Congr. S. Afr. Sug. Technol. Ass.* 276–285.

ROTINI O. T. (1931). Sopra il potere catalasico del terreno. *Ann. Labor Ric. Ferm.* *"Spallanzani"* 2, 333–351.

ROTINI O. T. (1933). La presenza e l'attività delle pirofosfatasi in alcuni substrati organici e nel terreno. *Atti Soc. Ital. Progr. Sci., 21. Riun. Rome, 1932* 2, 1–11.

ROTINI O. T. (1935a). Enzymatic transformation of urea in soil. *Ann. Labor. Ferment. L. "Spallanzani", Milan* 3, 173–184.

ROTINI O. T. (1935b). The catalytic transformation of cyanamide into urea. *Chim. Ind.* 17, 14–20.

ROTINI O. T. (1951). Relationships of fertilizers with cultivated soils and their transformations. *Ann. Fac. Agr. Univ. Pisa* 12, 159–178.

ROTINI O. T. (1956). Urea, biuret and cyanic acid in the fertilizing of agricultural crops. *Ann. Fac. Agr. Univ. Pisa* 17, 1–25.

ROTINI O. T. (1963). The effect of urea fertilizers on plants. *Agrokém. Talajt.* 12, 501–516.

ROTINI O. T. and CARLONI L. (1953). The transformation of metaphosphates into orthophosphates promoted by agricultural soil. *Ann. Sper. Agrar., Pisa* 7, 1789–1799.

ROTINI O. T. and GALOPPINI C. (1967). Effect of synthetic detergents on soil and crops. *In* "Chemistry, Physics and Application of Surface Active Substances" (C. Paquot, Ed.), Vol. 3, pp. 451–460. Gordon and Breach, London.

ROTINI O. T. and SESSA F. (1943). The ureolytic action of cultivated soils. *Chim. Ind. Milano* 25, 3–6.

RUSSELL E. J. (1921). *Soil Conditions and Plant Growth.* 4th ed. Longmans, Green and Co., London.

RUSSEL S. and KOBUS J. (1974). Dehydrogenase activity of different soils. *Agrártud. Közl.* 33, 161–168.

RYŠAVÝ P. and MACURA J. (1972a). The assay of β-galactosidase in soil. *Folia Microbiol.* 17, 370–374.

RYŠAVÝ P. and MACURA J. (1972b). The formation of β-galactosidase in soil. *Folia Microbiol.* 17, 375–380.

SARKISYAN S. S. and SHUR-BAGDASARYAN E. F. (1967). Interaction of vegetation and soils of various mountain steppe pastures overgrazed and eroded to different extents. *Pochvovedenie.* (12), 37–44.

SATYANARAYANA T. and GETZIN L. W. (1973). Properties of a stable cell-free esterase from soil. *Biochemistry* 12, 1566–1572.

SCHAEFER R. (1963). Dehydrogenase activity as an index for total biological activity in soils. *Ann. Int. Pasteur, Paris* 105, 326–331.

SCHARRER K. (1927), Zur Kenntniss der Hydroperoxyd spaltenden Eigenschaft der Böden. *Biochem. Z.* 189, 125–149.

SCHARRER K. (1928a). Beiträge zur Kentniss der Wasserstoffperoxyd zersetzenden Eigenschaft des Bodens. *Landw. Versuchs-Stationen* 107, 143–187.

SCHARRER K. (1928b), Katalytische Eigenschaften der Böden. *Z. Pflanzenernähr. Düng. Bodenkd.* 12, 323–329.

SCHARRER K. (1933). Reaction kinetics of the hydrogen peroxide decomposing properties of soils. *Z. Pflanzenernähr. Düng. Bodenkd.* Teil A. 31, 27–36.

SCHARRER K. (1936). Catalytic characteristics of the soil. *Forschungsdient* 1, 824–831.

SCHEFFER F. and TWACHTMANN R. (1953). Erfahrungen mit der Enzym-methode nach Hofmann. *Z. Pflanzenernähr. Düng. Bodenkd.* 62, 158–171.

SCHREINER O. and SULLIVAN M. X. (1910). Studies in soil oxidation. *USDA, Bureau of soils, Bulletin No.* 73, Washington, D.C.

SEETHARAMAN K., RAMABADRAN R. and MAHADEVAN A. (1968). Enzymes in Tamilnad, soils, *Proceedings, First All-India Symposium on Agricultural Microbiology*, University of Agricultural Science, Hebbal, Bangalore, 56–59.

SEQUI P. (1974). Enzymes in soil. *Ital. Agric.* 111, 91–109.

SHCHERBAKOVA T. A., MAKSIMOVA V. P. and GALUSHKO N. A. (1970). Isolation of amylolytic soil enzyme complex. *Dokl. Akad. Nauk. USSR* 14, 661–663.

SHCHERBAKOVA T. A., MAKSIMOVA V. P. and GALUSHKO N. A. (1971). Isolation of an active enzymatic complex from soil and its separation on DEAE-cellulose. *Mikrobiol. Biokhim. Issled. Pochvy, Mater. Nauch. Konf. Kiev* 1969, 108–111.

SKUJIŅŠ J. (1967). Enzymes in soil. *In* "Soil Biochemistry" (A. D. McLaren and G. H. Peterson, Eds) pp. 371–414. Marcel Dekker, New York.

SKUJIŅŠ J. (1973). Dehydrogenase: an indicator of biological activities in arid soils. *Bull. Ecol. Res. Commun.* (Stockholm) 17, 235–241.

SKUJIŅŠ J. (1976). Extracellular enzymes in soil. *CRC Crit. Rev. Microbiol.* 4, 383–421.

SKUJIŅŠ J. J. and McLAREN A. D. (1968). Persistence of enzymatic activities in stored and geologically preserved soils. *Enzymologia* 34, 213–225.

SKUJIŅŠ J. J. and McLAREN A. D. (1969). Assay of urease activity using ^{14}C-urea in stored, geologically preserved, and in irradiated soils. *Soil Biol. Biochem.* 1, 88–99.

SKUJIŅŠ J. J., BRAAL L., and McLAREN A. D. (1962). Characterization of phosphatase in a terrestrial soil sterilized with an electron beam. *Enzymologia* 25, 125–133.

SMOLIK J. (1925). Hydrogen-peroxyd-Katalyse der mährischen Böden. *Mitt. Intern. Bodenkundl. Ges.* 1, 6–20.

SØRENSEN H. (1955). Xylanase in the soil and the rumen. *Nature* 176, 74.

SØRENSEN, (1957). Microbial decomposition of xylan. *Acta Agric. Scand.* Suppl. 1, 1–86.

STENINA T. A. (1968). Enzyme activity of some soils of the central taiga. *Pochvovedenie* (2), 109–113.

STEVENSON I. L. (1959). Dehydrogenase activity in soils. *Can. J. Microbiol.* 5, 229–235.

STOKLASA J. (1924). Methoden zur biochemischen Untersuchung des Bodens. *In* "Handbuch der biologischen Arbeitsmethoden" (E. Abderhalden, Ed) Berlin-Wien, Abt. XI, pp. 1–262. Teil 3, Heft 1.

SUBRAHMANYAN V. (1927). Biochemistry of water-logged soils. II. The presence of deaminase in water-logged soils and its role in the production of ammonia. *J. Agr. Sci.* 17, 449–467.

SULLIVAN M. X. and REID F. R. (1912). Studies in soil catalysis. *USDA, Bureau of Soils, Bulletin No.* 86, Washington, D.C.

SUMNER J.B. (1926). The isolation and crystallization of the enzyme urease. *J. Biol. Chem.* 69, 435–441.

SUTTON C. D., GUNARY D. and LARSEN S. (1966). Pyrophosphate as a source of

48 J. SKUJIŅŠ

phosphorus for plants: II. Hydrolysis and initial uptake by a barley crop. *Soil Sci.* **101**, 199–204.

TABATABAI M. A. (1973). Michaelis constants of urease in soils and soil fractions. *Soil Sci. Soc. Am. Proc.* **37**, 707–710.

TABATABAI M. A. and BREMNER J. M. (1969). Use of *p*-nitrophenyl phosphate for assay of soil phosphatase activity. *Soil Biol. Biochem.* **1**, 301–307.

TABATABAI M. A. and BREMNER J. M. (1970a). Arylsulfatase activity of soils. *Soil Sci. Soc. Am. Proc.* **34**, 225–229.

TABATABAI M. A. and BREMNER J. M. (1970b). Factors affecting soil arylsulfatase activity. *Soil Sci. Soc. Am. Proc.* **34**, 427–429.

TABATABAI M. A. and BREMNER J. M. (1971). Michaelis constants of soil enzymes. *Soil Biol. Biochem.* **3**, 317–323.

TABATABAI M. A. and BREMNER J. M. (1972). Assay of urease activity in soils. *Soil Biol. Biochem.* **4**, 479–487.

THENTE B. (1970). Effects of toluene and high-energy radiation on urease activity in soil. *Lantbrukshögsk. Annlr* **36**, 401–418.

THORNTON J. I. and MCLAREN A. D. (1975). Enzymatic characterization of soil evidence. *J. Forensic Sci.* **20**, 674–692.

TOMESCU E. (1970). Contribution to the methodology for determination of cellulase in soil. *Microbiol., Lucr. Conf. Nat. Microbiol. Gen. Appl., Bucharest,* 1968, Soc. Stţiinţe Biol. R. S. Romania, Bucharest, 509–513.

TURKOVÁ V. and SROGL M. (1960). The relationship between enzymatic and other soil-biological tests in the same habitat. *Rostl. Výroba* **33**, 1431–1438.

TYLER G. (1974). Heavy metal pollution and soil enzymatic activity. *Pl. Soil* **41**, 303–311.

TYLER G. (1976a). Influence of vanadium on soil phosphatase activity. *J. Env. Qual.* **5**, 216–217.

TYLER G. (1976b). Heavy metal pollution, phosphatase activity, and mineralization of organic phosphorus in forest soils. *Soil Biol. Biochem.* **8**, 327–332.

UKHTOMSKAYA F. I. (1952). Role of enzymes in self-purification of soil. *Gigiena i Sanitariya* (11), 46–54.

VÁLY F. (1937). Catalase activity in soils. *Mezograzd. Kutatasok* **10**, 195–203.

VASYUK L. F. (1961). Growth of microflora and enzyme activity along layers and horizons of arable soddy-podzolic and soddy carbonate soils in the Leningrad region. *Byul. Nauchn.-Tekhn. Inform. Selskokhoz. Mikrobiol.* (10), 14–18.

VASYUK L. F. (1964). Activity of enzymes and characteristics of a bio-organo-mineral complex of virgin and cultivated soils. *Trudy Vsesoyuz. Nauch.-issled. Inst. Sel'skohhoz. Mikrobiol.* **1963** (18), 23–38.

VELA G. R., WYSS O. (1962). The effect of gamma radiation on nitrogen transformation in soil. *Bact. Proc.* **62**, 24.

VELASCO J. R. and LEVY L. R. (1949). Fertile agricultural soils. I. The catalases. *Anal. Real Sol Españ. Fis. Quim. Ser. B,* **45**, 821–860.

VERONA P. L. (1960). Effect of seed on soil saccharase activity *Atti Ist. Bot. Lab. critt. Univ. Pavia* **17**, 184–187.

VERSTRAETE W., VOETS J. P. and VAN LANCKER P. (1976). Evaluation of some enzymatic methods to measure the bio-activity of aquatic environments. *Hydrobiol.* **49**, 257–266.

VLASYUK P. A. and LISOVAL A. P. (1964). Effect of fertilizers on phosphatase activity of soil. *Vestn. Sel'skokhoz. Nauki* **7**, 52–55.

VLASYUK P. A., DOBROTVORSKAYA K. M. and GORDIENKO S. A. (1956). Urease activity in the rhizosphere of agricultural crops. *Dokl. Vses. Akadl Skh. Nauk* **8**, 28–31.

VLASYUK P. A., DOBROTVORSKAYA K. M. and GORDIENKO S. A. (1957). The intensity of the activity of enzymes in the rhizosphere of individual crops. *Dokl Akad. Sel'skokhoz.* **3**, 14–19.

VLASYUK P. A., DOBROTVORSKAYA K. M. and GORDIENKO S. A. (1959). Activity of enzymes in the rhizosphere of agricultural plants. *Nauch. Trudy ukrain. Inst. Fiziol. Rast.* No. 20, 12–17.

VOETS J. P. and DEDEKEN M. (1964a). A new method for measuring the activity of proteolytic enzymes in soil. *Naturwissenschaften* **51**, 267–268.

VOETS J. P. and DEDEKEN M. (1964b). Studies on biological phenomena of proteolysis in soil. *Ann. Inst. Pasteur, Paris* **107**, Suppl. 3, 320–329.

VOETS J. P. and DEDEKEN M. (1965). Influence of high-frequency and gamma irradiation on the soil microflora and the soil enzymes. *Meded. Landbouwhogesch. Opozoekingsta. Staat Gent* **30**, 2037–2049.

VOETS J. P. and DEDEKEN M. (1966). Soil enzymes. *Meded. Rijkafac. Landb-Wetter-schappin Gent* **31**, 177–190.

VOETS J. P., DEDEKEN M. and BESSEMS E. (1965). The behaviour of some amino acids in gamma irradiated soils. *Naturwissenschaften* **16**, 476.

VOETS J. P., AGRIANTO G., and VERSTRAETE W. (1975). Étude écologique des activités microbiologiques et enzymatiques des sols dans une foret de feuillus. *Rev. Écol. Biol. Sol.* **12**, 543–555.

WACHTEL P. (1912). *Die Wasserstoffperoxydkatalyse durch Boden.* Dissertation, Jena—cited in Kiss, 1958a.

WAKSMAN S. A. (1927). "Principles of Soil Microbiology". Williams and Wilkins Co., Baltimore, Maryland.

WAKSMAN S. A. (1952). "Soil Microbiology". John Wiley and Sons, New York.

WAKSMAN S. A. and DUBOS R. J. (1926). Microbiological analysis as an index of soil fertility. X. The catalytic power of the soil. *Soil Sci.* **22**, 407–420.

WEETALL H. H., WELIKY N. and VANGO S. P. (1965). Detection of microorganisms in soil by their catalatic activity. *Nature* **206**, 1019–1021.

WOODS A. F. (1899). The destruction of chlorophyll by oxidizing enzymes. *Zentralbl. Bakteriol. Parasitenkd. Abt. 2* **5**, 745–754.

YAROSHEVICH I. V. (1966). Effect of fifty years' application of fertilizers in a rotation on the biological activity of a chernozem. *Agrokhimiya* **6**, 14–19.

2

Origin and Range of Enzymes in Soil

J. N. LADD

Division of Soils, CSIRO, Glen Osmond, South Australia, Australia

I. Introduction

Kiss *et al.* (1972b, 1975), in recent reviews of enzymic activities of soils, have sought to distinguish those activities due to accumulated enzymes from those due to microorganisms proliferating during the assay period. By definition, accumulated enzymes are those "present and active in a soil in which no microbial proliferation takes place". They thus include enzymes which:

1. Function extracellularly, either free in the soil solution or bound to inorganic and organic soil constituents
2. Are present in particulate cell debris
3. Are present in dead cells or in viable but non-proliferating cells.

Cells and cell fragments may be animal, plant or microbial. Reviews by Skujiņš (1967, 1976) have emphasized the desirability of assaying enzymes which only function extracellularly in soils and have focused attention on the states in which extracellular (abiontic) enzymes may exist and possibly may become stabilized in soils. Both concepts of soil enzymes are valid.

Ideally the activity of a soil should be attributable to enzymes of different origins, functioning according to their concentration and catalytic properties as expressed under the conditions of the soil micro-environment in which they are located. In practice however, it has been impossible to measure contributions to a soil's activity by enzymes of different origins, even when sources are categorized broadly into plant, animal and microbial components. Further, assays of abiontic enzymes are restricted to enzymes active against externally added substrates of high molecular weight. No compound is available which can inhibit enzymes selectively and completely within live cells in soils.

Accumulated enzymes are assayed under conditions which minimize or eliminate microbial proliferation, either by using sensitive assays with brief incubation periods, or by using sterilized soils or by treating soils with bacteriostatic compounds e.g. toluene. Even so, results should be interpreted cautiously. Diminution of activity by soil treatments may be due to the desired elimination of microbial growth, observed especially in long-term (1–2 days) assays. However, soil treatments may directly affect activities of accumulated enzymes themselves. For example, activities of soil dehydrogenases, assayed for periods of 1–2 h, are completely lost when soils are incubated with toluene. This effect of toluene may result from changes in the integral structure of live but not necessarily proliferating cells. Also the use of short-term assays has demonstrated that irradiation of soils at dosages minimal for sterilization often decreases enzyme activities, resulting probably from partial denaturation of accumulated enzymes. Irradiation or toluene-treatment of soils may actually increase the activities of some enzymes, e.g. urease, possibly by changing the permeability of cell membranes and facilitating enzyme–substrate interaction. Nevertheless the employment of sterilized or toluene-treated soils is widespread in studies of soil enzymes and it is the wider concept of accumulated enzymes that has been adopted to define the scope of the present review of the origin and range of enzymes in soils.

II. Origin of Enzymes in Soils

It is accepted that enzymes in soils originate from animal, plant and microbial sources. However, the presence in soils of enzymes, derived directly and specifically from animal and plant sources, has yet to be demonstrated conclusively.

A widespread, often tacit assumption, is that soil enzymes are derived primarily from microorganisms. In many investigations several considerations are unclear:

(1) In which season soils are sampled
(2) Whether soils are sampled fallow or under plant cover
(3) How long soils have been in fallow or whether plant cover is young or mature
(4) Whether the soil enzymic activities are also exhibited by roots of existing vegetation or by older, partly-decomposed plant debris isolated from the soil
(5) Whether recently-live plant roots or older plant residues are removed from the soil prior to assay.

The effectiveness of plant removal would vary depending upon the age and type of plant, soil type, pretreatment of soils after sampling, and separation methods used. Soils are inevitably disturbed on sampling and are frequently dried and ground before assay, causing death of some soil animals and microorganisms and favouring the persistence in soils of enzymes which are inherently stable or are stabilized by bonding to soil colloids. In some studies, dried soils are remoistened and incubated for several weeks prior to assay, resulting in increased activities of enzymes from microbial sources and probably some destruction of enzymes of plant and animal residues. Clearly the time of sampling and the conditions of pretreatment and storage of a soil before assay could markedly influence the relative extents to which enzymes of different origin might contribute to the activity of the accumulated enzymes present.

In some investigations, the effects of season, cultivation and the nature of the plant cover on soil enzyme activities are well described and have been offered as indirect evidence for the persistence of plant enzymes in soils.

1. Activities of soil enzymes *in situ*

A direct demonstration of the location of some enzymes in natural soils may be achieved at the ultrastructural level by a combination of histochemical and electron microscopic techniques. Foster (1977) has developed techniques for embedding soils *in situ* to prevent relative mechanical movement of the soil constituents during their removal from the soil profile and preparation for electron microscopy. At the same time, the soil is chemically fixed by methods which stabilize the enzymes within the soil fabric at their locations of natural occurrence without causing complete denaturation and loss of enzymic activity (Foster and Rovira, 1976). Standard methods for histochemical localization of the enzymes at the ultrastructural level (Pearse, 1972) are then applied to the whole soil including animal, plant and microbial cells. The sites of enzyme activity can be obtained precisely (to within 1 nm). The adequacy of the chemical fixation and the reliability of the histochemical tests can be judged from the quality of preservation of the fine structure of the soil microorganisms.

In a test for acid phosphatase in the rhizosphere of a South Australian field soil under wheat, the enzyme was shown to be located at the plasma-lemma and in the cell walls of soil bacteria and in the outer cortical cells of wheat roots as expected. The reaction product was also associated with small fragments of organic matter which may be finely comminuted plant cell wall material (Fig. 1).* The phosphatase activity was demonstrated after fixing the enzymes *in situ* with glutaraldehyde and subsequently incubating with sodium β-glycerophosphate in the presence of lead nitrate. Insoluble, electron-dense lead phosphate was deposited at the site of enzyme action. Controls were run, in which either the substrate was omitted or the complete reaction mixture was incubated with sodium fluoride as a phosphatase inhibitor.

In another study, a fallowed sandy top soil was sampled 11 months after the last accession of living plant tissue. Extensive decomposition of pasture residues had occurred during the warm, moist period of fallow (4 months) and during the summer when the soils were intermittently dried and remoistened. Moist soil samples were sieved to remove larger plant debris, then suspended in cold distilled water to yield, after centrifugation, a light fraction which contained phosphatases, as assayed by the technique of Tabatabai and Bremner (1969).

The phosphatases of the light fraction were associated with intact cell walls of plant tissue, with cell wall fragments and with amorphous organic material of unknown origin (Figs. 2 and 3). The results suggest that plant phosphatases may persist in soils within plant tissue, remaining accessible to added substrate and other reagents. To what extent is the stability of enzymes in soils due to their protection within dead intact cells and what is their half-life compared with clay-enzyme or humic-enzyme complexes? Some observations suggest that cells deep within decomposing plant tissue are less damaged and contain more sites of active phosphatases than cells nearer the tissue surface.

The results are preliminary and interpretations are incomplete. How-ever, the technique would appear to have potential for deducing the origin of soil enzymes from their location in intact cells and in partly decomposed but still recognizable cell fragments. At later stages of decomposition the origin of the enzymes would be obscured as cell fragments became indis-tinguishable. The persistence of plant phosphatases in cell debris in soils is indeterminate. The technique does not permit measurements of reaction rates and it is possible that insoluble products may accumulate in dispro-portionately high amounts at reaction sites of enzymes in low concentrations.

Apart from the present inability to quantify enzyme reaction rates, the

* The author wishes to express his gratitude to Dr. R. C. Foster, Division of Soils, CSIRO, Glen Osmond for his valuable comments on the techniques for demonstrating, *in situ*, the action of enzymes in soil, and for making available previously unpublished electron micrographs.

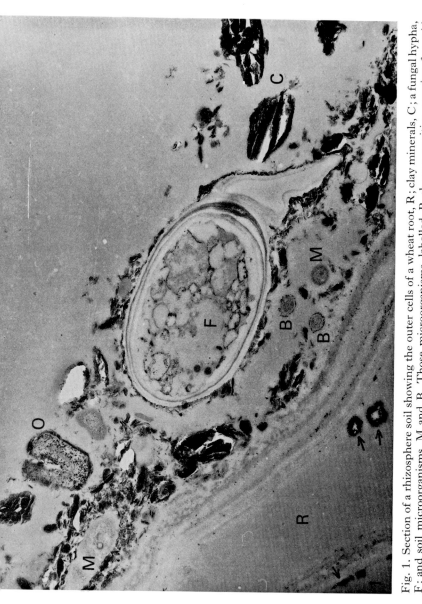

Fig. 1. Section of a rhizosphere soil showing the outer cells of a wheat root, R; clay minerals, C; a fungal hypha, F; and soil microorganisms, M and B. Those microorganisms labelled B show a positive reaction for acid phosphatase whilst those labelled M do not. Some reaction is evident in the root (arrows) and in particular organelles of the fungus. Note also an organic fragment, O, containing phosphatases.

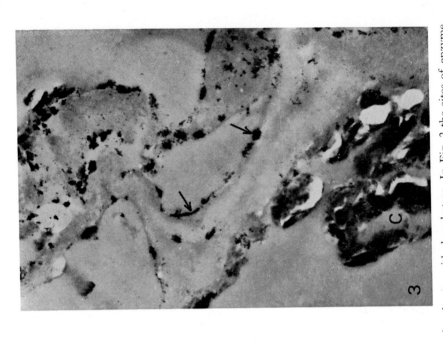

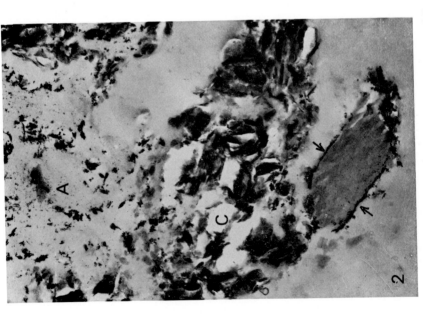

Figs 2 and 3. Sections of a light fraction from a fallow soil treated to locate acid phosphatase. In Fig. 2 the sites of enzyme ... located at the surface of a cell wall fragment (arrow) and throughout amorphous organic material, A.

technique is limited to those enzymes which are not denatured or completely inhibited by the fixation process and for which suitable substrates can be devised. Also, since soil inorganic components are electron-dense, the method is not always suitable for detecting enzymes bound to clay or clay-organic surfaces.

All other evidence relating to the origins of enzymes in soils is indirect. Such evidence rests partly on enzymic activities either of soils under fallow or different vegetations, of homogenates of fresh plants, plant exudates, or plant debris separated from soils or of specific microorganisms isolated from soils.

2. Influence of vegetation and plant residues

A number of investigations have shown that the activities of various enzymes in field or greenhouse soils are affected differentially by the nature of the plant cover (Ross, 1966, 1968a; Galstyan and Avundzhyan, 1967; Neal, 1973; Pancholy and Rice, 1973a, b; Blagoveshchenskaya and Danchenko, 1974; Cortez et al., 1975; Kiss et al., 1975; Spalding et al., 1975, and others). Pancholy and Rice (1973a, b) have demonstrated that activities of soil amylase, cellulase and invertase decreased and activities of soil dehydrogenase and urease increased under successional stages of revegetation to climax stands of prairie grass, oak forest and oak-pine forest. No correlation was found between soil enzymic activity and amounts of soil organic matter but the types of vegetation and hence organic matter added to the soil were concluded to be important in determining the activity gradients of the enzymes assayed.

Ross (1975a, b) has compared the invertase and amylase activities of green leaves, standing dead material, litter and roots of tussock grasses and of corresponding top soils at ten sites. On an organic carbon basis, activities of soil were generally higher than those of plant litter and roots. If the soil enzymes were of plant origin exclusively, then plant amylases and invertases would have been relatively stable in soil compared with other plant components. However, it is likely that the soil enzymes were due not only to an accumulation of plant enzymes but also to synthesis by soil microorganisms. Fractionation of the soils densimetrically showed that the enzyme activities of a light fraction were far greater than those of the heavy soil residue and, unlike the latter, were correlated significantly with environmental factors, especially temperature. Nevertheless most of the total activities of the soils were accounted for by the activities of the heavy fraction and were due to enzymes present mainly in a stable form.

The rates of hydrolysis of benzyloxycarbonyl phenylalanyl leucine (ZPL) and N-benzoyl arginine amide (BAA) by soil, by buffered soil extracts, extracted soil and plant debris separated from soil, have been determined (Ladd, 1972). Soil extracts, isolated plant fragments (and homogenates of recently-dried, washed plant roots—Ladd, unpublished) each hydrolysed ZPL more rapidly than they did BAA, whereas each substrate was hydrolysed by either soil or extracted soil at similar rates. Ladd (1972) also found that the activities of fresh soil extracts decreased during storage of the moistened soil at 20–25°C for several months, whereas activities of the complementary extracted soil sediments remained relatively constant. The results suggested that the extracted enzymes were derived from a source, possibly plant debris, which decomposed comparatively rapidly under the moist conditions favouring microbial growth.

The above results implicate plant enzymes as contributors to the enzymic activities of soils but in all cases the results are inconclusive. Soils under different types of vegetation exhibit changes, not only in certain enzymic activities but also in their populations of microorganisms and rates of respiration. Partly decomposed plant fragments, obtained from soils by hand-picking and by electrostatic attraction, or soil light fractions, separated densimetrically, are associated with microbial cells and products of microbial growth. The relative activities of plant and microbial enzymes in these preparations cannot be distinguished.

3. Persistence of enzymes in soils

Free enzymes, produced extracellularly by live cells or released from recently dead and disrupted cells, are anticipated to have high turnover rates in soils. In the absence of renewed synthesis, the concentrations of free enzymes in soils would rapidly decline, due either to their vulnerability to hydrolysis by microbial proteinases or to their reaction with soil colloids. The general acceptance that soil enzymes may be of plant, animal and microbial origin is based partly on the belief that in soils, opportunities exist for enzymes to become stabilized and hence to persist for long periods after the original source has been removed or destroyed.

Bartholomew (1965) has concluded that most, if not all of the organic-N of soils is of microbial origin. Supportive evidence included studies which demonstrated that during the decomposition of plant material in the presence of excess amounts of inorganic-^{15}N, the atom % enrichment of the organic-N and inorganic-N became nearly equal. It was deduced that equivalence of label would have been achieved if plant-N had been completely mineralized, the inorganic-N derived from the plant had completely equilibrated with the added inorganic-^{15}N and if the newly synthesized organic-^{15}N was due exclusively to immobilization from the inorganic-N pool as a result of microbial growth. However, since exact equivalence of label was not achieved, the possibility remains that not all of the plant-N was mineralized. Further, the decomposition was not carried out in soil, thus limiting the opportunities for stabilization of the plant protein.

Simonart and Mayaudon (1961) have compared the extents of decomposition in soil of a ^{14}C-labelled plant protein preparation and a protein hydrolysate. Based on $^{14}CO_2$ release after 30 days' incubation when mineralization rates of both substrates were low, they calculated that 37% of the plant protein originally added became directly stabilized, free from attack by the soil microflora. Their conclusions were further supported by analyses of ^{14}C-amino acids of soil hydrolysates obtained before and after incubation of the protein and protein hydrolysates. Simonart and Mayaudon (1961) suggested that protein stabilization in soil was due to the adsorption of the proteins on clays or by chemical bonding of proteins to humic acids or lignin.

Clays and complex organic heterocondensates are known to bond to proteins, generally slowing the rates of degradation of the latter by microorganisms or added proteinases (Ensminger and Gieseking, 1942; Pinck and Allison, 1951; Zittle, 1953; McLaren, 1954; Estermann et al., 1959; Handley, 1961; Bremner, 1965; McLaren and Skujiņš, 1967; Mayaudon, 1968; Verma et al., 1975; and others). Reactions of enzymes with clays or organic polymers usually decrease enzyme activities but may stabilize the enzymes towards proteinases or denaturing agents (Aomine and Kobayashi, 1964; Goldstein and Swain, 1965; Ambrož, 1966c; Kobayashi and Aomine, 1967; Ladd and Butler, 1969a, b, 1975; Sørensen, 1969; Rowell et al., 1973; and others).

The stability of many enzymes in soils has been amply demonstrated. Whether their stability is due to the formation in soils of enzyme–clay or enzyme–organic polymer complexes, analogous to those in model experiments, has yet to be proven conclusively. However enzymes have been extracted from soils under relatively mild conditions (Bartha and Bordeleau, 1969b; Chalvignac and Mayaudon, 1971; Burns et al., 1972a, b; Ladd, 1972; McLaren et al., 1975; Nannipieri et al., 1975; Cacco and Maggioni, 1976). In some cases the enzymes have been partly purified and shown to be complexed with soil humic material (Burns et al., 1972a, b; 1975; Nannipieri et al., 1975; Cacco and Maggioni, 1976). McLaren and his colleagues have demonstrated that a urease–organic matter complex from soils is more resistant to attack by a proteinase than is a pure urease preparation.

Other evidence suggesting that soil enzymes occur partly as adsorbed extracellular complexes is based on studies of kinetic properties (Paulson and Kurtz, 1970) and of activation energies (Dalal, 1975) and on numerous studies relating changes in enzyme activity to change in microbial populations in soils. There is no doubt that mechanisms exist in soil for stabilizing enzymes from a variety of sources. Some stability may be conferred within autolysing dead cells before rupture of the cell membrane or cell wall.

Mechanisms for protecting enzymes from the destructive action of micro-bial proteinases may also limit the protected enzymes to a range of available substrates of comparatively low molecular weight (Burns *et al.*, 1972a, b; Ladd and Butler, 1975).

III. Range of Enzymes in Soils

Since, by definition, accumulated enzymes in soils include those of living organisms, the range of enzymes present clearly must be far wider than the approximately fifty enzymes whose activities have been indicated to date. Even so, the observed activity of many of these cannot be conclusively ascribed to the action of a particular enzyme or system of enzymes. This restriction applies obviously to soil dehydrogenases, which are assayed by measuring the formation of a reduced dye resulting from the oxidation of generally unknown endogenous substrates present in unknown and possibly rate-limiting concentrations. However the reservation applies also to other soil enzymes e.g. proteolytic enzymes and nucleases, whose activities have been measured without specific identification of the bonds hydrolysed or of all products formed.

Only a few enzymes have been extracted from soils. Partial purification of extracted enzymes has been achieved by some investigators but not enough either to permit comparisons of the properties of an enzyme from different sources in the soil or to define the nature of the bonding of soil enzymes to humic colloids in soils. Thus the range of soil enzymes, listed below, is not derived from studies of isolated enzymes catalysing clearly defined reactions, but includes enzymes which reasonably but not conclu-sively may have contributed to the activity measured. Where an involve-ment of a specific enzyme is more clearly indicated both trivial and syste-matic nomenclature is given.

Soil enzymes most frequently studied are oxidoreductases and hydro-lases. Several investigations of transferase and lyase activities have been reported but isomerase and ligase activities not at all. Studies of dehydro-genase, catalase, invertase, protease, phosphatase and urease activities in soils account for most publications on soil enzymes. References to these studies are selective, especially since several of these enzymes are reviewed in more detail elsewhere in this book. Other reviews covering the range and properties of enzymes in soils include Durand (1965), Voets and Dedeken (1966), Skujinš (1967, 1976), Ramírez-Martínez (1968), Kuprevich and Shcherbakova (1971b), Sequi (1974) and Kiss *et al.* (1975).

1. Oxidoreductases

1.1. Dehydrogenases

Dehydrogenase activities of soils are determined from the rates of reduction of 2, 3, 5-triphenyltetrazolium chloride to triphenylformazan which is extracted and measured spectrophotometrically (Lenhard, 1956, 1966; Stevenson, 1959; Hirte, 1963; Schaefer, 1963; Casida et al., 1964; Galstyan, 1964; Kiss and Boaru, 1965; Kozlov and Mikhailova, 1965; Peterson, 1965, 1967; Roth, 1965; Thalmann, 1966; Klein et al., 1971; Ross, 1971; Bremner and Tabatabai, 1973; Ladd and Paul, 1973; Skujiņš, 1973). A variety of assay conditions have been described. Soil suspensions may be buffered (pH 7–8) or amended with calcium carbonate. In some assays, succinate, glucose and other sugars may be added, whereas in other assays dye formation is dependent upon oxidation of endogenous substrates. Incubations may be anaerobic, and may be for 1–2 h (Ross, 1971; Ladd and Paul, 1973) or for 24 h or longer.

Since treatment of soils with toluene or chloroform destroys dehydrogenase activity, bacteriostatic and bactericidal compounds are omitted from reaction mixtures. Thus the contribution to the observed activity by proliferating microorganisms is unknown but can be anticipated to be substantial in those assays especially where soils are amended with carbon substrates and incubated for 24 h. Even in short-term assays, where microbial growth may be minimal, dehydrogenase activity may be influenced not only by enzyme concentration but also by the nature and concentration of added and endogenous C substrates and of alternative electron acceptors. Bremner and Tabatabai (1973) have shown that added Fe_2O_3, MnO_2, SO_4^{2-}, PO_4^{3-} and Cl^- stimulated soil dehydrogenase activity whereas NO_3^-, NO_2^- and Fe^{3+} appeared to inhibit. The apparent inhibition may have been due to these latter compounds acting as alternative electron acceptors. Bremner and Tabatabai (1973) have suggested that the effect of NO_3^- may explain the results of Stevenson (1959), who found that leaching of soils with water increased their dehydrogenase activities and that addition of concentrated leachates to the leached soils decreased their activities to those of unleached soils.

Active dehydrogenases are considered to exist in soils as integral parts of intact cells and dehydrogenase activities are thought to reflect the total range of oxidative activities of the soil microflora. Nevertheless dehydrogenase activities do not consistently correlate with numbers of microorganisms in soils or with rates of oxygen consumption or carbon dioxide evolution by soils (Stevenson, 1959, 1962; Hirte, 1963; Schaefer, 1963; Casida et al., 1964; Roth 1965; Gilot and Dommergues, 1967; Raguotis,

1967; Bauzon et al., 1968, 1969; Rawald et al., 1968; Galstyan and Avundz-hyan, 1970; Ross and Roberts, 1970; Laugesen and Mikkelsen, 1973; Ross, 1973a, b; Skujiņš, 1973; Russel and Kobus, 1974). Dehydrogenase activi-ties increase with increasing microbial populations following amendment of soils with nutrients (Ladd and Paul, 1973) and decrease when soils are dried (Cerná, 1969; Ross, 1970; Pancholy and Rice, 1972; Ahrens, 1975) or when treated with triazine derivatives (Spiridonov and Spiridonova, 1973), methyl parathion (Naumann, 1970) or chloramphenicol (Cerná, 1973). Dehydrogenase activities are related to soil organic matter contents (Galstyan and Avundzhyan, 1970; Laugesen and Mikkelsen, 1973) but vary with the season (Rawald et al., 1968; Ross and Roberts, 1970; Ross, 1973a) and decrease with depth of sample from the soil profile (Musa and Mukhtar, 1969). Activities are best retained by storing undried soil samples in the cold (Ross, 1970).

1.2. Glucose Oxidase

Buffered, toluene-treated soil suspensions slowly oxidize glucose when incubated at 37°C for 1–2 days. Gluconate and 2-ketogluconate have been identified as metabolic products, indicative of glucose oxidase and gluconate dehydrogenase activities. Acid production and oxygen consumption were approximately equivalent after incubation for 24 h but not for longer periods (Ross, 1968b; 1974). Rates of glucose oxidation by toluene-treated soils were sufficiently low to suggest that activities of various soil carbo-hydrases, assayed by measuring glucose formation in the presence of toluene, would not be seriously underestimated.

Ladd and Paul (1973) have described a more sensitive assay for glucose oxidation, similar to that of Hobbie and Crawford (1969), in which glu-cose-^{14}C is partly converted to $^{14}CO_2$ after incubation with soil suspensions for 1–2 h in the absence of toluene or other bacteriostatic agents. Glucose oxidation by an air-dried soil was readily detectable. Activities reflected changes in viable bacterial numbers following addition of nutrients to the soil (Ladd and Paul, 1973).

1.3. Aldehyde Oxidase

Libbert and Paetow (1962) have demonstrated the formation of indole-3-carboxylate after incubating washed soil suspensions with indole-3-aldehyde for 1–2 h.

1.4. Urate Oxidase, Uricase (1.7.3.3. Urate : oxygen oxidoreductase)

Uricase activities of toluene-treated soils have been calculated from the recoveries of added uric acid after incubation and correction for adsorption of substrate by soil particles (Durand, 1961a, b, 1965, 1966). Buffered extracts of soils also exhibited uricase activity (Martin-Smith, 1963). Uricases extracted at pH 7·0 were more sensitive to inhibition by metal chelating compounds and less sensitive to inhibition by sulphydryl group reagents than uricases extracted at pH 8·4.

1.5. Catechol Oxidase, Polyphenol Oxidase, Tyrosinase (1.10.3.1. o-Diphenol: oxygen oxidoreductase)*

Soils or soil extracts may oxidize various o-diphenols (catechol, catechin), benzenetriols (phloroglucinol, pyrogallol) and monophenols (tyrosine). Soil catecholase activity generally has been indicated using catechol as substrate (Galstyan, 1958; Kozlov, 1964; Kuprevich and Shcherbakova, 1971a; Ross and McNeilly, 1973). In some assays, activities have been based on rates of oxygen consumption, without demonstration of the formation of o-quinone. However Galstyan (1958) has shown that soil extracts readily catalysed the oxidation of pyrogallol to purpurogallin and Mayaudon et al. (1973b) have demonstrated that soil extracts, partially purified to remove soluble humic compounds, oxidized D-catechin, catechol and DL-3, 4-dihydroxyphenylalanine to quinone derivatives.

Ross and McNeilly (1973) have demonstrated that catechol-oxidizing activities of soils under beech forest were inversely related to the catechol phenolic content of the soils but were not correlated with soil pH, moisture content, organic-C or total content of polyphenols. The catechol-oxidizing enzymes appeared to be of microbial origin and not derived from beech leaf litter.

1.6. p-Diphenol Oxidase, Laccase (1.10.3.2. p-Diphenol: oxygen oxidoreductase)*

Oxidation of p-diphenols or related compounds by soils or soil extracts has been demonstrated using hydroquinone (Kuprevich and Shcherbakova, 1971a; Mayaudon et al., 1973b) and p-phenylenediamine (Mayaudon et al., 1973b; Mayaudon and Sarkar, 1974a, b, 1975).

1.7. Ascorbate Oxidase (1.10.3.3. L-Ascorbate: oxygen oxidoreductase)

Galstyan and Marukyan (1973) have demonstrated that soils oxidized ascorbate to dehydroascorbate after incubation for 1 h at 30°C. Activities of untreated soils generally exceeded those of heat-sterilized control soils indicative of the participation of ascorbate oxidase.

* Both now classified as monophend, monooxygenases (1.14.18.1.)

1.8. Catalase (1.11.1.6. Hydrogen peroxide: hydrogen peroxide oxidoreductase)

Catalase activities of soils are based on the rates of release of oxygen from added hydrogen peroxide (Kuprevich and Shcherbakova, 1956, 1971a; Weetall *et al.*, 1965; Beck, 1971) or on the recoveries of hydrogen peroxide (Baroccio, 1958; Johnson and Temple, 1964). Assay periods range from 2–60 min. Autoclaved control soils also catalyse hydrogen peroxide decomposition due to the presence of inorganic manganese and iron compounds (Kuprevich and Shcherbakova, 1971b). Activities of heated control soils may range from 5–65% of those of untreated soils (Sharova, 1953; Baroccio, 1958; Johnson and Temple, 1964; Kuprevich and Shcherbakova, 1971b).

Radulescu and Kiss (1971) have demonstrated catalase activity in lake sediments up to 13000 years old. In field soils, catalase activity varied with the season (Shumakov, 1960; Slavnina and Sorokina, 1964; Latỳpova, 1965; Raguotis, 1967) and was influenced by the nature of the vegetative cover (Shumakov, 1960; Slavnina and Sorokina, 1964; Latỳpova, 1965; Khan, 1970). Catalase activity was related to soil organic matter content and decreased with depth of sample in the soil profile but was unrelated to the number of microorganisms in soils (Nizova, 1960; Beck, 1971; Roizin and Egorov, 1972) or to the surface area of rendzina aggregates of different size (Cerná, 1966).

Catalase activities may be both increased and decreased by treating soils with herbicides (Zinchenko and Osinskaya, 1969), insecticides (Tsirkov, 1970) or nematocides (Abdel'Yussif *et al.*, 1976). Tannic acid and other aromatic acids inhibit soil catalases, the inhibition increasing with decreasing phenolic content of the acid applied (Kunze, 1970, 1971; Gnittke and Kunze, 1975).

1.9. Peroxidase (1.11.1.7. Donor: hydrogen peroxide oxidoreductase)

Peroxidase activities in soils and soil extracts have been inferred from the oxidation of pyrogallol (Galstyan, 1958; Shchatsmar and Kalikina, 1972) catechol (Kozlov, 1964) *p*-dianisidine (Bordeleau and Bartha, 1969) and *o*-dianisidine (Bartha and Bordeleau, 1969a, b; Bordeleau and Bartha, 1972) in the presence of hydrogen peroxide.

Peroxidatic activities towards *o*-dianisidine by buffered cell-free extracts of soils have been correlated with the capacities of the soils to transform herbicide-derived chloroanilines (Bartha and Bordeleau, 1969a, b). Techniques for the isolation of peroxidase-synthesizing microorganisms from soils have also been described (Bordeleau and Bartha, 1969, 1972). Burge (1973), using the same enzyme assay, found no correlation between soil peroxidase activity, the number of peroxidase-synthesizing microorganisms and the ability of soils to convert 3, 4-dichloroaniline to 3, 3′, 4, 4′tetrachloroazobenzene. Although soil extract peroxidase and horseradish peroxidase were similarly inhibited by potassium cyanide and sodium azide, the enzymes differed in that the peroxidase from soil was unable to convert 3, 4-dichloroaniline to 3, 3′, 4, 4′-tetrachloroazobenzene. The results suggest that different soils may contain peroxidases with different catalytic properties, but also emphasize the necessity to purify soil enzymes from soil constituents which possibly may affect their catalytic activities differentially.

2. Transferases

2.1. Dextransucrase (2.4.1.5. α-1, 6-Glucan : D-fructose 2-glucosyltransferase)

Dragan-Bularda and Kiss (1972b) have demonstrated dextran synthesis in a soil incubated with sucrose for several days in the presence of toluene.

2.2. Levansucrase (2.4.1.10. β-2, 6-Fructan : D-glucose-6-fructosyltransferase)

Levan formation has been demonstrated in toluene treated and γ-irradiated soils after incubation with sucrose for several days (Kiss and Dragan-Bularda, 1968, 1970, 1972).

2.3. Aminotransferase (2.6.1.–)

α-Alanine was formed by incubating a toluene-treated soil with pyruvate and either leucine, valine, glutamate or aspartate (Hoffmann, 1959, 1963). α-Ketoisovalerate and oxaloacetate were formed in those soils to which leucine and aspartate respectively, had been added.

2.4. Rhodanese (2.8.1.1. Thiosulphate : cyanide sulphurtransferase)

Buffered soil suspensions, incubated for 1 h at 37°C with thiosulphate and cyanide, formed thiocyanate, indicative of rhodanese activity (Tabatabai and Singh, 1976). The assay was rapid and precise. Activities were increased by the addition of SO_4^{2-} and especially Cl^- to soils, were unaffected by treatment of soils with toluene and were eliminated by treatment with formaldehyde. Activation energies of rhodanese activity in several soils ranged from 21 672 to 34 062 J mole^{-1} and were similar to that (33 180 J mole^{-1}) of purified rhodanese from animal sources.

3. Hydrolases

3.1. Carboxylesterase (3.1.1.1. Carboxylic ester hydrolase)

Pancholy and Lynd (1972, 1973) have described a sensitive assay based on the hydrolysis of 7-hydroxy 4-methylcoumarin butyrate to a fluorescent product, 7-hydroxy 4-methyl umbelliferone. Esterase activity of extracts from a loamy sand was increased by addition of Ca^{2+}, Mg^{2+}, K^+ and Na^+ and was decreased by Cu^{2+}, S^{2-}, Fe^{2+} and EDTA.

An esterase which hydrolyses malathion to its mono-acid has been extracted from a clay loam and partially purified (Getzin and Rosefield, 1968, 1971; Satyanarayana and Getzin, 1973). The enzyme was comparatively stable, withstanding extraction in $0\cdot2$ M NaOH. Denaturation occurred when solutions were kept for 24 h at pHs $< 2\cdot0$ or $> 10\cdot0$, or for 15 min at above 70°C. Activities were retained when solutions of the purified esterase were stored at 4°C or −10°C, or when incubated with proteolytic enzymes, but were partly lost after lyophilization. The enzyme, possibly a glycoprotein, persisted in soils when added and incubated for eight weeks.

3.2. Arylesterase (3.1.1.2. Aryl ester hydrolase)

Phenyl acetate and phenyl butyrate were hydrolysed by toluene-treated soils, the esterase being mainly associated with clay particles (Haig, 1955). Cacco and Maggioni (1976) obtained extracts of an Alpine podzol which hydrolysed naphthyl acetate. Electro-focusing in acrylamide gel showed that esterase activity was associated with several fractions, which also contained humic compounds. The soil extracts may contain different proteins possessing esterase activity. Alternatively, the properties of a soil esterase may be influenced by bonding of the enzyme to different humic components. Compared with a naphthyl acetate esterase preparation from wheat kernels, the esterase from soil was more stable to heat and was resistant to proteolysis by the enzyme pronase. The soil and plant enzymes also differed in their pH-activity profiles and Michaelis constants. Such differences could be due possibly to bonding of the soil esterase in a humic–enzyme complex.

3.3. Lipase (3.1.1.3. Glycerol ester hydrolase)

Pokorná (1964) has shown that tributyrin is hydrolysed by toluene-treated soils to glycerol and butyrate. After incubation for 72 h, butyrate concentrations in soils with and without toluene differed by approximately 15%. Tributyrin hydrolysis was more rapid in peats than in muds.

3.4. Phosphatase

Various organic orthophosphoric mono–and di-esters are hydrolysed when incubated with either untreated, toluene-treated or irradiated soils, indicative of phosphatase activity. Assays have been based on the rates of release of orthophosphate (Rogers, 1942, Drobnikova, 1961; Stefanic, 1971) or of the organic moiety e.g. phenol (Kroll and Krámer, 1955; Krámer, 1957; Krámer and Erdei, 1958; Halstead, 1964; Hoffman, 1968a, b; Khaziev, 1972b) phenolphthalein (Krasil'nikov and Kotelev, 1959) glycerol (Kiss and Péterfi, 1961b) β-naphthol (Ramírez-Martínez, 1965; Ramírez-Martínez and McLaren, 1966a) and p-nitrophenol (Tabatabai and Bremner, 1969). Assay periods generally were less than 2–3 h.

The effect of substrate concentrations on phosphatase reaction rates has been studied (Ramírez-Martínez, 1966; Tabatabai and Bremner, 1971; Cervelli et al., 1973; Brams and McLaren, 1974; Irving and Cosgrove, 1976). K_m values were determined for systems in which soil crumbs were perfused with substrate (Brams and McLaren, 1974) and in a batch-type system in which substrate adsorption was taken into account (Cervelli et al., 1973). Irving and Cosgrove (1976) have demonstrated that the hydrolysis of p-nitrophenyl phosphate by a kraznozem soil deviated from Michaelis-Menten kinetics.

Phosphatase activities of field soils showed little (Ramírez-Martínez and McLaren, 1966b) or no (Keilling et al., 1960) variation with season but were influenced by the cropping system and nature of the plant cover (Khan, 1970; Neal, 1973; Blagoveshchenskaya and Danchenko, 1974). Activities generally increased after the addition of inorganic and organic fertilizers to soils (Vlasyuk et al., 1957; Chunderova, 1964; Dubovenko, 1964; Vlasyuk and Lisoval, 1964; Burangulova and Khaziev, 1965a; Halstead and Sowden, 1968; Kudzin et al., 1968; Muresanu and Goian, 1969; Khan, 1970; Rankov and Dimitrov, 1971; Stefanic et al., 1971). However, inhibition of phosphatases by phosphatic fertilizers (Afanasyeva and Gerus, 1964; Burangulova and Khaziev, 1965a; Muresanu and Goian, 1969) and by liming of soils (Halstead, 1964) has been reported. Phosphatase activities were directly related to the organic-P of soils (Garilova et al., 1974) but were not consistently related to available-P contents (Krámer, 1957; Hofmann and Kesseba, 1962; Burangulova and Khaziev, 1965a; Hoffmann and Elias-Azar, 1965).

Phosphatase activities generally were not correlated with numbers and respiratory activities of microorganisms in soils (Krámer and Yerdei, 1960; Ramírez-Martínez and McLaren, 1966b; Vukhrer and Shamshieva, 1968; Roizin and Egorov, 1972) although activities increased markedly during rapid microbial growth (Ramírez-Martínez and McLaren, 1966b; Ladd and Paul, 1973).

Irradiation of soils did not affect (Voets *et al.*, 1965) or slightly decreased (McLaren *et al.*, 1962; Skujiņš *et al.*, 1962) soil phosphatase activity. Activities were also unaffected by toluene (Suciu, 1970) but were decreased by air-drying (Ramírez-Martínez and McLaren, 1966b; Speir and Ross, 1975) and heating (Halstead, 1964) of soils, by the addition of nematocides (Abdel'Yussif *et al.*, 1976) and by increasing concentrations of heavy metals (Tyler, 1974). Activities were detected in soils at −20°C (Bremner and Zantua, 1975) and in dried samples of aged, permafrost peat (Skujiņš and McLaren, 1968) and ancient lake sediments (Radulescu and Kiss, 1971).

3.5. Nucleases and Nucleotidases

Toluene-treated soils, incubated with nucleic acid at pH 7 at 60°C, released inorganic orthophosphate, indicative of nuclease and nucleotidase activity (Rogers, 1942). Agre *et al.* (1969) demonstrated nuclease production in liquid culture of thermophilic actinomycetes, isolated from soil.

Soil nuclease activities were low when assayed at 28°C (Nilsson, 1957), decreased with depth of sample in the profile (but were unrelated to soil organic matter contents) and were higher in calcareous chernozems than leached chernozems (Burangulova and Khaziev, 1965b). Nuclease activity increased when urea and ammonium carbonate were added to soils, but decreased after addition of NO_3^- (Khaziev, 1966).

3.6. Phytase

Soil phytase activity has been inferred from the release of inorganic phosphate in toluene-treated soils, incubated with sodium phytate at 45°C for 20 h (Jackman and Black, 1952). Skujiņš (1976) has warned that commercial phytates contain lower phosphate esters as impurities, which possibly may have been hydrolysed by phosphatases unable to react with phytate.

3.7. Arylsulphatase (3.1.6.1. Aryl-sulphate sulphohydrolase)

Arylsulphatase activities of soils have been assayed rapidly and precisely from the rates of hydrolysis of *p*-nitrophenyl sulphate to *p*-nitrophenol by buffered soil suspensions in the presence of toluene (Tabatabai and Bremner, 1970a). Activities decreased markedly with sampling depth and were correlated with soil organic matter content but not with pH or percentage of S, N, clay or sand contents (Tabatabai and Bremner, 1970b). Michaelis constants of arysulphatases in different soils were similar but differed substantially from K_m values of commercial sulphatases from microbial and animal sources (Tabatabai and Bremner, 1971).

Arylsulphatase activities were detectable in soils incubated at −20°C, but not at −30°C (Bremner and Zantua, 1975). Intermittent drying and wetting of soils decreased activities (Cooper, 1972). Activities in Nigerian field soils changed with season, increasing during continuously-moist rainy periods and decreasing as soils dried out. Tabatabai and Bremner (1970b) have reported that arylsulphatase activities increased when field-moist soils were dried at 22–24°C.

3.8. Amylase (3.2.1.–)

Buffered soil suspensions and soil extracts hydrolyse starch under conditions limiting microbial proliferation, indicative of accumulated α- and β-amylases (Drobník, 1955; Hofmann and Hoffmann, 1955; Galstyan, 1965b, 1974; Ross, 1965; Shcherbakova et al., 1970, 1971; Pancholy and Rice, 1973a; and others). β-Amylases were more active than α-amylases (Hofmann and Hoffmann, 1955). Optimum pH for assay was approximately 5–6 (Markosyan and Galstyan, 1963; Küster and Gardiner, 1968).

Soil amylase activity was inducible (Drobník, 1955) and decreased with sample depth and organic matter content (Hofmann and Hoffmann, 1955; Ross, 1968a). Activities were influenced by the season (Galstyan, 1965; Ross and Roberts, 1970; Cortez et al., 1972) and by the nature of the vegetative cover (Ross, 1966, 1968a; Pancholy and Rice, 1973a, b; Cortez et al., 1975).

Amylase activities decreased after air-drying of soils (Ross, 1965) and after storing soils at −20°C (Ross, 1965) or at 21°C (Pancholy and Rice, 1972). Nevertheless amylase activities of air-dried soils were still detectable after dispersion by ultrasonic vibration in an organic solvent ("Nemagon"). The average amylase activity of light fractions (sp. gr. < 2·06) of several soils was, on a weight basis, 18 times that of the heavy fractions (Ross, 1975c).

3.9. Cellulase (3.2.1.4. β-Glucan 4-glucanohydrolase)

Soils treated with toluene, petroleum ether or merthiolate to prevent microbial proliferation, hydrolyse cellophane (Márkus, 1955; Kislitsina, 1965, 1968) cellulose powder (Rawald, 1968, 1970; Mereshko, 1969; Benefield, 1971; Kong et al., 1971; Drozdowicz, 1971; Kong and Dommergues, 1972) or carboxymethyl cellulose (Kong and Dommergues, 1970, 1972; Tomescu, 1970; Drozdowicz, 1971; Kislitsina, 1971; Kong et al., 1971; Ambroz, 1973; Pancholy and Rice, 1973a, b). Cellulase activity was indicated by disappearance of substrates, measured gravimetrically or viscosimetrically, or by appearance of reducing sugars.

Cellulase activities of soils were increased by prior incubation with cellulose (Kong and Dommergues, 1972; Ambroz, 1973) and were decreased by addition of paraquat (Giardina et al., 1970) or purified wattle tannin preparations (Benoit and Starkey, 1968). Pancholy and Rice (1973a, · b) have measured cellulase activities in soils in two successional stages and in a climax stand of three vegetation types. Activities were not correlated with total organic matter content in the soil but were related to the type of organic matter added to the soil during the succession.

3.10. Laminarinase, Lichenase (3.2.1.6. β-1, 3-Glucan 4-glucanohydrolase)

Hydrolysis of β-1, 3-glucan bonds by soils has been demonstrated by Jones and Webley (1968). Laminarin was decomposed by toluene-treated soil aggregates which had been incubated previously without toluene but with fungal cell walls containing β-1, 3-glucan polysaccharide. Kiss et al. (1962) have shown that soils hydrolysed lichenin in the presence of toluene.

3.11. Inulase (3.2.1.7. Inulin 1-fructanohydrolase)

Hoffmann (1959) and Kiss and Péterfi (1961a) have shown that toluene-treated soils hydrolysed the β-1, 2-fructan bonds of added inulin.

3.12. Xylanase (3.2.1.8 Xylan 4-xylanohydrolase)

Sørensen (1955, 1957) has shown that γ-irradiated and toluene-treated soils, when incubated as buffered (pH 6·5) suspensions with added xylan for 24 h at 37°C, released reducing sugar.

3.13. Dextranase (3.2.1.11. α-1, 6-Glucan 6-glucanohydrolase)

Incubation of irradiated or toluene-treated soils with added dextran resulted in glucose formation (Drăgan-Bularda and Kiss, 1972a). Dextranase was inducible.

3.14. Levanase

Kiss et al. (1965) and Kiss and Drăgan-Bularda (1968) have shown that irradiated and toluene-treated soils hydrolysed added levan.

3.15. Polygalacturonase, Pectinase (3.2.1.15. Polygalacturonide glycanohydrolase)

Kaiser and Monzon de Asconegui (1971) have shown that buffered soil suspensions hydrolyse added pectin when incubated in the presence of toluene. Polygalacturonase activity was inhibited by purified tannin. Benoit and Starkey (1968) have suggested that the effect of tannins on extracellular enzyme activity may be to retard the decomposition in soils of high molecular weight compounds of plant origin.

3.16. α-Glucosidase, Maltase (3.2.1.20. α-D-Glucoside gluco-hydrolase)

Phenyl-α-D-glucoside (Hofmann and Hoffmann, 1954) and maltose (Drobník, 1955; Hofmann and Hoffmann, 1955; Kiss, 1958b; Kiss and Péterfi, 1960a, b) were hydrolysed by toluene-treated soils, indicative of α-glucosidase activity. Activities were increased in soils previously incubated with maltose without toluene, and were partly decreased by addition of silver nitrate or mercuric chloride.

3.17. β-Glucosidase, Cellobiase, Gentiobiase, Emulsin (3.2.1.21. β-D-Glucoside glucohydrolase)

β-Glucosidase activities in soils have been based on the hydrolysis of p-hydroxyphenyl-β-D-glucoside (Hofmann and Hoffmann, 1953; Galstyan, 1965a, b), cellobiose (Hoffmann, 1959), salicin (Hoffmann and Dedeken, 1965) and p-nitrophenyl-β-D-glucoside (Hayano, 1973; Hayano and Shiojima, 1974). Assays using the latter substrate are based on the rates of formation of p-nitrophenol and are rapid and precise.

Using different assays and soils, β-glucosidase activity was optimal at pH 5·9–6·2 (Markosyan and Galstyan, 1963) and at pH 4·8 (Hayano, 1973). Activities decreased with depth of sample (Berger-Landefelt, 1965) and with drying of soils (Cerná, 1969), they varied seasonally in the field (Galstyan, 1965a) and were correlated with the surface areas of crushed and uncrushed soil aggregates (Cerná, 1968, 1970).

3.18. α-Galactosidase, Melibiase (3.2.1.22. α-D-Galactoside galac-tohydrolase)

Toluene-treated soils hydrolysed phenyl-α-D-galactoside (Hofmann and Hoffmann, 1953, 1954) and melibiose (Hoffmann, 1959) indicative of α-galactosidase.

3.19. β-Galactosidase, Lactase (3.2.1.23. β-D-Galactoside galacto-hydrolase)

β-Galactosidase activities of soils have been demonstrated using phenyl-D-galactoside (Hofmann and Hoffmann, 1953, 1954), lactose (Hoffmann, 1959) and o-nitrophenyl-β-galactoside (Rysavý and Macura, 1972a, b) as substrates. The enzyme was inducible, attaining maximum activity after prior incubation of the soils with lactose for 6–24 h.

3.20. Invertase, Saccharase, Sucrase, β-Fructofuranosidase (3.2.1.26. β-D-Fructofuranoside fructohydrolase)

Irradiated or toluene-treated soils hydrolyse sucrose to glucose and fructose, indicative of accumulated invertase activity. Incubation periods ranged from a few hours to several days (Hofmann and Seegerer, 1951; Kuprevich and Shcherbakova, 1956, 1971a; Kiss, 1957, 1958a, b; Kleinert, 1962; Peterson and Astafyeva, 1962; Galstyan, 1965a; Hoffmann and Pallauf, 1965; Voets and Dedeken, 1965; Voets et al., 1965; Kiss et al., 1971, 1972a; Kunze and Richart, 1973). Optimum pH for assaying soil invertases was approximately 4.2–5.0 (Markosyan and Galstyan, 1963; Küster and Gardiner, 1968; Chunderova, 1970).

Invertase activities of field soils varied seasonally (Nizova, 1960; Freytag, 1965; Galstyan, 1965b; Ross, 1966; Raguotis, 1967; Cortez et al., 1972) and were influenced by the vegetative cover (Balicka and Sochacka, 1959; Shumakov, 1960; Nowak, 1964; Freytag, 1965; Latỳpova, 1965; Ross, 1966, 1968a; Raguotis, 1967; Khan, 1970; Pancholy and Rice, 1973a, b; Blagoveshchenskaya and Danchenko, 1974). Invertase activities were correlated with surface area of aggregates (Cerná, 1966, 1970) and with soil organic matter contents (Berger-Landefelt, 1965; Ross, 1966, 1968a, 1975a; Kudryachev, 1968; Ross and Roberts, 1968, 1970; Vukhrer and Shamshieva, 1968; Kiss et al., 1971; Roizin and Egorov, 1972). However, because activities differed with time of sampling, organic-C concentrations alone did not accurately reflect invertase activities. Invertase activities of soils were not consistently related to numbers of microorganisms present (Balicka and Trzebinski, 1956; Nowak, 1964; Raguotis, 1967; Roizin and Egorov, 1972; Ross and Roberts, 1973; Ross, 1975b, 1976) or to rates of oxygen consumption or carbon dioxide evolution by soils (Nizova, 1960; Turková and Srogl, 1960; Moureaux, 1965; Ross, 1973b). Relationships varied with different plant associations on the same soil type; plant roots with microorganisms may contribute to the total assayed invertase activities of soils.

Soil invertase was stimulated by addition of nematocides (Abdel'Yussif et al., 1976). Activities decreased when soils were dried (Latỳpova, 1965; Cerná, 1969) but showed little or no further decrease when air-dried soils were stored at laboratory temperatures (Nizova, 1969; Ross, 1965; Pancholy and Rice, 1972). Invertase activities of dried soils were partly lost (average decrease, 23%) after dispersion of the soils by ultrasonic vibration in an organic solvent, sp. gr. 2·06. The mean invertase activity of a light fraction of various soils was 6·7 times that of the heavy fraction.

3.21. Proteinases, Peptidases

When amino compounds of low molecular weight are formed in soils incubated with proteins soil proteinase activity is indicated. Protein substrates include ovalbumin (Ambroz, 1965), casein (Kuprevich and Shcherbakova, 1961, 1971b; Ambroz, 1965, 1966a, b, 1971; Ladd, 1972; Ladd and Butler, 1972; Mayaudon et al., 1975), azocasein (Macura and Vágnerová, 1969), haemoglobin (Antoniani et al. 1954; Ladd, 1972; Ladd and Butler, 1972) and especially gelatine (Fermi, 1910; Hofmann and Niggemann, 1953; Hoffmann and Teicher, 1957; Voets and Dedeken, 1964, 1965; Voets et al., 1965; Ambroz, 1965, 1966a, b, 1971). Assays may be of short duration (1–2 h) without addition of bacteriostatic compounds (Ladd and Butler, 1972), or more commonly, for 1–2 and up to 16 days in the presence of toluene to minimize microbial growth and utilization of protein hydrolytic products. Activities are based on the release of amino compounds determined with a ninhydrin (Ladd and Butler, 1972) or a copper reagent (Hoffmann and Teicher, 1957), or on the formation of coloured products (Macura and Vágnerová, 1969), TCA-soluble tyrosyl derivatives (Ladd and Butler, 1972) or specifically, arginine (Voets and Dedeken, 1964).

Toluene-treated soils hydrolysed gelatine without NH_4^+ release (Voets and Dedeken, 1964) and did not deaminate added amino acids (Voets et al., 1965). However, Ambroz (1966b) has shown that NH_4^+ was present in toluene-treated soils incubated with gelatine for 1–2 days, which suggested deaminase activity. Concentrations of NH_4^+ in soils with the highest amounts of added toluene were about 20% of those without toluene. Unfortunately concentrations of NH_4^+ at the beginning of incubation and in corresponding soil controls without gelatine were not shown.

Dedeken and Voets (1965) have identified amino acids formed in toluene-treated soils incubated without added protein substrates. Ladd and Amato (unpublished data) have found that unamended soils, held under chloroform for 24 h at 25°C, released ninhydrin-reactive compounds, part of which was accounted for as NH_4^+. During extended periods of fumigation, the amounts of ninhydrin-reactive compounds increased, due almost exclusively to increased NH_4^+ formation. Fumigation with chloroform drastically decreased the numbers of viable bacteria in the soils and eliminated soil dehydrogenase and glucose-oxidizing activities. The results support those of Ambroz (1965) and indicate that soil deaminases may be active for several days at least, in the presence of organic solvents acting as bacteriostatic and bactericidal agents. Nevertheless deaminase activities may be sufficiently low in the presence of such agents not to pose a serious problem in long-term proteinase assays, based on the general release of amino compounds.

Soils differed in their relative activities towards casein and gelatine (Ambroz, 1965, 1966a, 1970). Also rates of casein hydrolysis assayed at pH 8·5, were decreased more by heating and drying of soils, then rates of gelatine hydrolysis, assayed at pH 6·0 (Ambroz, 1966a, b, 1968).

Casein hydrolysing activities of soils did not reflect the capacities of the soils to mineralize N (Ross and McNeilly, 1975).

Proteinase activities decreased with depth of sampling (Hoffmann and Teicher, 1957), surface area of rendzina aggregates (Cerná, 1970) and soil organic matter content (Bei-Bienko, 1970; Franz, 1973). In the field, activities varied with season but were not correlated with changes in microbial populations (Franz, 1973; Ladd et al., 1976). Nevertheless, marked increases in proteinase activities have occurred in soils previously incubated under conditions to promote microbial growth (Ladd and Paul, 1973; Mayaudon et al., 1975). Ladd and Paul (1973) have found that the most rapid increases in casein-hydrolysing activities took place during a period of relatively rapid oxidation of microbial metabolites and when numbers of viable bacteria were declining markedly. The activities of the newly-formed proteinases declined subsequently as microbial activity decreased. Ladd and Butler (1975) have suggested that extracellular enzymes such as proteinases, active against added high molecular weight substrates, would be short-lived in soils. Either the enzyme protein would be vulnerable to autolysis or attack by other proteinases, or the enzyme would be stabilized by complexing with or adsorption to soil colloids. Bonding of the enzyme to soil colloids might protect the enzyme protein but might also render it inaccessible and thus inactive towards substrates of high molecular weight (Rowell et al. 1973). Thus activities of proteinases would decline rapidly unless free active enzymes were continually resynthesized, dependent upon the availability of energy sources.

Proteinase activities of field soils decreased temporarily after addition of paraquat (Giardina et al., 1970; Namdeo and Dube, 1973a) dalapon (Namdeo and Dube, 1973b) and simazine (Spiridonov and Spiridonova, 1973) and by fumigation with chloropicrin or methyl bromide (Ladd et al., 1976). Casein hydrolysing activities were not appreciably affected by fumigation with chloroform.

γ-Irradiated (2 Mrad) soils and toluene-treated soils hydrolysed gelatine to similar extents (Voets and Dedeken, 1965; Voets et al., 1965). Ladd et al. (unpublished data) have found that in short-term assays in which soils were incubated in the absence of toluene, γ-irradiation (2·6 Mrad) decreased activities towards casein by approximately 60%. An increased irradiation dose (10·2 Mrad) completely destroyed activities in most soils.

Proteinases have been extracted from soils (Ladd and Paul, 1973; Mayaudon et al., 1975). The activities of the extracts were positively correlated with soil organic matter contents and negatively correlated with soil clay contents (Mayaudon et al., 1975).

Soils hydrolyse dipeptide derivatives (Ladd and Butler, 1972). Having but one peptide bond, the use of such substrates introduces specificity into the reaction and more easily allows the establishment of a linear relationship between their rates of hydrolysis and soil concentrations. Soils also hydrolysed the amides N-benzoyl L-arginine amide (BAA) (McLaren et al., 1957; Ladd and Butler, 1972) and propanil (Burge, 1973). Although several proteinases have been conveniently assayed using BAA or specific dipeptide derivatives as substrates, in soils their hydrolyses may possibly be catalysed also by dipeptidases. Ladd and Butler (1972) have demonstrated that soils which ranged widely in their pH, clay content and organic matter content were consistent in their preferential hydrolysis of several anionic dipeptide derivatives containing amino acids with hydrophobic side chains. Benzyloxycarbonyl phenylalanyl leucine (ZPL) was hydrolysed most rapidly. Rates of hydrolysis of ZPL were highly correlated with clay contents of the soils but not with organic matter contents nor with the rates of hydrolysis of the cationic substrate, BAA.

Compared with casein-hydrolysing activities, rates of hydrolysis of ZPL and BAA by soils were less affected by seasonal changes (Ladd et al., 1976), heating or drying of soils or conditions causing marked fluctuations in microbial populations for example, intermittent drying and wetting of soils or incubation after amendment with an energy source (Ladd and Paul, 1973). Enzymes hydrolysing ZPL and BAA were generally less sensitive to inactivation by fumigation with chloropicrin or methyl bromide, or by γ-irradiation of soils. Irradiating soils with 5·1 Mrad, decreased average activities towards ZPL, BAA and casein by 51, 70 and 77% respectively (Ladd et al. unpublished data). In contrast, McLaren et al. (1957) have found that BAA-hydrolysing activity was completely destroyed by irradiating a soil with 5·1 Mrad.

Enzymes hydrolysing ZPL were extracted from dry soils and from soil fractions of different particle sizes (Ladd, 1972). Activities of the extracts were moderately well correlated with the clay and organic matter contents of the soils but poorly correlated with the amounts of soluble humic compounds, the majority of which were readily separated by precipitation with 0·1 M CaCl$_2$ without loss of enzyme activity. Extracted enzymes hydrolysing ZPL were optimally active near 55°C and pH 7·5 and were stable after storage at 25°C for at least 10 days and after incubation with microbial proteinases. γ-Irradiation of freeze-dried extracts partially destroyed ZPL-hydrolysing activities. Extracts from stored, moistened soils were less active than those from dried soils. Evidence was presented suggesting that plant residues may have been the source of the extracted enzymes (Ladd, 1972).

3.22. Asparaginase (3.5.1.1. L-Asparagine amidohydrolase)

Toluene-treated soils, incubated with asparagine for 2–3 days (Drobník, 1956; Beck and Poschenrieder, 1963) or for 1–5 h (Mouraret, 1655) released NH_4^+, indicative of asparaginase activity. Asparaginase may be bound to cell constituents (Mouraret, 1965). There was no obvious correlation between activities and numbers of microorganisms in cultivated and virgin soils (Roizin and Egorov, 1972).

3.23. Glutaminase (3.5.1.2. L-Glutamine amidohydrolase)

Toluene-treated soils hydrolysed glutamine to glutamate and NH_4^+ (Galstyan, 1973; Galstyan and Saakyan, 1973).

3.24. Urease (3.5.1.5. Urea amidohydrolase)

Assays for soil urease have been based on the rates of utilization of added urea (Kuprevich, 1951; Van Niekerk, 1964; Porter, 1965; Simpson, 1968) or on the rates of formation o^f carbon dioxide from urea (Conrad, 1942; Skujiņš and McLaren, 1969; Norstad et al., 1973) or on the rates of formation of NH_4^+ from urea (Hofmann and Schmidt, 1953; Stojanovic, 1959; Hoffmann and Teicher, 1961; Kuprevich et al., 1966; McGarity and Myers, 1967; Thente, 1970; Kozlovskaya et al., 1972; Tabatabai and Bremner, 1972; and others). Most assays use buffered soil suspensions incubated for periods less than 1–2 h. Soils may be untreated, treated with toluene or sterilized by irradiation.

Urease activities of soils may be either increased or decreased by addition of toluene (Rotini, 1935; Roberge, 1968; Bhavanandan and Fernando, 1970; Thente, 1970; Tabatabai and Bremner, 1972; Norstadt et al., 1973), or by irradiation of soils (McLaren et al., 1957; Vela and Wyss, 1962; Van Niekerk, 1964; Roberge, 1968; Roberge and Knowles, 1968a, b; Skujiņš and McLaren, 1969; Thente, 1970). The effects varied with the soil, the duration of storage and moisture contents of treated soils, radiation dose, conditions of assay etc. Rupture of cell membranes or changes in their permeability may facilitate increased activity of ureases of formerly live cells towards added urea. Such increases may, under some conditions, more than offset losses of activity of soil ureases due to enzyme denaturation caused by the soil treatments. The results emphasize the necessity for cautious interpretation of data from assays in which sterilized soils are employed or where bacteriostatic agents are added to prevent microbial growth.

The effect of urea concentration on soil urease activity has been investigated (Paulson and Kurtz, 1970; Gould *et al.*, 1973; Tabatabai, 1973; Ardakani *et al.*, 1975). Assuming the applicability of Michaelis-Menten kinetics to the soil urease systems, calculated Michaelis constants (K_m) varied with the soil and conditions of assay (Tabatabai, 1973). Variations may be due partly to differences in the proportions of enzymes present in live cells or adsorbed to soil colloids (Paulson and Kurtz, 1970).

Urease activities of field soils varied seasonally and were influenced by the vegetation (Stojanovic, 1959; Galstyan 1965b; Raguotis, 1967; Ishizawa and Tanabe, 1968; Stenina, 1968; Khan, 1970; Pancholy and Rice, 1973a; Blagoveshchenskaya and Danchenko, 1974). Urease activities decreased with depth of sample in profiles and were correlated with soil organic matter contents (Myers and McGarity, 1968; Vukhrer and Shamshieva, 1968; Musa and Mukhtar, 1969; Laugesen, 1972; Roizin and Egorov, 1972; Franz, 1973). Activities increased after addition of inorganic and especially organic fertilizers to soils (Filev *et al.*, 1968; Alekseeva *et al.*, 1969; Bei-Bienko, 1970; Nikiforenko, 1971; Stefanic *et al.*, 1971; Laugesen, 1972).

Urease activities of soils are considered to be due mainly to enzymes located extracellularly. However, the proportions vary according to the effects of soil conditions on previous microbial growth (Paulson and Kurtz, 1969; Lloyd and Sheaffe, 1973; Zantua and Bremner, 1976). Urease activity increased with surface area of soil aggregates (Cerná, 1966). Activities were not affected by freeze-drying, air-drying (22°C, 48 h) or oven-drying (55°C, 24 h) of soils (Zantua and Bremner, 1975). Activities of air-dried soils were constant during storage for 1 year; activities of moist soils did not change during incubation in closed containers for times up to 3 months (Pancholy and Rice, 1972; Zantua and Bremner, 1975). Urease activity has been detected in ancient (9550 years) permafrost soil samples (Skujiņš and McLaren, 1969), in soils assayed at −20°C (Bremner and Zantua, 1975) and at −33°C (Tagliabue, 1958) and in soils assayed air-dry at 80% relative humidity (Skujiņš and McLaren, 1967).

The stability of soil urease may be due to its existence mainly as a complex with soil organic constituents. Bonding of the enzyme in an organic colloidal matrix may render it inaccessible to destruction by soil proteinases and may confer some stability against denaturation by heating and drying, yet may not exclude small molecular weight substrates and products. This view is supported by studies of the properties of an extracted active urease–organic matter complex from soils (Burns *et al.*, 1972a, b; McLaren *et al.*, 1975; Nannipieri *et al.*, 1975).

Effects of insecticides (Tsirkov, 1970) herbicides (Zinchenko and Osinskaya, 1969; Giardina *et al.*, 1970; Namdeo and Dube, 1973a, b) nematocides (Abdel'Yussif *et al.*, 1976) and of heavy metals (Tyler, 1974) on urease activities have been determined. Both inhibition and stimulation of activity by the various amendments have been observed. Quinone derivatives effectively inhibit soil ureases (Bremner and Douglas, 1971, 1973; Bundy and Bremner, 1973).

3.25. Inorganic Pyrophosphatase (3.6.1.1. Pyrophosphate phosphohydrolase)

Soil pyrophosphatases have been assayed using toluene-treated (Rotini, 1933; Khaziev, 1972a) and untreated (Douglas et al., 1976) soils. Activities decreased with depth of sample, were correlated with organic matter contents and were mainly eliminated when the soils were autoclaved.

3.26. Polymetaphosphatase

Rotini and Carloni (1953) have demonstrated that added metaphosphate was hydrolysed to orthophosphate by a heat-sterilized and a toluene-treated soil. Activities were partly due to accumulated enzymes but mainly to catalysis by inorganic soil constituents, especially hydrated manganese dioxide.

4. Lyases

4.1. Aspartate Decarboxylase (4.1.1.11. L-Aspartate 1-carboxy-lase)*

Drobník (1956, 1961) has shown that β-alanine was formed after the incubation of toluene-treated soils with either aspartate or asparagine.

4.2. Glutamate decarboxylase (4.1.1.15. L-Glutamate 1-carboxylyase)

γ-Aminobutyrate was formed by soils incubated with glutamate which indicated glutamate-1-decarboxylase activity (Kuprevich and Shcherbakova, 1971a; Umarov and Aseeva, 1970). The latter authors have described a sensitive assay in which $^{14}CO_2$ formation from glutamate -5-^{14}C was measured after incubation with soils for 2 h.

4.3. Aromatic Amino Acid Decarboxylases

Mayaudon et al. (1973a) have shown that neutral, cell-free extracts of fresh soils catalysed the release of $^{14}CO_2$ from DL-3, 4-dihydroxyphenylalanine-1-^{14}C (DOPA), DL-tyrosine-1-^{14}C and DL-tryptophan-1-^{14}C at the relative rates 10:3:1. DL-Phenylalanine-1-^{14}C was not decarboxylated.

* Now classified with glutamate decarboxylase (4.1.1.15)

DOPA was thought to be oxidized to the o-quinone derivative prior to decarboxylation.

Tryptophan was converted by soils and soil extracts to β-indoleacetate (Chalvignac, 1968, 1971; Pilet and Chalvignac, 1970; Chalvignac and Mayaudon, 1971). The amounts of indoleacetate formed by toluene-treated, buffered soil suspensions, incubated at 40°C for 48 h, slightly exceeded those by soils subjected to a preliminary heat treatment (130°C, 15 min). By comparison, the activities of soil extracts were decreased more by heat treatment (100°C, 10 min) than by the presence of toluene, suggesting that in soils the tryptophan-decomposing enzymes may be stabilized against thermal denaturation, possibly by bonding in a clay–humus complex.

Enzymically active extracts from a forest mull also contained humic compounds. Under the conditions of assay, optimal activity occurred at 65°C, only the L-isomer of tryptophan being degraded to form primarily, indoleacetamide and secondarily, β-indoleacetate.

Indoleacetate was also formed from the hydrolysis of indole-3-aceto-nitrile, after incubation with washed soil suspensions for 1–2 h (Libbert and Paetow, 1962).

IV. Conclusions

While it is reasonable to accept that enzymes from cells and cell debris of all kinds contribute to the total activity of accumulated enzymes in soils, the relative activities of enzymes of different origin are unknown. It is anticipated that in many cases the contributions may vary, not only seasonally and with the type and age of plant cover but also with the treatment of soils between sampling and assay. Available evidence for the persistence of plant enzymes in soils is mainly indirect since activities of enzymes derived from the soil microflora, necessarily present, cannot be assessed. Also, attempts to attribute a soil activity to enzymes of particular microorganisms cannot be based on the relative activities of cultures of the isolated microorganisms.

Techniques for directly observing, at the ultrastructure level, the sites of enzyme action in soils are potentially very useful for demonstrating the presence of active plant, animal and microbial enzymes. However these methods do not quantitatively determine enzyme reaction rates and, in respect to the origin of the enzymes, are limited to those enzymes associated with intact cells or with large, recognizable cell fragments.

The range of accumulated enzymes in soils could be extended by devising sensitive assays to measure activities of enzymes present in low concentrations, possibly confined to non-proliferating, live cells. Of greater value would be the development of assays which distinguished activities of enzymes in live and dead cells, in cell debris or adsorbed to clays or organic residues. The long-expressed need to determine the catalytic properties of enzymes in different states in the natural soil has yet to be met. The use of established histochemical and microscopic techniques for demonstrating enzyme reactions *in situ* may prove to be of considerable value, especially if the methods are developed to employ suitable specific substrates which differ in their net charge or molecular weight range.

At best, activities of accumulated enzymes in soils are measured under standardized conditions. Nevertheless, assay conditions are chosen arbitrarily and differ markedly from those of the natural soil environment. Frequently soils are incubated as buffered suspensions with natural or synthetic substrates, which, because of kinetic considerations, are added in abnormally high concentrations. In some assays, the methods used to eliminate microbial proliferation either increase or decrease the activities of enzymes already present. Accordingly, it is doubtful whether present methods can be used to assess the relative importance of accumulated enzymes and of enzymes of proliferating microorganisms in the transformations of compounds under conditions normally encountered in soil.

References

ABDEL'YUSSIF R. M., ZINCHENKO V. A. and GRUZDEV G. S. (1976). The effect of nematocides on the biological activity of soil. *Izv. Timiryazev. Sel'skokhoz. Akad.* (1), 206–214.

AFANAS'EVA A. L. and GERUS O. I. (1964). Effect of fertilizers on biological activity of soil. *Dokl. sibir. Pochv. Novosibirsk. Sibir. Otd. Akad. Nauk* 85–95.

AGRE N. S., TATARSKAYA R. I. and DEMBO G. V. (1969). Nuclease activity of thermophilic soil actinomycetes. *Mikrobiologiya* **38**, 1085–1090.

AHRENS E. (1975). The comparability of dehydrogenase activity in air-dry and moist soil samples. *Landw. Forschung* **28**, 310–316.

ALEKSEEVA N. S., GOLOVKO E. A. and PEREVERZEV V. N. (1969). Nitrogen regime and biological activity of peat-bog soil of the Kola peninsula. *Agrokhimiya* **8**, 3–13.

AMBROŽ Z. (1965). The proteolytic complex decomposing proteins in soil. *Rostl. Výroba* **11**, 161–170.

AMBROŽ Z. (1966a). Some reasons for differences in the occurrence of proteases in soils. *Rostl. Výroba* **12**, 1203–1210.

AMBROŽ Z. (1966b). Some notes on the determination of activities of certain proteases in the soil. *Sb. Vys. Šk. Zeměd. Brně* A1, 57–62.

AMBROŽ Z. (1966c). Adsorption of two proteases occurring frequently in soil. *Sb. Vys. Šk. Zeměd. Brně* A2, 161–167.

AMBROŽ Z. (1968). Investigation of the effects of soil structure and associated factors on the activity of proteases. *Rostl. Výroba* **14**, 201–208.

AMBROŽ Z. (1970). Factors influencing distribution of some proteolytic enzymes in soil. *Zbl. Bakt. Parasitenkd. Abt. II* **125**, 433–437.

AMBROŽ Z. (1971). The determination of activities of gelatinase and caseinase in the soil. *Biol. Sol.* **13**, 28–29.

AMBROŽ Z. (1973). Study of the cellulase complex in soil. *Rostl. Výroba* **19**, 207–212.

ANTONIANI C., MONTANARI T. and CAMORIANO A. (1954). Soil enzymology. I. Cathepsin-like activity. A preliminary note. *Ann. Fac. Agric. Univ. Milano* **3**, 99–101.

AOMINE S. and KOBAYASHI Y. (1964). Effects of allophane on the enzymatic activity of a protease. *Soil Pl. Fd., Tokyo* **10**, 28–32.

ARDAKANI M. S., VOLZ, M. G. and MCLAREN A. D. (1975). Consecutive steady-state reactions of urea, ammonium and nitrite nitrogen in soil. *Can. J. Soil Sci.* **55**, 83–91.

BALICKA N. and SOCHACKA Z. (1959). Biological activity in light soils. *Zesz. Problem. Postęp. Naukrol.* **21**, 257–265.

BALICKA N. and TRZEBINSKI M. (1956). Enzyme activity and the presence of vitamin B_2 in soils. *Acta Microbiol Polon.* **5**, 377–384.

BAROCCIO A. (1958). Catalase activity of soil as a biopedological index of fertility. *Agrochimica* **2**, 243–257.

BARTHA R. and BORDELEAU L. M. (1969a). Transformation of herbicide-derived chloroanilines by cell free peroxidases in soil. *Bact. Proc. 4.* A26.

BARTHA R. and BORDELEAU L. M. (1969b). Cell-free peroxidases in soil. *Soil Biol. Biochem.* **1**, 139–143.

BARTHOLOMEW W. V. (1965). Mineralization and immobilization of nitrogen in the decomposition of plant and animal residues. *In* "Soil Nitrogen" (W. V. Bartholomew and F. E. Clark, Eds) Vol. 10, 285–306, Amer. Soc. Agron., Inc., Madison.

BAUZON D., VAN DEN DRIESSCHE R. and DOMMERGUES Y. (1968). Respirometric and enzymatic characterization of surface horizons of forest soils. *Sci. Sol.* **2**, 55–77.

BAUZON D., VAN DEN DRIESSCHE R. and DOMMERGUES Y. (1969) Litter effect 1. The *in situ* influence of forest litters on several biological characteristics of soils. *Oecol. Plant.* **4**, 99–122.

BECK T. (1971). Determination of the catalase activity of soils. *Z. Pflanzenernähr. Düng. Bodenkd.* **130**, 68–81.

BECK T. and POSCHENRIEDER H. (1963). Experiments on the effect of toluene on the soil microflora. *Pl. Soil* **18**, 346–357.

BEI-BIENKO N. V. (1970). Effect of mineral nitrogen fertilizers on enzyme activity in soil. *Pochvovedenie* (2), 87–93.

BENEFIELD C. B. (1971). A rapid method for measuring cellulase activity in soils. *Soil Biol. Biochem.* **3**, 325–329.

BENOIT R. E. and STARKEY R. L. (1968). Enzyme inactivation as a factor in the inhibition of decomposition of organic matter by tannins. *Soil Sci.* **105**, 203–208.

BERGER-LANDEFELDT U. (1965). Activity of some soil enzymes under different plant associations. *Flora, Jena* **155**, 452–473.

BHAVANANDAN V. P. and FERNANDO V. (1970). Studies on the use of urea as a fertilizer for tea in Ceylon. 2. Urease activity in tea soils. *Tea Q.* **41**, 94–106.

BLAGOVESHCHENSKAYA Z. K. and DANCHENKO N. A. (1974). The activity of soil enzymes with prolonged application of fertilizers to maize in monoculture and to crops of a farm rotation. *Pochvovedenie* (10), 124–130.

BORDELEAU L. M. and BARTHA R. (1969). Rapid technique for enumeration and isolation of peroxidase-producing microorganisms. *Appl. Microbiol.* **18**, 274–275.

BORDELEAU, L. M. and BARTHA, R. (1972). Biochemical transformations of herbicide-derived anilines in culture medium and in soil. *Can. J. Microbiol.* **18**, 1857–1864.

BRAMS W. H. and MCLAREN A. D. (1974). Phosphatase reactions in a column of soil. *Soil Biol. Biochem.* **6**, 183–189.

BREMNER J. M. (1965). Organic nitrogen in soils. *In* "Soil Nitrogen" (W. V. Bartholomew and F. E. Clark, Eds) Vol. 10, pp. 93–149, Amer. Soc. Agron., Inc., Madison.

BREMNER J. M. and DOUGLAS L. A. (1971). Inhibition of urease activity in soils. *Soil Biol. Biochem.* **3**, 297–307.

BREMNER J. M. and DOUGLAS L. A. (1973). Effect of some urease inhibitors on urea hydrolysis in soils. *Soil Sci. Soc. Am. Proc.* **37**, 225–226.

BREMNER J. M. and TABATABAI M. A. (1973). Effect of some inorganic substances on TTC assay of dehydrogenase activity in soils. *Soil Biol. Biochem.* **5**, 385–386.

BREMNER J. M. and ZANTUA M. I. (1975). Enzyme activity in soils at subzero temperatures. *Soil Biol. Biochem.* **7**, 383–387.

BUNDY L. G. and BREMNER J. M. (1973). Effects of substituted p-benzoquinones on urease activity in soils. *Soil Biol. Biochem.* **5**, 847–853.

BURANGULOVA M. N. and KHAZIEV F. Kh. (1965a). Effect of mineral fertilizers on the phosphatase activity in soil. *Agrokém. Talajt.* **14**, 101–110.

BURANGULOVA M. N. and KHAZIEV F. Kh. (1965b). The nuclease activity of soils *Biol. Nauki* **3**, 198–201.

BURGE W. D. (1973). Transformation of propanil-derived 3, 4-dichloroaniline in soil to 3, 3′, 4, 4′-tetrachloroazobenzene as related to soil peroxidase activity. *Soil Sci. Soc. Am. Proc.* **37**, 392–395.

BURNS R. G., EL-SAYED M. H. and MCLAREN A. D. (1972a). Extraction of a urease-active organo-complex from soil. *Soil Biol. Biochem.* **4**, 107–108.

BURNS R. G., PUKITE A. H. and MCLAREN A. D. (1972b). Concerning the location and persistence of soil urease. *Soil Sci. Soc. Am. Proc.* **36**, 308–311.

CACCO G. and MAGGIONI A. (1976). Multiple forms of acetyl-naphthyl-esterase activity in soil organic matter. *Soil Biol. Biochem.* **8**, 321–325.

CASIDA L. E., KLEIN D. A. and SANTORO T. (1964). Soil dehydrogenase activity. *Soil Sci.* **98**, 371–376.

CERNÁ S. (1966). Enzymatic activity of soil in relation to its structure. *Acta Univ. Carol. Biol.* **1**, 83–89.

CERNÁ S. (1968). β-Glucosidase activity in structural soil. *Acta Univ. Carol. Bot.* **1967**, 267–272.

CERNÁ S. (1969). The influence of desiccation of structural soil upon the intensity of biochemical reactions. *Acta Univ. Carol. Biol.* **1968**, 285–288.

CERNÁ S. (1970). The influence of crushing of structural aggregates on the activity of hydrolytic enzymes in the soil. *Acta Univ. Carol. Biol.* **6**, 461–466.

CERNÁ S. (1973). Inhibition of dehydrogenase activity in the soil. *Rostl. Výoba* **19**, 51–58.

CERVELLI S., NANNIPIERI P., CECCANTI B. and SEQUI P. (1973). Michaelis constants of soil acid phosphatase. *Soil Biol. Biochem.* **5**, 841–845.

CHALVIGNAC M. A. (1968). Evidence and preliminary study of a soil enzyme capable of decomposing tryptophane in the absence of active bacteria. *C. R. Acad. Sci. Ser. D* **266**, 637–639.

CHALVIGNAC M. A. (1971). Stability and activity of a tryptophane-degrading enzyme system in a variety of soil types. *Soil Biol. Biochem.* **3**, 1–7.

CHALVIGNAC, M. A. and MAYAUDON J. (1971). Extraction and study of soil enzyme metabolizing tryptophan. *Pl. Soil* **34**, 25–31.

CHUNDEROVA A. I. (1964). Phosphatase activity of corn rhizosphere microflora. *Mikrobiologiya* **33**, 89–92.

CHUNDEROVA A. I. (1970). Activity of enzymes of soil and its pH. *Agrokhimiya* (5), 71–77.

CONRAD J. P. (1942). The occurrence and origin of urease-like activities in soils. *Soil Sci.* **54**, 367–380.

COOPER P. J. M. (1972). Arylsulphatase activity in Northern Nigerian soils. *Soil Biol. Biochem.* **4**, 333–337.

CORTEZ, J., LOSSAINT, P. and BILLÉS G. (1972). Soil biological activity in Mediterranean ecosystems. III. Enzymatic activity. *Rev. Écol. Biol. Sol.* **9**, 1–19.

CORTEZ J., BILLÉS G. and LOSSAINT P. (1975). A comparative study of the biological activity of soils under shrub and herbaceous vegetation of the Mediterranean garrigue. II. Enzymatic activities. *Rev. Écol. Biol. Sol.* **12**, 141–156.

DALAL R. C. (1975). Effect of toluene on the energy barriers in urease activity of soils. *Soil Sci.* **120**, 256–260.

DEDEKEN M. and VOETS J. P. (1965). Studies on the metabolism of amino acids in soil. I. Metabolism of glycine, alanine, aspartic acid and glutamic acid. *Suppl. Ann. Inst. Pasteur., Paris* **109**, 103–111.

DOUGLAS L. A., RIAZI-HAMADANI A. and FIELD, J. F. B. (1976). Assay of pyrophosphatase activity in soil. *Soil Biol. Biochem.* **8**, 391–393.

DRĂGAN-BULARDA M. and KISS S. (1972a). Dextranase activity in soil. *Soil Biol. Biochem.* **4**, 413–416.

DRĂGAN-BULARDA M. and KISS S. (1972b). Occurrence of dextransucrase in soil. *3rd Symp. Soil Biol.*, National Society of Soil Scientists, Bucharest, 119–128.

DROBNÍK J. (1955). Degradation of starch by the enzyme complex of soils. *Folia Biol.* **1**, 29–40.

DROBNÍK J. (1956). Degradation of asparagine by the enzyme complex of soils. *Cesk. Mikrobiol.* **1**, 47.

DROBNÍK J. (1961). On the role of toluene in the measurement of the activity of soil enzymes. *Pl. Soil* **14**, 94–95.

DROBNIKOVA V. (1961). Factors influencing the determination of phosphatases in soil. *Folia Microbiol.* **6**, 260–267.

DROZDOWICZ A. (1971). The behaviour of cellulase in soil. *Rev. Microbiol.* **2**, 17–23.

DUBOVENKO E. K. (1964). Phosphatase activity of different soils. *Zhivlennya ta Udobr. Sil.-gospod. Kul't., Ukr. Nauk. Dosl. Inst. Zemberobstva*, pp. 29–32.

DURAND G. (1961a). Sur la Degradation Aérobic de L'Acide Urique par le Sol. Thèse doct., Université de Toulouse.

DURAND G. (1961b). Degradation of purine and pyrimidine bases in soil: aerobic degradation of uric acid. *C. R. Acad. Sci.* **252**, 1687–1689.

DURAND G. (1965). Enzymes in soil. *Rev. Écol. Biol. Sol.* **2**, 141–205.

DURAND G. (1966). Sc.D. Thesis, University of Toulouse.

ENSMINGER L. E. and GIESEKING J. E. (1942). Resistance of clay-adsorbed proteins to proteolytic hydrolysis. *Soil Sci.* **53**, 205–209.

ESTERMANN E. F., PETERSON G. H. and MCLAREN A. D. (1959). Digestion of clay-protein, lignin-protein and silica-protein by enzymes and bacteria. *Soil Sci. Soc. Am. Proc.* **23**, 31–36.

FERMI C. (1910). On the presence of enzymes in soil, water and dust. *Zbl. Bakt. Parasitenkd., Abt. II* **26**, 330–334.

FILEV D. S., PASHOVA V. T. and YAROSHEVICH I. V. (1968). Effect of a single application of high rates of fertilizers on the productivity of maize and soil properties. *Dokl. Vses. Akad. Sel's kokhoz. Nauk* **12**, 2–4.

FOSTER R. C. (1977). Ultramicromorphology of some South Australian soils. Commission 1 Meeting, I.S.S.S. Adelaide 1976, in press.

FOSTER R. C. and ROVIRA A. D. (1976). Ultrastructure of wheat rhizosphere. *New Phytol.* **76**, 343–352.

FRANZ G. (1973). Comparative investigations on the enzyme activity of some soils in Nordrhein-Westfalen and Rheinland-Pfalz. *Pedobiologia* **13**, 423–436.

FREYTAG H. E. (1965). The saccharase activity in soil. *Albrecht Thaer. Arch.* **9**, 47–66.

GALSTYAN A. S. (1958). Determination of the comparative activity of peroxidase and polyphenoloxidase in soil. *Dokl. Akad. Nauk Arm. SSR* **26**, 285–288.

GALSTYAN A. S. (1964). Dehydrogenases of Soil. *Dokl. Akad. Nauk* **156**, 166–167.

GALSTYAN A. S. (1965a). Dynamics of enzymatic processes of soils. *Dokl. Akad. Nauk Arm. SSR* **40**, 39–42.

GALSTYAN A. S. (1965b). Method of determining activities of hydrolytic enzymes of soil. *Pochvovedenie* (2), 68–74.

GALSTYAN A. S. (1973). Formation of easily hydrolysable nitrogen in the soil. *Biol. Zh. Arm.* (26), 15–19.

GALSTYAN A. S. (1974). Fermentativnaya Aktivnost' Pochv. Armenii. Ayastan, Erevan.

GALSTYAN A. S. and AVUNDZHYAN Z. S. (1967). Enzyme activity of fine earth under lichens and mosses on rocks. *Dokl. Akad. Nauk Arm. SSR* **45**, 78–81.

GALSTYAN A. S. and AVUNDZHYAN Z. S. (1970). Dehydrogenases of the clay fraction of soil. *Dokl. Akad. Nauk SSSR* **195**, 707–709.

GALSTYAN A. S. and MARUKYAN L. G. (1973). Determination of ascorbate oxidase activity in the soil. *Dokl. Akad. Nauk Arm. SSR* **57**, 100–102.

GALSTYAN A. S. and SAAKYAN E. G. (1973). Determination of soil glutaminase activity. *Dokl. Akad. Nauk SSSR* **209**, 1201–1202.

GAVRILOVA A. N., SAVCHENKO N. I. and SHYMKO N. A. (1974). Forms of phosphorus and the phosphatase activity of the principal soil types in the BSSR. *Trans. 10th Int. Congr. Soil Sci.* **4**, 281–289.

GETZIN L. W. and ROSEFIELD I. (1968). Organophosphorus insecticide degradation by heat-labile substances in soil. *J. Agric. Fd Chem.* **16**, 598–601.

GETZIN L.W. and ROSEFIELD I. (1971). Partial purification and properties of a soil enzyme that degrades the insecticide malathion. *Biochim. Biophys. Acta* **235**, 442–453.

GIARDINA M. C., TOMATI U. and PIETROSANTI W. (1970). Hydrolytic activities of the soil treated with paraquat. *Meded. Rijksafac. Landbouwett., Gent* **35**, 615–626.

GILOT J. C. and DOMMERGUES Y. (1967). Note on the calcareous stony mor of the subalpine station of R.C.P. 40. *Rev. Écol. Biol. Sol.* **4**, 357–383.

GNITTKE J. and KUNZE C. (1975). The influence of tannic acid on catalase activity of soil samples. *Zbl. Bakt. Parasitenkd. Abt.* **130**, 37–40.

GOLDSTEIN J. L. and SWAIN T. (1965). The inhibition of enzymes by tannins. *Phytochemistry* **4**, 185–192.

GOULD W. D., COOK F. D. and WEBSTER G. R. (1973). Factors affecting urea hydrolysis in several Alberta soils. *Pl. Soil* **38**, 393–401.

HAIG A. D. (1955). Some characteristics of esterase- and urease-like activity in the soil. Ph.D. Dissertation, Univ. California, Davis.

HALSTEAD R. L. (1964). Phosphatase activity of soils as influenced by lime and other treatments. *Can. J. Soil Sci.* **44**, 137–144.

HALSTEAD R. L. and SOWDEN F. J. (1968). Effect of long-term additions of organic matter on crop yields and soil properties. *Can. J. Soil Sci.* **48**, 341–348.

HANDLEY W. R. C. (1961). Further evidence for the importance of residual leaf protein complexes in litter decomposition and the supply of nitrogen for plant growth. *Pl. Soil* **15**, 37–73.

HAYANO K. (1973). A method for the determination of β-glucosidase activity in soil. *Soil Sci. Pl. Nutr.* **19**, 103–108.

HAYANO K. and SHIOJIMA M. (1974). Estimation of β-glucosidase activity in soils. *Trans. 10th Int. Congr. Soil Sci.* **3**, 136–142.

HIRATE W. (1963). Notes on the determination of dehydrogenase activity in soil. *Zentrabl. Bakteriol. Parasitenkd., Abt. II* **116**, 478–484.

HOBBIE J. E. and CRAWFORD C. C. (1969). Respiration correction for bacterial uptake of dissolved organic compounds in natural waters. *Limnol. Oceanogr.* **14**, 528–532.

HOFFMANN G. (1959). Investigations on the synthetic effect of enzymes in soil. *Z. Pflanzenernähr. Düng. Bodenkd.* **85**, 193–201.

HOFFMANN G. (1963). Synthetic effects of soil enzymes. Recent Progress in Microbiology. *Symposium 8th Int. Congr. Microbiol.,* pp. 230–234. Montreal.

HOFFMANN G. (1968a). Phosphatases in the enzyme system of cultivated soils (in Germany) and possibilities of determining their activity. *Z. Pflanzenernähr. Düng. Bodenkd.* **118**, 153–160.

HOFFMANN G. (1968b). A photometric determination of phosphatase activity in soil. *Z. Pflanzenernähr. Düng. Bodenkd.* **118**, 161–172.

HOFFMANN G. and DEDEKEN M. (1965). A colorimetric method for determining β-glucosidase activity in soils. *Z. Pflanzenernähr. Düng. Bodenkd.* **108**, 193–198.

HOFFMANN G. and ELIAS-AZAR K. (1965). Various factors in soil fertility of Northern Iran soils and their relationship to the activities of hydrolytic enzymes. *Z. Pflanzenernähr. Düng. Bodenkd.* **108**, 199–217.

HOFFMANN G. and PALLAUF J. (1965). A colorimetric method for determining saccharase activity in soils. *Z. Pflanzenernähr, Düng. Bodenkd.* **110**, 193–201.

HOFFMANN G. and TEICHER K. (1957). The enzyme system of our arable soils. VII. Proteases. *Z. Pflanzenernähr. Düng. Bodenkd.* **77**, 243–251.

HOFFMANN G. and TEICHER K. (1961). A colorimetric technique for the determination of urease activity in soils. *Z. Pflanzenernähr. Düng. Bodenkd.* **95**, 55–63.

HOFMANN E. and HOFFMANN G. (1953). Occurrence of a- and β-glycosidases in the soil. *Naturwissenschaften* **40**, 511.

HOFMANN E. and HOFFMANN G. (1954). The enzyme system of our arable soils. V. α- and β-Galactosidase and α-glucosidase. *Biochem. Z.* **325**, 329–332.

HOFMANN E. and HOFFMANN G. (1955). The enzyme system of our arable soils. IV. Amylase. *Z. Pflanzenernähr. Düng. Bodenkd.* **70**, 97–104.

HOFMANN E. and KESSEBA A. (1962). Investigations of enzymes in Egyptian soils. *Z. Pflanzenernähr. Düng. Bodenkd.* **99**, 9–20.

HOFMANN E. and NIGGEMANN J. (1953). The enzyme system of our arable soils. III. Proteinase. *Biochem. Z.* **324**, 308–310.

HOFMANN E. and SCHMIDT W. (1953). The enzyme system of our arable soils. II. Urease. *Biochem. Z.* **324**, 125–127.

HOFMANN E. and SEEGERER A. (1951). The enzyme system of our arable soils. I. Saccharase. *Biochem. Z.* **322**, 174–179.

IRVING G. C. and COSGROVE D. J. (1976). The kinetics of soil acid phosphatase. *Soil Biol. Biochem.* **8**, 335–340.

ISHIZAWA S. and TANABE I. (1968). Urease activity of Japanese soils. *In* "Progress in Soil Biodynamics and Soil Productivity" (A. Primavesi, Ed.) 105 Pallotti, Santa Maria, Brazil.

JACKMAN R. H. and BLACK C. A. (1952). Phytase activity in soils. *Soil Sci.* **73**, 167–171.

JOHNSON J. L. and TEMPLE K. L. (1964). Some variables affecting the measurement of "catalase activity" in soil. *Soil Sci. Soc. Am. Proc.* **28**, 207–209.

JONES D. and WEBLEY D. M. (1968). A new enrichment technique for studying lysis of fungal cell walls in soil. *Pl. Soil* **28**, 147–157.

KAISER P. and MONZON DE ASCONEGUI M. S. (1971). Measurement of pectinolytic enzyme activity in soil. *Biol. Sol* **14**, 16–19.

KEILLING J., CAMUS A., SAVIGNAC G. DAUCHEZ P., BOITEL M. and PLANET (1960). Contribution à l'étude de la biologie des sols. *C. R. Acad. Agric. Fr.* **46**, 647–652.

KHAN S. U. (1970). Enzymatic activity in a grey wooded soil as influenced by cropping systems and fertilizers. *Soil Biol. Biochem.* **2**, 137–139.

KHAZIEV F. KH. (1966). Dependence of the nuclease activity of soil on the pH and influence of various substances on it. *Biol. Nauki* (2), 223–227.

KHAZIEV F. KH. (1972a). Determination of pyrophosphatase activity of soil. *Pochvovedenie* (2), 67–73.

KHAZIEV F. KH. (1972b). Determination of phosphatase activity in soils. *Biol. Sol* **16**, 22–23.

KISLITSINA V. P. (1965). Cellulase activity in soils of some East Siberian environments depending on conditions of the geographic area. *Dokl. Inst. Geogr. Sibir. Daln. Vostoka* **10**, 40–44.

KISLITSINA V. P. (1968). Cellulase activity of East Siberian forest and steppe soils. *Les Pochva, Tr. Vses. Nauch. Konf. Les. Pochvoved.* **1965**, 417–422.

KISLITSINA V. P. (1971). Method for the determination of cellulase activity in soils. *Mikrobiol. Biokhim. Issled. Pochv. Mater Nauch. Konf.* **1969**, 111–115.

KISS S. (1957). The invertase activity of earthworm casts and soils from ant hills. *Agrokém. Talajtan* **6**, 65–68.

KISS S. (1958a). Talajenzimek. *In* "Talajtan" (M. J. Csapó, Ed.) pp. 491–622 Agro-Silvica, Bucharest.

KISS S. (1958b). New data regarding the identity of soil saccharase and soil α-glucosidase (maltase). *Stud. Univ. Babes-Bolyai Sér. Biol.* **2**, 51–55.

KISS S. and BOARU, M. (1965). Methods for the determination of dehydrogenase activity in soil. *Symposium on Methods in Soil Biology, Bucharest*, 137–143.

KISS S. and DRĂGAN-BULARDA M. (1968). Levan-sucrase activity in soil under conditions unfavourable for the growth of microorganisms. *Rev. Roum. Biol. Ser. Bot.* **13,** 435–438.

KISS S. and DRĂGAN-BULARDA M. (1970). Levan formation and decomposition in soil. *Microbiol. Lúcr. Conf. Nat. Microbiol. Gen. Appl., Bucharest* **1968,** 483–486.

KISS S. and DRĂGAN-BULARDA M. (1972). Persistence of levansucrase activity in soil in the presence of chloromycetin. *Stud. Univ. Babes-Bolyai Ser. Biol.* **2,** 139–144.

KISS S. and PÉTERFI S. (1960a). Importance of substrates in determining and comparing maltase (α-glucosidase) and lactase (β-glucosidase) activities in soil. *Stud. Univ. Babes-Bolyai Ser. Biol.* **2,** 275–276.

KISS S. and PÉTERFI S. (1960b). Inhibition of the activity of soil maltase. *Pochvovedenie* (8), 84–86.

KISS S. and PÉTERFI S. (1961a). Presence of carbohydrases in the peat of a Salicea community. *Stud. Cercet. Biol., Cluj* **12,** 209–216.

KISS S. and PÉTERFI S. (1961b). A chromatographic method for determining phosphomonoesterases in soil. *Stud. Univ. Babes-Bolyai, Ser. Biol.* **2,** 292–296.

KISS S., BOSICA I. and POP M. (1962). On the enzymatic degradation of lichenin in soil. *Contrib. Bot., Cluj* 335–340.

KISS S., BOARU M. and CONSTANTINESCU L. (1965). Levanase activity in soil. *Symposium on Methods in Soil Biology, Bucharest*, 129–136.

KISS S., NEMES M. and DRĂGAN-BULARDA M. (1971). The saccharase activity of some soils of the Vladeasa Mountains (Rumania). *Stiinta Sol* **9,** 17–25.

KISS S., DRĂGAN-BULARDA M. and KHAZIEV F. K. (1972a). Influence of chloromycetin on the activities of some oligases of soils. *Lucr. Conf. Nat. Stiinta Solului* **1970,** 451–462.

KISS S., DRĂGAN-BULARDA M. and RADULESCU D. (1972b). Biological significance of the enzymes accumulated in soil. *3rd Symp. Soil Biology, Bucharest*, 19–78.

KISS S., STEFANIC G. and DRĂGAN-BULARDA M. (1974). Soil enzymology in Romania. II. *Contrib. Bot., Cluj* 197–207.

KISS S., DRĂGAN-BULARDA M. and RADULESCU D. (1975). Biological significance of enzymes accumulated in soil. *Adv. Agron.* **27,** 25–87.

KLEIN D. A. LOH T. C. and GOULDING R. L. (1971). A rapid procedure to evaluate the dehydrogenase activity of soils low in organic matter. *Soil Biol. Biochem.* **3,** 385–387.

KLEINERT H. (1962). Determination of the invertase activity of soil and the decrease of activity under constant conditions of storage. *Albrecht Thaer. Arch.* **6,** 477–484.

KOBAYASHI Y. and AOMINE S. (1967). Mechanism of inhibitory effect of allophane and montmorillonite on some enzymes. *Soil Sci. Pl. Nutr., Tokyo* **13,** 189–194.

KONG K. T., and DOMMERGUES Y. (1970). Limitation of cellulolysis in organic soils. 1. Respirometric study. *Rev. Écol. Biol. Sol* **7,** 441–456.

KONG K. T. and DOMMERGUES Y. (1972). Limitation of cellulolysis in organic soils. II. Study of the enzymes in soil. *Rev. Écol. Biol. Sol* **9,** 629–640.

KONG K. T., BALANDREAU, J. and DOMMERGUES Y. (1971). Measurement of cellulase activity in organic soils. *Biol. Sol.* **13,** 26–27.

KOZLOV K. (1964). Enzymatic activity of the rhizosphere and soils in the East Siberia area. *Folia Microbiol.* **9**, 145–149.

KOZLOV K. A. and MIKHAILOVA E. N. (1965). Dehydrogenase activity of some soils of eastern Siberia. *Pochvovedenie* (2), 58–63.

KOZLOVSKAYA N. A., NIKITINA G. D. and RUNKOV S. V. (1972). A rapid micro-diffusion method for the determination of the urease activity of peat-bog soils. *Agrokhimiya* **5**, 144–149.

KRÁMER M. (1957). Phosphatase activity as an indicator of biologically useful phosphorus in soil. *Naturwissenschaften* **44**, 13.

KRÁMER M. and ERDEI S. (1958). Investigation of the phosphatase activity of soils by means of disodium phenyl-phosphate. I. Method. *Agrokém. Talajt.* **7**, 361–366.

KRÁMER M. and YERDEI G. (1960). Application of the method of phosphatase activity determination in agricultural chemistry. *Sov. Soil Sci.* (9), 1100–1103.

KRASIL'NIKOV N. A. and KOTOLEV V. V. (1959). Adsorption of phosphatases of soil microorganisms by corn roots. *Mikrobiologiya* **28**, 515–517.

KROLL, L. and KRÁMER M. (1955). The influence of clay minerals on the activity of soil phosphatase. *Naturwissenschaften* **42**, 157–158.

KUDZIN Y. K., YAROSHEVICH I. V. and GUBENKO V. A. (1968). Mobilization of phosphorus reserves in chernozem during prolonged application of nitrogen and phosphorus fertilizers. *Dokl. Vses. Akad. Sel'skokhoz. Nauk* **12**, 6–8.

KUNZE C. (1970). The effect of streptomycin and of aromatic carboxylic acids on catalase activity of soil samples. *Zentralbl. Bakteriol. Parasitenkd., Abt. II* **124**, 658–661.

KUNZE C. (1971). Catalase activity in soil samples as influenced by tannin, gallic acid and *p*-hydroxybenzoic acid. *Oecol. Plant.* **6**, 197–201.

KUNZE C. and RICKART H. (1973). Determination of saccharase activity in soil samples. *Experientia* **29**, 641–642.

KUDRYACHEV A. I. (1968). Effect of peat on the enzymatic activity of light soils. *Vestsi Akad. Navuk Belarusk. SSR. Ser. Biyal. Navuk* **1**, 49–52.

KUPREVICH V. F. (1951). The biological activity of soil and methods for its determination. *Dokl. Akad. Nauk SSSR* **79**, 863–866.

KUPREVICH V. F. and SHCHERBAKOVA T. A. (1956). Determination of invertase and catalase activity of soils. *Vestsi Akad. Navuk Belarusk. SSR, Ser. Biyal.* **2**, 115–116.

KUPREVICH V. F. and SHCHERBAKOVA T. A. (1961). Method for determining proteolytic activity of the soil. *Dokl. Akad. Nauk Belorus. SSR* **3**, 133–136.

KUPREVICH V. F. and SHCHERBAKOVA T. A. (1971a). *Pochvennaya Enzimologiya* Nauka Tekh. Minsk, 1966 (transl. from Russian). Soil Enzymes. US Department of Commerce, National Technical Information Service, Springfield Virginia.

KUPREVICH V. F. and SHCHERBAKOVA T. A. (1971b). Comparative enzymatic activity in diverse types of soil. *In* "Soil Biology" Vol. 2 (A. D. McLaren and J. J. Skujiņš, Eds) pp. 167–201, Marcel Dekker, Inc., New York.

KUPREVICH V. F., SHCHERBAKOVA T. A. and TSYUPA G. P. (1966). Soil urease activity. *Dokl. Akad. Nauk Belorus SSR* **10**, 336–338.

KÜSTER E. and GARDINER J. J. (1968). Influence of fertilizers on microbial activities of peatland. *Proc. 3rd Int. Peat Congr.* 314–317.

LADD J. N. (1972). Properties of proteolytic enzymes extracted from soil. *Soil Biol. Biochem.* **4**, 227–237.

LADD J. N. and BUTLER J. H A. (1969a). Inhibitory effect of soil humic compounds on the proteolytic enzyme, Pronase. *Aust. J. Soil Res.* **7**, 241–251.

LADD J. N. and BUTLER J. H. A. (1969b). Inhibition and stimulation of proteolytic enzyme activities by soil humic acids. *Aust. J. Soil Res.* **7**, 253–261.

LADD J. N. and BUTLER J. H. A. (1972). Short-term assays of soil proteolytic enzyme activities using proteins and dipeptide derivatives as substrates. *Soil Biol. Biochem.* **4**, 19–30.

LADD J. N. and BUTLER J. H. A. (1975). Humus-enzyme systems and synthetic organic polymer-enzyme analogs. In "Soil Biochemistry" Vol. 4 (E. A. Paul, and A. D. McLaren, Eds), pp. 143–194, Marcel Dekker, New York.

LADD J. N. and PAUL E. A. (1973). Changes in enzymic activity and distribution of acid-soluble, amino acid nitrogen in soil during nitrogen immobilization and mineralization. *Soil Biol. Biochem.* **5**, 825–840.

LADD J. N., BRISBANE P. G., BUTLER J. H. A. and AMATO M. (1976). Studies on soil fumigation. III. Effects on enzyme activities, bacterial numbers and extractable ninhydrin reactive compounds. *Soil Biol. Biochem.* **8**, 255–260.

LATÝPOVA R. M. (1965). Effect of environmental conditions on activity of soil enzymes. *Trudý Beloruss. Sel'Skokhoz. Akad.* **37**, 60–65.

LAUGESEN K. (1972). Urease activity in Danish soils. *Tidsskr. Pl.* **76**, 221–229.

LAUGESEN K. and MIKKELSEN J. P. (1973). Dehydrogenase activity in Danish soils. *Tidsskr. Pl.* **77**, 516–520.

LENHARD G. (1956). The dehydrogenase activity in soil as a measure of the activity of soil microorganisms. *Z. Pflanzenernähr. Düng. Bodenkd.* **73**, 1–11.

LENHARD G. (1966). The dehydrogenase activity for the study of soils and river deposits. *Soil Sci.* **101**, 400–402.

LLOYD A. B. and SHEAFFE M. J. (1973). Urease activity in soils. *Pl. Soil* **39**, 71–80.

LIBBERT E. and PAETOW W. (1962). Untersuchungen über die enzymatische Hydrolyse von Indol-3-acetonitril und Oxydation von Indol-3-aldehyd in Erdboden und in roher Milch. *Flora, Jena* **152**, 540–544.

MACURA J. and VÁGNEROVÁ K. (1969). Colorimetric determination of the activity of proteolytic enzymes in soil. *Rostl. Výroba* **15**, 173–180.

MARKOSYAN L. V. and GALSTYAN A. H. (1963). Optimum pH of some hydrolases of soil. *Isv. Akad. Nauk Arm. SSR Biol. Nauki* **16**, 45–52.

MÁRKUS L. (1955). Determination of carbohydrates from plant materials with anthrone reagent. II. Assay of cellulase activity in soil and farmyard manure. *Agrokem. Talajt.* **4**, 207–216.

MARTIN-SMITH M. (1963). Uricolytic enzymes in soil. *Nature* **197**, 361–362.

MAYAUDON J. (1968). Stabilization biologique des protéines ^{14}C dans le sol. *In* "Isotopes and radiation in soil organic matter studies", pp. 177–188, International Atomic Energy Agency, Vienna.

MAYAUDON J. and SARKAR J. M. (1974a). Etude des diphenol oxydases extraites d'une litiere de foret. *Soil Biol. Biochem.* **6**, 269–274.

MAYAUDON J. and SARKAR J. M. (1974b). Chromatographie et purification des diphenol oxydases du sol. *Soil Biol. Biochem.* **6**, 275–285.

MAYAUDON J. and SARKAR J. M. (1975). Laccases of *Polyporus versicolor* in soil and litter. *Soil Biol. Biochem.* **7**, 31–34.

90   J. N. LADD

MAYAUDON J., EL-HALFAWI M. and BELLINCK C. (1973a). Decarboxylation of aromatic 1-¹⁴C amino acids by soil extracts. *Soil Biol. Biochem.* **5**, 355–367.

MAYAUDON J., EL-HALFAWI M. and CHALVIGNAC M. A. (1973b). Properties of diphenol oxidases extracted from soils. *Soil Biol. Biochem.* **5**, 369–383.

MAYAUDON J., BATISTIC, L. and SARKAR J. M. (1975). Properties of proteolytically active extracts from fresh soils. *Soil Biol. Biochem.* **7**, 281–286.

MCGARITY J. W. and MYERS M. G. (1967). A survey of urease activity in soils of northern New South Wales. *Pl. Soil* **27**, 217–238.

MCLAREN A. D. (1954). The adsorption and reaction of enzymes and proteins on kaolinite. II. The action of chymotrypsin on lysozyme. *Soil Sci. Soc. Am. Proc.* **18**, 170–174.

MCLAREN A. D. and SKUJIŅŠ J. J. (1967). The physical environment of microorganisms in soils. In "The Ecology of Soil Bacteria" (T. R. G. Gray and D. Parkinson, Eds), pp. 3–24, Liverpool Univ. Press.

MCLAREN A. D. RESHETKO L. and HUBER W. (1957). Sterilization of soil by irradiation with an electron beam and some observations on soil enzyme activity. *Soil Sci.* **83**, 497–502.

MCLAREN A. D. LUSE R. A. and SKUJIŅŠ J. J. (1962). Sterilization of soil by irradiation and some further observations on soil enzyme activity. *Soil Sci. Soc. Am. Proc.* **26**, 371–377.

MCLAREN A. D., PUKITE A. H. and BARSHAD I. (1975). Isolation of humus with enzymatic activity from soil. *Soil Sci.* **119**, 178–180.

MERESHKO M. Y. (1969). Effect of herbicides on biological activity of microflora in chernozem soils. *Mikroobiol. Zh, Kiev* **31**, 525–529.

MOURARET M. (1965). Contribution à l'étude de l'activité des enzymes du sol: L'asparaginase. ORSTOM, Paris.

MOUREAUX C. (1965). Glycolysis and total microbiological activity in some West African soils. *Cah. Pédol.* ORSTOM **3**, 43–78.

MURESANU P. L. and GOIAN M. (1969). Investigation of the phosphomonoesterase activity in fertilized and limed soils. *Agrokém. Talajt.* **18**, 102–106.

MUSA M. M. and MUKHTAR N. O. (1969). Enzymatic activity of a soil profile in the Sudan Gezera. *Pl. Soil* **30**, 153–155.

MYERS M. G. and MCGARITY J. W. (1968). The urease activity in profiles of five great soil groups from northern New South Wales. *Pl. Soil* **28**, 25–37.

NAMDEO K. N. and DUBE J. N. (1973a). Proteinase enzyme as influenced by urea and herbicides applied to grassland oxisol. *Ind. J. Exper. Biol.* **11**, 117–119.

NAMDEO K. N. and DUBE J. N. (1973b). Residual effect of urea and herbicides on hexosamine content and urease and proteinase activities in a grassland soil. *Soil Biol. Biochem.* **5**, 855–859.

NANNIPIERI P., CERVELLI S. and PEDRAZZINI F. (1975). Concerning the extraction of enzymatically active organic matter from soil. *Experientia* **31**, 513–515.

NAUMANN K. (1970). Dynamics of the soil microflora following application of insecticides. IV. Investigations on the effect of parathion-methyl on soil respiration and dehydrogenase activity. *Zentralbl. Bakteriol. Parasitenkd., Abt. II* **125**, 119–133.

NEAL J. L. (1973). Influence of selected grasses and forbs on soil phosphatase activity. *Can. J. Soil Sci.* **53**, 119–121.

NIKIFORENKO L. I. (1971). Effect of fertilizers on total content and mobility of nitrogen in grey podzolized soil. *Agrokhimiya* **5**, 10–14.

NILSSON P. E. (1957). Influence of crop on biological activities in soil. *K. Lantbrukshögsk. Annlv* **23**, 175–218.

NIZOVA A. A. (1960). The question of the biological activity of soils. *Pochvovedenie* (10), 96–101.

NIZOVA A. A. (1969). Saccharase activity in derno-podzolic heavy loamy soil. *Mikrobiologiya* **38**, 336–339.

NORSTADT F. A., FREY C. R. and SIGG H. (1973). Soil urease: paucity in the presence of the fairy ring fungus *Marasmius oreades* (*Bolt.*) *Soil Sci. Soc. Am. Proc.* **37**, 880–885.

NOWAK W. (1964). Comparative studies on the bacterial population and saccharase activity of the soil. *Pl. Soil* **20**, 302–318.

PANCHOLY S. K. and LYND J. Q. (1972). Quantitative fluorescence analysis of soil lipase activity. *Soil Biol. Biochem.* **4**, 257–259.

PANCHOLY S. K. and LYND J. Q. (1973). Interactions with soil lipase activation and inhibition. *Soil Sci. Soc. Am. Proc.* **37**, 51–52.

PANCHOLY S. K. and RICE E. L. (1972). Effect of storage conditions on activities of urease, invertase, amylase and dehydrogenase in soil. *Soil Sci. Soc. Am. Proc.* **36**, 536–537.

PANCHOLY S. K. and RICE E. L. (1973a). Soil enzymes in relation to old field succession: amylase, cellulase, invertase, dehydrogenase and urease. *Soil Sci. Soc. Am. Proc.* **37**, 47–50.

PANCHOLY S. K. and RICE E. L. (1973b). Carbohydrases in soil as affected by successional stages of revegetation. *Soil Sci. Soc. Am. Proc.* **37**, 227–229.

PAULSON K. N. and KURTZ L. T. (1969). Locus of urease activity in soil. *Soil Sci. Soc. Am. Proc.* **33**, 897–901.

PAULSON K. N. and KURTZ L. T. (1970). Michaelis constant of soil urease. *Soil Sci. Soc. Am. Proc.* **34**, 70–72.

PEARSE A. G. E. (1972). "Histochemistry" Vol. 2, pp. 1260–1302, Churchill Livingstone, London.

PETERSON N. V. (1965). Modification of the method of determining dehydrogenase activity in soil. *Mỳkrobiol Zh.* **27**, 18–22.

PETERSON N. V. (1967). Dehydrogenase activity in soil as an index of its microbial activity. *Mikrobiologiya* **36**, 518–525.

PETERSON N. V. and ASTAFYEVA E. V. (1962). Method of determining saccharase activity of soil. *Mikrobiologiya* **31**, 918–922.

PILET P. E. and CHALVIGNAC M–A. (1970). Effect on an enzyme system extracted from soil on growth and auxin catabolism of *Lens culinaris* roots. *Ann. Inst. Pasteur, Paris* **118**, 349–355.

PINCK L. A. and ALLISON F. E. (1951). Resistance of a protein-montmorillonite complex to decomposition by soil microorganisms. *Science* **114**, 130–131.

POKORNÁ V. (1964). Methods for determining lipolytic activity of raised bog and fen peats and muds. *Pochvovedenie* (1), 106–109.

PORTER L. K. (1965). Enzymes. *In* "Methods of Soil Analysis" Part 2 (C. A. Black, D. D. Evans, J. L. White, L. E. Ensminger, F. E. Clark, Eds), p. 1536, American Society of Agronomy, Madison, Wisconsin.

RADULESCU D. and KISS S. (1971). Enzyme activities in the sediments of the Zanoaga and Zanoguta lakes (Retezat mountain mass). *In* "Progress in Roumanian Palinology" (Editura Academie i Republicii Socialiste Romania) pp. 243–248. Bucharest.

RAGUOTIS A. D. (1967). Biological activity of sod-podzolic forest soils on the Lithuanian SSR. *Pochvovedenie* (6), 51–57.

RAMÍREZ-MARTÍNEZ, J. R. (1965). A method for the determination of soil phosphatase activity using Na-β-naphthylphosphate. *Soil Biol.* 4, 12–13.

RAMÍREZ-MARTÍNEZ J. R. (1966). Characterization and localization of phosphatase activity in soils. Ph.D. Thesis, University of California, Berkeley, USA.

RAMÍREZ-MARTÍNEZ J. R. (1968). Organic phosphorus mineralization and phosphatase activity in soil. *Folia Microbiol.* 13, 161–174.

RAMÍREZ-MARTÍNEZ J. R. and MCLAREN A. D. (1966a). Determination of soil phosphatase activity by a fluorimetric technique. *Enzymologia* 30, 243–253.

RAMÍREZ-MARTÍNEZ J. R. and MCLAREN A. D. (1966b). Some factors influencing the determination of phosphatase activity of native soils and in soil sterilized by irradiation. *Enzymologia* 31, 23–38.

RANKOV V. and DIMITROV G. (1971). Soil phosphatase activity resulting from manuring of tomato and cabbage. *Pochvozn.; Agrokhim.* 6, 93–98.

RAWALD W. (1968). The intensity of cellulose decomposition, occurrence of cellulose-decomposing microorganisms, and cellulase activity in some forest and arable soils. *Tagungsber., Deut. Akad. Landw, Berlin* 98, 171–179.

RAWALD W. (1970). Enzyme activity in soil as a component of the biological activity in soil with particular reference to the evaluation of soil fertility status, and aspects of the tasks of soil enzymological research. *Zentralbl. Bakteriol. Parasitenkd., Abt. II* 125, 363–384.

RAWALD W., DOMKE K. and STOHR G. (1968). Studies on the relations between humus quality and microflora of the soil. *Pedobiologia* 7, 375–380.

ROBERGE M. R. (1968). Effects of toluene on microflora and hydrolysis of urea in a black spruce humus. *Can. J. Microbiol.* 14, 999–1003.

ROBERGE M. R. and KNOWLES R. (1968a). Factors effecting urease activity in a black spruce humus sterilized by gamma radiation. *Can. J. Soil Sci.* 48, 355–361.

ROBERGE M. R. and KNOWLES R. (1968b). Urease activity in a black spruce humus sterilized by gamma radiation. *Soil Sci. Soc. Am. Proc.* 32, 518–521.

ROGERS H. T. (1942). Dephosphorylation of organic phosphorus compounds by soil catalysts. *Soil Sci.* 54, 439–446.

ROIZIN M. B. and EGOROV V. I. (1972). Biological activity of podzolic soils of the Kola Peninsula. *Pochvovedenie* (3), 106–114.

ROSS D. J. (1965). Effects of air-dry, refrigerated and frozen storage on activities of enzymes hydrolysing sucrose and starch in soils. *J. Soil Sci.* 16, 86–94.

ROSS D. J. (1966). A survey of activities of enzymes hydrolysing sucrose and starch in soils under pasture. *J. Soil Sci.* 17, 1–15.

ROSS D. J. (1968a). Activities of enzymes hydrolysing sucrose and starch in some grassland soils. *Trans 9th Int. Congr. Soil Sci., Adelaide* 3, 299–308.

ROSS D. J. (1968b). Some observations on the oxidation of glucose by enzymes in soil in the presence of toluene. *Pl. Soil* 28, 1–11.

ROSS D. J. (1970). Effects of storage on dehydrogenase activities of soils. *Soil Biol. Biochem.* 2, 55–61.

ROSS D. J. (1971). Some factors influencing the estimation of dehydrogenase activities of some soils under pasture. *Soil Biol. Biochem.* **3**, 97–110.

ROSS D. J. (1973a). Biochemical activities in a soil profile under hard beech forest. 2. Some factors influencing oxygen uptakes and dehydrogenase activities. *N.Z. Jl Sci.* **16**, 225–240.

ROSS D. J. (1973b). Some enzyme and respiratory activities of tropical soils from New Hebrides. *Soil Biol. Biochem.* **5**, 559–567.

ROSS D. J. (1974). Glucose oxidase activity in soil and its possible interference in assays of cellulase activity. *Soil Biol. Biochem.* **6**, 303–306.

ROSS D. J. (1975a). Studies on a climosequence of soils in tussock grasslands. 5. Invertase and amylase activities of topsoils and their relationships with other properties. *N.Z. Jl Sci.* **18**, 511–518.

ROSS D. J. (1975b). Studies on a climosequence of soils in tussock grasslands. 6. Invertase and amylase activities of tussock plant materials and of soil. *N.Z. Jl Sci.* **18**, 519–526.

ROSS D. J. (1975c). Studies on a climosequence of soils in tussock grasslands. 7. Distribution of invertase and amylase activities in soil fractions. *N.Z. Jl Sci.* **18**, 527–534.

ROSS D. J. (1976). Invertase and amylase activities in ryegrass and white clover plants and their relationships with activities in soils under pasture. *Soil Biol. Biochem.* **8**, 351–356.

ROSS D. J. and MCNEILLY B. A. (1973). Biochemical activities in a soil profile under hard beech forest. 3. Some factors influencing the activities of polyphenol-oxidizing enzymes. *N.Z. Jl Sci.* **16**, 241–257.

ROSS D. J. and MCNEILLY B. A. (1975). Studies of a climosequence of soils in tussock grasslands. 3. Nitrogen mineralization and protease activity. *N.Z. Jl Sci.* **18**, 361–375.

ROSS D. J. and ROBERTS H. S. (1968). A study of activities of enzymes hydrolysing sucrose and starch and of oxygen uptake in a sequence of soils under tussock grassland. *J. Soil Sci.* **19**, 186–196.

ROSS D. J. and ROBERTS H. S. (1970). Enzyme activities and oxygen uptakes of soils under pasture in temperature and rainfall sequences. *J. Soil Sci.* **21**, 368–381.

ROSS D. J. and ROBERTS H. S. (1973). Biochemical activities in a soil profile under hard beech forest. 1. Invertase and amylase activities and relationships with other properties. *N.Z. Jl Sci.* **16**, 209–224.

ROTH G. (1965). The biochemical activity of soils. A new method of assessment. *Proc. 39th Congr. S. Afr. Sug. Technol. Ass.* 276–285.

ROTINI O. T. (1933). La presenza e l'attività delle pirofosfatasi in alcuni substrati organici e nel terreno. *Atti Soc. Ital. Progr. Sci.* **2**, 78–86.

ROTINI O. T. (1935). Enzymatic transformation of urea in soil. *Ann. Labor. Ric. Fer. Spallanzani* **3**, 179–192.

ROTINI O. T. and CARLONI L. (1953). The transformation of metaphosphates into orthophosphates promoted by agricultural soil. *Ann. Sper. Agrar., Pisa* **7**, 1789–1799.

ROWELL M. J., LADD J. N. and PAUL E. A. (1973). Enzymatically active complexes of proteases and humic acid analogues. *Soil Biol. Biochem.* **5**, 699–703.

RUSSEL S. and KOBUS J. (1974). Dehydrogenase activity of different soils. *Agrártudományi Közlemények* **33**, 161–168.

RYSAVÝ P. and MACURA J. (1972a). The assay of β-galactosidase in soil. *Folia Microbiol.* **17**, 370–374.

RYSAVÝ P. and MACURA J. (1972b). The formation of β-galactosidase in soil. *Folia Microbiol.* **17**, 375–380.

SATYANARAYANA T. and GETZIN L. W. (1973). Properties of a stable cell-free esterase from soil. *Biochemistry* **12**, 1566–1572.

SCHAEFER R. (1963). Dehydrogenase activity as an index for total biological activity in soils. *Ann. Inst. Pasteur, Paris* **105**, 326–331.

SEQUI P. (1974). Enzymes in soil. *Ital. Agric.* **111**, 91–109.

SHAROVA A. S. (1953). Biological activity of Latvian, Lithuanian and Estonian soils. *Latv. PSR Zinàt. Akad. Vèstis* **1**, 107–142.

SHCHATSMAR L. I. and KALIKINA T. F. (1972). Determination of the oxidase activities of soils. *Dokl. Vses. Akad. Sel'skokhoz. Nauk* **4**, 15–16.

SHCHERBAKOVA T. A., MAKSIMOVA V. P. and GALUSHKO N. A. (1970). Isolation from soil of amylolytic enzymes. *Dokl. Akad. Nauk Belorus. SSR* **14**, 661–663.

SHCHERBAKOVA T. A., MAKSIMOVA V. P. and GALUSHKO N. A. (1971). Isolation of an active enzymatic complex form soil and its separation on DEAE-cellulose. *Mikrobiol. Biokhim. Issled. Pochvy, Mater. Nauch. Konf., Kiev.* **1969**, 108–111.

SHUMAKOV V. S. (1960). Biochemical activity of dark grey wooded steppe soil under different plantations. *Pochvovedenie* (10), 47–54.

SIMONART P. and MAYAUDON J. (1961). Humification des protéines ^{14}C dans le sol. *Pédologie. Symp. Int. 2, Appl. sc. nucl. péd.*, 91–103.

SIMPSON J. R. (1968). Losses of urea nitrogen from the surface of pasture soils. *Trans. 9th Int. Congr. Soil Sci., Adelaide* **2**, 459–466.

SKUJIŅŠ J. (1967). Enzymes in soil. *In* "Soil Biochemistry". (A. D. McLaren and G. H. Peterson, Eds) pp. 371–414, Marcel Dekker, New York.

SKUJIŅŠ J. (1973). Dehydrogenase: an indication of biological activities in arid soils. *In* "Modern methods in the study of microbial ecology". Bulletins from the Ecological Research Committee, Sweden **17**, 235–241.

SKUJIŅŠ J. (1976). Extracellular enzymes in soil. *CRC Crit. Rev. Microbiol.* **4,** 383–421.

SKUJIŅŠ J. J. and MCLAREN A. D. (1967). Enzyme reaction rates at limited water activities. *Science* **158**, 1569–1570.

SKUJIŅŠ J. J. and MCLAREN A. D. (1968). Persistence of enzymatic activities in stored and geologically preserved soils. *Enzymologia* **34**, 213–225.

SKUJIŅŠ J. J. and MCLAREN A. D. (1969). Assay of urease activity using ^{14}C-urea in stored, geologically preserved, and in irradiated soils. *Soil Biol. Biochem.* **1**, 89–99.

SKUKIŅŠ J. J., BRAAL L. and MCLAREN A. D. (1962). Characterization of phosphatase in a terrestrial soil sterilized with an electron beam. *Enzymologia* **25**, 125–133.

SLAVNINA T. P. and SOROKINA G. E. (1964). Biological activity of grey forest soils in the rhizosphere of some cultivated plants. *Trudy Tomsk. Gos. Univ.* **172**, 65–72.

SØRENSEN H. (1955). Xylanase in the soil and the rumen. *Nature* **176**, 74.

SØRENSEN H. (1957). Microbial decomposition of xylan. *Acta Agric. Scand. Suppl.* **1**, 1–86.

SØRENSEN L. H. (1969). Fixation of enzyme protein in soil by the clay mineral montmorillonite. *Experientia* **25**, 20–21.

SPALDING B., DUXBURY J. M. and STONE E. L. (1975). *Lycopodium* fairy rings: effect on soil respiration and enzymatic activities. *Soil Sci. Soc. Am. Proc.* **39**, 65–70.

SPEIR T. W. and ROSS D. J. (1975). Effects of storage on the activities of protease, urease, phosphatase and sulphatase in three soils under pasture. *N.Z. Jl Sci.* **18**, 231–237.

SPIRIDONOV Y. Y. and SPIRIDONOVA G. S. (1973). Effect of repeated application of *sym.* triazines on biological activity of soil. *Agrokhimiya* **3**, 122–131.

STEFANIC G. (1971). Total soil phosphatasic capacity. *Biol. Sol.* **14**, 10–11.

STEFANIC G., BOERIU I. and DUMITRU L. (1971). The effect of fertilizing and rates of liming on the total microflora and enzyme activity in a clay-illuvial podzolic soil. *Stiinta Sol* **9**, 45–54.

STENINA T. A. (1968). Enzyme activity of some soils of the central taiga. *Pochvovedenie* (2), 109–113.

STEVENSON I. L. (1959). Dehydrogenase activity in soils. *Can. J. Microbiol.* **5**, 229–235.

STEVENSON I. L. (1962). The effect of decomposition of various crop plants on the metabolic activity of the soil microflora. *Can. J. Microbiol.* **8**, 501–509.

STOJANOVIC B. J. (1959). Hydrolysis of urea in soil as affected by season and by added urease. *Soil Sci.* **88**, 251–255.

SUCIU M. (1970). Graduate Thesis, Babes-Bolyai University, Cluj.

TABATABAI M. A. (1973). Michaelis constants of urease in soils and soil fractions. *Soil Sci. Soc. Am. Proc.* **37**, 707–710.

TABATABAI M. A. and BREMNER J. M. (1969). Use of p-nitrophenyl phosphate for assay of soil phosphatase activity. *Soil. Biol. Biochem.* **1**, 301–307.

TABATABAI M. A. and BREMNER J. M. (1970a). Arylsulfatase activity of soils. *Soil Sci. Soc. Am. Proc.* **34**, 225–229.

TABATABAI M. A. and BREMNER J. M. (1970b). Factors affecting soil arylsulfatase activity. *Soil Sci. Soc. Am. Proc.* **34**, 427–429.

TABATABAI M. A. and BREMNER J. M. (1971). Michaelis constants of soil enzymes. *Soil Biol. Biochem.* **3**, 317–323.

TABATABAI M. A. and BREMNER J. M. (1972). Assay of urease activity in soils. *Soil Biol. Biochem.* **4**, 479–487.

TABATABAI M. A. and SINGH B. B. (1976). Rhodanese activity of soils. *Soil Sci. Soc. Am. J.* **40**, 381–385.

TAGLIABUE L. (1958). Cryoenzymological research on urease in soils. *Chimica Milano* **24**, 488–491.

THALMANN A. (1966). The determination of dehydrogenase activity in soil by means of TTC (triphenyltetrazolium chloride). *Soil Biol.* **6**, 46–47.

THENTE B. (1970). Effects of toluene and high energy radiation on urease activity in soil. *Lantbrukshögsk. Annlr* **36**, 401–418.

TOMESCU E. (1970). Contribution to the methodology for determination of cellulase in soil. *Microbiol., Lucr. Conf. Nat. Microbiol. Gen. Appl., Bucharest* **1968**, 509–513.

TSIRKOV I. (1970). Effect of the organic chlorine insecticides hexachlorane, heptachlor, lindane and dieldrin on activity of some soil enzymes. *Pochvozn. Agrokhim.* **4**, 85–88.

TURKOVÁ V. and SROGL M. (1960). The relationship between enzymatic and other soil biological tests in the same habitat. *Rostl. Výroba* **33**, 1431–1438.

TYLER G. (1974). Heavy metal pollution and soil enzymatic activity. *Pl. Soil* **41**, 303–311.

UMAROV M. M. and ASEEVA I. V. (1970). Decarboxylation and chelation of glutamic acid in soil. *Pochvovedenie* (1), 95–98.

VAN NIEKERK P. E. LE R. (1964). A modified method for the determination of urease activity in the soil. *S. Afr. J. Agric. Sci.* **7**, 131–134.

VELA G. R. and WYSS O. (1962). The effect of gamma radiation in nitrogen trans- formation in soil. *Bact. Proc.* **62**, 24.

VERMA L., MARTIN J. P. and HAIDER K. (1975). Decomposition of carbon-14-labelled proteins, peptides, and amino acids; free and complexed with humic polymers. *Soil Sci. Soc. Am. Proc.* **39**, 279–284.

VLASYUK P. A. and LISOVAL A. P. (1964). Effect of fertilizers on phosphatase activity of soil. *Vestn. Sel'skokhoz. Nauki, Vses. Akad. Sel'skokhoz Nauk* **9**, 52–55.

VLASYUK P. A. DOBROTVORSKAYA K. M. and GORDIENKO S. A. (1957). The intensity of the activity of enzymes in the rhizosphere of individual crops. *Dokl. Akad. Sel'skokhoz. Nauk* **3**, 14–19.

VOETS J. P. and DEDEKEN M. (1964) Studies on biological phenomena of proteolysis in soil. *Ann. Inst. Pasteur, Paris* **107**, Suppl. 3, 320–329.

VOETS J. P. and DEDEKEN M. (1965). Influence of high-frequency and gamma irradiation on the soil microflora and the soil enzymes. *Meded. Landbouwhogesch. Opzoekingsta. Staat Gent* **30**, 2037–2049.

VOETS J. P. and DEDEKEN M. (1966). Soil enzymes. *Meded. Rijksafac. Landbou- wett., Gent* **31**, 177–190.

VOETS J. P., DEDEKEN M. and BESSEMS E. (1965). Behaviour of some amino acids in gamma irradiated soils. *Naturwissenshaften* **52**, 476.

VUKHRER E. G. and SHAMSHIEVA K. T. (1968). Activity of some enzymes in soils of central Tien Shan. *Pochvovedenie* (3), 94–100.

WEETALL H. H., WELIKY N. and VANGO S. P. (1965). Detection of microorganisms in soil by their catalatic activity. *Nature* **206**, 1019–1021.

ZANTUA M. I. and BREMNER J. M. (1975). Preservation of soil samples for assay of urease activity. *Soil Biol. Biochem.* **7**, 297–299.

ZANTUA M. I. and BREMNER J. M. (1976). Production and persistence of urease activity in soils. *Soil Biol. Biochem.* **8**, 369–374.

ZINCHENKO V. A. and OSINSKAYA T. V. (1969). Alteration of biological activity of soil during incubation with herbicides. *Agrokhimiya* **9**, 94–101.

ZITTLE C. A. (1953). Adsorption studies of enzymes and other proteins. *Adv. Enzymol.* **14**, 319–374.

3

Kinetics and Consecutive Reactions of Soil Enzymes

A. D. MCLAREN

Department of Soils and Plant Nutrition, University of California, Berkeley, California, USA

I. Introduction

We may think of soil as a system of humus and clay immobilized enzymes (McLaren, 1975a). This is a useful viewpoint (Ladd and Butler, 1975a) and timely because of a parallel story in enzyme technology. Many enzymes have been complexed with synthetic polymers, such as maleic acid–ethylene co-polymer, and become more stable as a result (Zaborsky, 1973). For example, trypsin molecules in a polymer network cannot collide and undergo self digestion; natural humus–enzyme systems seem to have similar stability (Mayaudon *et al.*, 1975), trapped enzymes may be protected from added proteinases by a humus matrix (Burns *et al.*, 1972).

For small substrate molecules, reaction rates should not be much reduced simply because an enzyme is immobilized; most of the diffusion mobility of the system resides in the substrate (Schurr, 1964). With large substrate molecules, the enzyme carrier, polymer network or a clay, can also offer a diffusion impediment to substrate molecules. Electrostatic properties of a carrier of the enzyme can also influence reaction rates for a variety of reasons (McLaren and Packer, 1970).

Before considering enzyme kinetics in soil, a brief description of classical kinetics is helpful.

1. Classical enzyme kinetics

Most studies have involved very dilute aqueous solutions of highly purified enzymes catalysing chemical changes in highly purified substrates. The latter may also be diluted or dilute by virtue of being insoluble with only surface molecules in contact with water and enzyme molecules. For example, at one extreme we have the pair urease and urea; at the other amylase and starch grains or lipase and oil droplets.

The usual scheme is as follows. Let substrate (S) react with hydrated enzyme (E) to give a reactive intermediate $(E \cdot S)$, which decomposes to products (P) and regenerated enzyme:

$$S + E \underset{1}{\overset{2}{\rightleftharpoons}} E \cdot S \overset{3}{\longrightarrow} P + E \tag{1}$$

It can be shown that with soluble substrate in excess, the rate of reaction, i.e., the decrease in concentration of substrate with time, is given by

$$-dS/dt = k_3 \frac{E_0 S}{K_m + S} \tag{2}$$

$K_m = \dfrac{k_2 + k_3}{k_1}$; k_3 is a rate constant for the third step in equation (1) and k_1 and k_2 are rate constant for the reversible steps in equation (1). K_m is an affinity constant that characterizes the (usually) transient intermediate $E \cdot S$. It is numerically equal to the substrate concentration that gives a rate equal to one-half the maximum rate for a given total concentration of enzyme (E_0) in the system.

With soluble enzyme in excess, compared to an insoluble substrate, our equation is (McLaren and Packer, 1970)

$$-dS/dt = k_3 \frac{E_0 S}{K_m + E_0} \tag{3}$$

Although these equations are basic, it must be kept in mind that pH, ionic strength, temperature, and many other factors influence the values of k_1, k_2, k_3 (Irving and Cosgrove, 1976). Some of these are covered adequately in many texts of biochemistry. Others, particularly as related to heterogeneous systems, have not reached such textbooks and will be identified as needed.

For some values of K_m for soil enzymes, see Table 1. A composite of several K_m values for a soil may have forensic value: they are independent of sample size provided the soil sample is uniform. A composite is characteristic for each soil and cropping practice (Thornton and McLaren, 1975).

Comparison of reaction rates in dilute solution with those found for the same enzyme and substrate in an insoluble state shows a change in K_m as expected. The action of trypsin on gelatin in both sol and gel states revealed that k_1 is smaller if gelatin is a gel. Diffusion of enzyme to peptide bonds is clearly inhibited by gel formation and K_m is increased (McLaren and Packer, 1970).

On the other hand, if an enzyme is in a gel matrix of an inert, charged polymer the apparent value of K_m can be greater or less than with a soluble enzyme, depending on whether the matrix and substrate molecules carry like or unlike charges respectively (Engasser and Horvath, 1975). A negatively charged matrix can attract hydrogen ions and the bound enzyme will therefore show a shift of pH optimum to larger values. If the enzyme depends on a free SH group for activity, a preferential attraction of R^+–SS–R^+ over R^+–SH by a charged, clay-bound enzyme in a redox, buffered system will result in decreased activity as compared with soluble enzyme in the corresponding solution system (Benesi and McLaren, 1976).

2. Enzyme reactions in soil: theory

One way to study a soil enzyme is to add substrate to a sample and then to measure changes in amounts of S and/or P. Another procedure is to pass a substrate through a soil column. The amount of P can be measured at the exit to the column or as a function of depth in the column. The two procedures are fundamentally different in some ways.

2.1. Batch studies

In a beaker of soil to which substrate is added uniformly, one may observe changes in S or P with time, i.e., dS/dt or dP/dt. Such changes may be measured in time intervals too short for microbial growth to interfere or else growth can be prevented by irradiation (Cawse, 1975), by adding chemical inhibitors, by removing oxygen, etc., depending on the specificity desired (Powlson, 1975). However, at any time dS/dP is the same throughout the soil sample and so is the ratio P/S in the bulk sense. Microsites are ignored, i.e., any variations from point to point are averaged.

2.2. Column studies

In contrast, in a soil column of length X to which S is added at the top in a flowing solution, P and S are removed at the bottom at $S = L$ (the total length of the column). In the column for any overall time of flow, the ratio P/S increases with depth and dS/dX and dP/dX are vector quantities.

Furthermore, if microorganisms multiply in response to the added S, with an increase in biomass, enzymes and perhaps excreted enzymes as well, their numbers also tend to vary with depth, depending on the delay time τ (time after flowing substrate has passed microbes at depth X) they have had to grow. Tortuosity of path length, and hence time of flow of a volume element containing S and P from $X = 0$ to $X = L$ will also influence the ratios of S/P throughout and must be included in a complete treatment (see below; Brams *et al.*, 1975).

Let us consider the uniformly packed soil column. It is first infiltrated with buffer until saturated. At $t = 0$, S in buffer is passed into the column. After a time the front will be at a distance $= f \cdot t$, where f is flow rate in the column for idealized plug or piston flow. If f is kept constant and the enzyme distribution in the soil is both uniform and constant, equation (2) may be rewritten as $(k = k_3)$

$$-dS/dX = \frac{k}{f} \cdot \frac{E_0 S}{(K + S)} \qquad (4)$$

In words the vector dS/dX is reduced if f is increased because substrate is in contact with soil enzyme for a shorter time in a path length X.

If S is much less than K_m everywhere in the column we are left with an equation that is first order in S:

$$-dS/dX = \frac{k}{f} \cdot \frac{E_0 S}{K_m} = \frac{k'}{f} S \qquad (4a)$$

Here k' is taken as constant and equals $\dfrac{kE_0}{K_m}$. Upon integration we have

$$S = S_0 \, e^{-k'X/f} \qquad (5)$$

S_0 is the entering concentration at $X = 0$. Clearly S has a fixed value at any designated depth and dS/dt at this X is zero. In other words, during a constant rate of flow with a constant S_0, the column is in a steady state.

If, on the other hand, the time rate of change at X is not zero, due to a change in catalyst concentration, or other perturbation (e.g. entrapment of substrate) equation (4) becomes

$$\left(\frac{\delta S}{\delta t}\right)_X + f \left(\frac{\delta S}{\delta X}\right)_t = -k'S \qquad (6)$$

Partial derivatives must be used because S is dependent on both X and t. Finally, because of tortuosity and diffusion between adjacent volume elements of solution with different values of dS/dX, equation (6) can be generalized as (Brams *et al.*, 1975)

$$\frac{\delta S}{\delta t} + f \frac{\delta S}{\delta X} + \frac{D \delta^2 S}{\delta X^2} = -k'S \tag{7}$$

An evaluation of the dispersion term D, which is an inseparable dispersion (due to tortuosity) and diffusion parameter, requires an independent measurement in a non-reacting system (Cho, 1971), i.e., $k' = 0$. With charged substrates a term for ion exchange may also be added to the materials balance equation (Equation 7) (Cho, 1971). Most studies with enzyme columns have been performed under steady state conditions and the first term drops out.

By neglecting dispersion, one is left with equation (6); Saunders and Bazin (1973) have shown that this is equivalent to

$$f \left(\frac{\delta S}{\delta X'} \right)_\tau = -k'S \tag{8}$$

Here $\delta S/\delta X'$ means differentiation with respect to $X (X \equiv X')$ with

$$\tau = t - X/f \tag{9}$$

held constant. The quantity τ is the delay time, the time at which the substrate in solution front has passed depth X. This notion becomes important if biomass-enzyme changes with time at X due to growth of an organism during a time τ that it has been in contact with nutrient-substrate.

2.3. Consecutive reactions

Situations may arise in which the product of one reaction, B, may be substrate for another as in

$$A \xrightarrow{\ 1\ } B \xrightarrow{\ 2\ } P \tag{10}$$

Both reactions 1 and 2 may proceed by steps such as in equation (1).

If we assume that each reaction is first order we have

$$-\frac{dA}{dX} = \frac{k_1'A}{f} \tag{4}$$

$$-\frac{dB}{dX} = \frac{k_2'B}{f} - \frac{k_1'A}{f} \tag{11}$$

and

$$P = A_0 - A - B \tag{12}$$

Note that $k_1' = \dfrac{k_1 E_{0_1}}{K_{m_1}}$ and $k_2' = \dfrac{k_2 E_{0_2}}{K_{m_2}}$ and E_0 may be extracellular or within the biomasses, whichever is responsible for the consecutive reactions (Capellos and Bielski, 1972).

II. Applications

Having outlined a general approach we proceed to examples. As mentioned, water, pH, surface charge of carrier, temperature, etc. are involved and, with heterogeneous systems, these parameters cannot always be included analytically in rate equations. Strictly empirical equations tend to lack insight and will be ignored.

Enzymes are active only in the presence of water, although the activity of water can be low as in soils dried by dessication. Urease is active in soil at a water activity of around 0·5 at room temperature (Skujiņš and McLaren, 1969) and even in soils at -10 to $-20°C$. Activity in frozen soils may depend on unfrozen water on clay surfaces (Bremner and Zantua, 1975) at similar minimal water activities (Skujiņš and McLaren, 1971).

In using equations such as (2) it is assumed that water is in great excess and that the pH is stabilized, usually at the optimum for enzyme activity. In the following sections it will be assumed that soil enzyme reactions have been conducted in a system saturated with water at an activity of nearly one (100% relative humidity). Further, the temperature must be carefully controlled (*vide infra*).

1. Evaluation of rates with enzymes constant

By equation (2), the reaction rate $r = dS/dt$ can have a maximum value $R = k_3 E_0$ with large concentrations of substrate (i.e., with $S >> K_m$). The value of R was found to increase slightly for urease activity if the soil suspension is shaken (Tabatabai, 1973). Presumably shaking exposes a little more crumb matrix urease to substrate. K_m for Thurman soil (Table 1) and others, decreases to about half with shaking, which suggests that k_1 increases, i.e., access of substrate to enzyme is facilitated during agitation.

By equation (2), r increases with the amount of soil present (i.e., E_0 increases with this amount) but, as expected is soil-specific (Tabatabai and Bremer, 1972). R and r increase with temperature. Fig. 1 gives an example (Dalal, 1975). It is suggested that urease activity in the presence of toluene is due to extracellular enzyme only.

Use of equation (2) however, gives only apparent values of K_m. In heterogeneous systems rates must be studied in relation to molecular environments. Not all of the substrate may be available to react with

TABLE 1

Some values of K_m for soils[a]

Substrate	Enzyme Activity	Soil	$K_m{}^b$ X10³M
Nitrophenylphosphate	Phosphatase	Columbia	2·0–
		(Sandy loam)	4·2
		Hanford	2·3–
		(Sandy loam)	4·4
		Dublin	
		(Clay loam)	3–4
Nitrophenylsulphate	Sulphatase	Columbia	20–
			40
		Hanford	3–4
Urea	Urease	Thurman	2·2
		(Sand)	(31ppm)
Nitrite	*Nitrobacter*	Hanford	1·6
			(23ppm)
Ammonium	*Nitrosomonas*	Hanford	0·7
			(8ppm)

[a] Thornton and McLaren, 1975; Tabatabai, 1973; McLaren, 1976b.
[b] ppm of N where indicated.

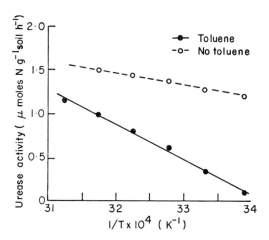

Fig. 1. Plots of urease activities, k_3 with and without toluene, as a function of reciprocal temperature for a Trinidad soil (Dalal, 1975).

enzyme at a given time. The substrate nitrophenyl phosphate is adsorbed according to Freundlich's equation (Equation 13).

$$S_a = KS_e^n \tag{13}$$

Here S_a is the amount of substrate adsorbed per g soil in equilibrium with dissolved substrate, S_e, and K and n are constants. For some soils K_m uncorrected for adsorption of substrate may be too large by an order of magnitude (Cervelli *et al.*, 1973).

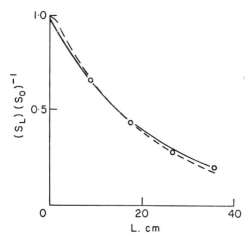

Fig. 2. Plots of equations 4 (solid line) and 7 (dashed line) for phosphatase activity in a soil column of length L = 36 cm and with a flow rate of 1·1 cm per minute (Brams *et al.*, 1975).

The influence of stirring, see above, is not similar to that of passing solution through soil in a column: K_m for phosphatase activity is over two times greater in column than in batch experiments (Brams and McLaren, 1974). A comparison of equations (4) and (7) under steady state conditions showed that a correction of k' for dispersion, $(0·96 \text{ cm}^2\text{min}^{-1} = D)$ is small, about 10%, Fig. 2. Because soil and substrate are both negatively charged, substrate (nitrophenylphosphate) concentrations near soil particles are less than in bulk solution; the apparent value of k' is therefore too large in this steady state, flowing system; a correction made for electrostatic repulsion with Dublin clay loam (Brams *et al.*, 1975), shows that the true value is lower when measured directly. In the experiment of Cervelli *et al.* (1973), a batch study, some of the substrate added to the soil was adsorbed and unavailable to immobilized enzyme. In the column (Brams *et al.*, 1975), a constant fraction of the adsorption sites are covered and substrate at a constant concentration is offered to the enzyme by a flowing solution, however only a fraction of this soluble substrate is, in a steady state, "seen" by enzyme, because of ionic repulsion. Thus in both the static and the dynamic reactors the enzyme-substrate contacts are less than expected with a concentration S_0 but for different reasons.

One practical implication of this work involves soil phosphates. Applications of soluble organic phosphate fertilizers during irrigation can lead to a liberation of phosphate at greater depths and concentration than can be achieved by percolating inorganic phosphate (Rolston et al., 1975).

2. Reaction rates with microbial growth

The previous sections have discussed soil enzyme activity as if E_0 were constant. The E_0 might be associated with biomass, either intercellular or in the outer membrane or cell wall, as with yeast invertase (McLaren and Packer, 1970). Otherwise the enzyme may be in humus (Lloyd, 1975; Ramírez-Martínez and McLaren, 1966; Mayaudon, et al., 1975; Ladd and Butler, 1975a) or in, or on clays, which is less probable (Rotini and Sequi, 1974; McLaren, 1975a). At least humus–enzyme complexes have been isolated from soil but humus-free clays with enzyme activity have not, nor have clays with expanded lattices containing protein been found in soil. Proteins complexed with clays in the laboratory are easily removed by microbial enzyme activity (McLaren and Barshad, 1976).

When treated with substrate, soil microbes that respond may or may not secrete extracellular enzymes. *Nitrobacter* grow with added nitrite but no extracellular nitrite oxidizing catalyst appears. With added urea, urease activity may (Ardakani et al., 1975; Larson and Kallio, 1954) or may not increase. It may increase with an increase of urea decomposers if sugar is also added as an energy source for growth (Paulson and Kurtz, 1969). Thus far simultaneous kinetics of reactions of a substrate with both a biomass and its corresponding extracellular enzyme component have not been treated as separate entities in a soil system as no data are available.

Perhaps the most simple case of one microbe acting on one substrate in an otherwise sterile soil is the oxidation of added nitrite by *Nitrobacter* and perhaps the most precise data is that of Schmidt (1974). Schmidt added nitrite and *Nitrobacter* to a sterile soil and followed both nitrate production and growth, i.e., increase in bacterial counts. Such growth tends to have a somewhat higher pH optimum in soil than in liquid medium (McLaren and Skujiņš, 1963). Growth is also slower on beads or in soil (Skujiņš, 1963; Morrill and Dawson, 1962) perhaps because only one side of each cell is in contact with substrate with cells on surfaces.

Lees and Quastel (1946) related substrate consumption and product formation to microbial growth (cf. Hattori, 1973) and concluded that a maximum population of nitrifiers (not counted, however) might be obtained; ultimately a constant rate of nitrification was observed. Nishio and Furusaka (1971), under similar circumstances, observed that a constant population may not be achieved even though nitrification takes place at a constant rate. (They offer an ingenious explanation for this seeming paradox but it has not been checked experimentally and will not be discussed here.) The works of Quastel and of Furusaka were performed with a reperfusion system which, in effect, is analogous to an hypothetical study of a continuously well stirred, very wet soil paste with substrate in a beaker. In either set-up, diffusion of substrate and oxygen should not be rate limiting because of a continuous agitating of the liquid phase; otherwise the kinetics should not differ much in kind with that of a corresponding system consisting of a stationary, saturated soil, with substrate, provided that rates of change again are not diffusion limited. From rates of change of substrate and of growth of nitrifiers (Schmidt, 1974; Morrill and Dawson, 1962; Ardakani *et al.*, 1973, 1974; Nishio and Furusaka, 1971; and others) we may conclude that growth conditions have not differed very much in any of these studies. Generation times are fairly consistent and are considerably longer than found with nitrifiers in liquid suspension (Gray and Williams, 1971). The fact that they are about the same in a beaker of soil and in a flowing column of soil argues against a diffusion limited reaction. Geometric considerations, topological in nature, may play a role in limiting access of substrates to adsorbed cells; surface charge of soil particle may be important (Hattori, 1973) but more cannot be written on this point at present. The notion of Lees and Quastel (1946) that adsorbed substrate (NH_4^+) is oxidized preferentially over solution substrate cannot be justified from any data available (Doner and McLaren, 1976).

For the oxidation of ammonium to nitrate, $NH_4^+ \xrightarrow{\;\;1\;\;} NO_2^- \xrightarrow{\;\;2\;\;} NO_3^-$ by nitrifiers, with no other nutrients limiting (such as carbon dioxide, oxygen, micronutrients) we may use equation (14) and (15) [cf. equations (4a) and (11)]:

$$-\frac{d(NH_4^+)}{dt} = k_1'(NH_4^+) \tag{14}$$

$$-\frac{d(NO_2^-)}{dt} = k_2'(NO_2^-) - k_1'(NH_4^+) \tag{15}$$

provided that concentrations are low compared to K_{m_1} and K_{m_2} for the reactions (McLaren, 1970) and that enzyme concentrations, i.e., corresponding numbers of nitrifiers, are constant. Also,

$$d(NO_3^-)/dt = k_2'(NO_2^-) \tag{16}$$

Note from equations (14) and (15), that if $d\mathrm{NO_2^-}/dt$ is small, i.e., $(\mathrm{NO_2^-})$ is small throughout the course of the reactions,

$$k_2'(\mathrm{NO_2^-}) \simeq k_1'(\mathrm{NH_4^+}) \tag{17}$$

or that

$$-d(\mathrm{NH_4^+})/dt = d(\mathrm{NO_3^-})/dt \tag{18}$$

Lees and Quastel (1946) assumed that the rate of nitrate production is proportional to growth, i.e.,

$$d(\mathrm{NO_3^-})/dt \; \alpha \; dm/dt \tag{19}$$

Here $dm/dt = \gamma_1 m$, γ_1 is growth rate constant, where m is the number of ammonium oxidizers. When soil is perfused with nitrite instead of ammonium a similar assumption was used except that the growth constant γ_2 was different and was characteristic of *Nitrobacter* spp. in their system.

2.1. Batch studies

There are difficulties in accepting equation (19) as it stands. First, as Lees and Quastel (1946) noted, nitrification can continue even if it is postulated that growth has ceased. Schmidt (1974) showed this to be true in batch, soil culture. Secondly, if we rewrite equation (19) as

$$-dS/dt = A \; dm/dt \tag{20}$$

we note that $A = -dS/dm$ which is substrate consumed per cell during growth. This is misleading for the nitrifiers because only a small part of substrate nitrogen becomes biomass (McCarty and Haug, 1971) and therefore $1/A$ is not the familiar growth yield (Hattori, 1973). A way out of these problems is to add a term B for both maintenance and wasted metabolism, inclusive, to give equation (21):

$$-dS/dt = A \frac{dm}{dt} + Bm \tag{21}$$

B is a function of substrate concentration (see below). In other words, the nitrifiers are inefficient and only a part of the energy of oxidation is used to maintain cells that are growing or have ceased growing but are still living (McLaren, 1970). Substituting for A dm/dt the quantity $A\gamma m$, we can say that the B term is greater by far than the Aγ term; B is the rate of nitrite consumption per *Nitrobacter* organisms, for example, in ways that lead to nitrate production.

The advantage of using dm/dt, instead of $A\gamma m$, in equation (21) is that if and when $dm/dt = 0$, substrate utilization is still expressed explicitly in terms of the prevailing numbers m. This number may be maximal for the system and can be designated as m_{max}. Instead of writing $dm/dt = \gamma m$ for growth, we can write a "logistic" equation

$$dm/dt = \gamma m \left(1 - \frac{m}{m_{max}}\right) \tag{22}$$

for which $dm/dt \longrightarrow 0$ as $m \longrightarrow m_{max}$. On second thought, that some such an equation must apply follows from the finite volume of the sample system. The number m becomes, e.g. a number per cm^3 of soil (Saunders and Bazin, 1973).

The integrated form of equation (22) is

$$m = \frac{m_{max}}{1 + \left(\dfrac{m_{max} - m_0}{m_0}\right) \exp(-\gamma t)} \tag{23}$$

which reproduces precisely the data of Schmidt for growth, Fig. 3.

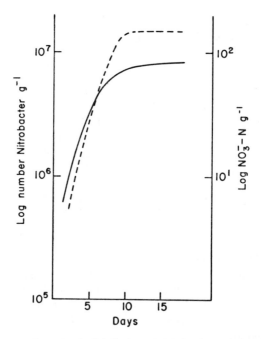

Fig. 3. Plot of equations 23 (solid line) and 24 (broken line) for simultaneous growth of and nitrate production by *Nitrobacter*. After Schmidt (1974). For purpose of plotting, equation 25 must be integrated (McLaren, 1976a).

The rate of nitrate production can be found as follows. We neglect the first term in equation (21) to obtain

$$-dS/dt \cong Bm \qquad (21a)$$

Next we assume that B has the form

$$B = \frac{B_\infty S}{K_m + S} \qquad (21b)$$

since an enzyme reaction is involved (McLaren, 1970). Here B_∞ is a maximum value, achieved with substrate concentrations large as compared to K_m. Product formation in Fig. 3 was actually calculated from a composite of equations (21a) and (23), namely,

$$-d(NO_2^-)/dt = \cfrac{B_\infty m_{max}(NO_2^-)}{\left[1 + \cfrac{m_{max} - m_0}{m_0}\right] e^{-t\gamma}\left[K_m + (NO_2^-)\right]} \qquad (24)$$

In summary, use of the B term provides for nitrate production after growth ceases, Fig. 3 (McLaren, 1976a). The work of Nishio and Furusaka, on continued growth during a constant rate of nitrification, does not fit this scheme and their topological explanation of growth within capillaries but with a constant amount of substrate or organisms-solution interfaces at the openings, deserves detailed study.

It may be noted that both the B and the m terms in equation (21a) involve very large numbers of substrate ions and organisms per cm^3 of soil respectively. Thus equation (24) is of the form of a classical chemical equation.

2.2. Column studies

Macura (1966) has revived continuous percolation (perfusion or infiltration are alternate descriptive words) as a technique. His work with nitrification in soil columns was the basis for the kinetic treatment that follows.

Consider a short column of soil (e.g. 3 cm) through which nitrite has passed long enough to build up a maximum population of *Nitrobacter* throughout. From equation (21a) and equation (21b) we have (compare equation 4) in the steady state

$$-\frac{d(NO_2^-)}{dX} = \frac{B_\infty m_{max}(NO_2^-)}{f\left(K_m + (NO_2)\right)} \qquad (25)$$

$$= \frac{K_f(NO_2^-)}{K_m + (NO_2^-)}$$

$K_f = B_\infty \, m_{max}/f$ and is a constant for a given flow rate. It varies experimentally with m and f as expected (McLaren, 1976b). The apparent $K_m = 23$ ppm NO_2–N for Hanford soil (Table 1). The integrated form of equation (25) has been applied successfully to data obtained in a column of soil diluted with sand (Ardakani $et\ al.$, 1973). K_m becomes 5 ppm if corrected for dispersion [cf. equation (7) (McLaren, 1976)].

For a long column the population of nitrifiers can decrease with depth and, if the times for collecting concentrations of products of reactions are short compared to growth rates, a quasi-steady state is achieved (Ardakani $et\ al.$, 1975; Kirda $et\ al.$, 1974). This expected growth phenomenon can be seen by replacing t by X/f in the logistic equation (McLaren, 1971).

For a relatively short time of flow during which an exponential decline in numbers of $Nitrosomonas$ with depth develops, we expect that

$$m_x = m_{x=0} \, e^{-rX} \tag{26}$$

where $m_{x=0}$ is the population at the top of the column after a few days of flow (several doubling times of about a day and half for each). In the experiment of Kirda $et\ al.$ (1974) $r = 0 \cdot 15$ cm^{-1}, an empirical constant.

From equation (21)

$$-d(NH_2^+)/dt = Bm \tag{21a}$$

since $(dm/dt)_x = 0$ if the times of passage in the column and collection are fast compared to growth rate. Substitution of equation (21b) and of equation (26) into equation (21a) and conversion to a depth function gives $-d(NH_4^+)/dX = B_\infty m_{x=0} \dfrac{e^{-rX}}{f}$ under $zero\ order$ conditions. This equation describes the data of Kirda $et\ al.$ (1974) satisfactorily (McLaren, 1975b). These authors discussed their data in terms of a more complicated expression under their impression that (NH_4^+) was much less than K_m whereas the opposite was true.

A theoretical approach to exponential growth in a column of length L allows us to deduce an equation analogous to equation (26) (McLaren, 1971). For growth at a depth X (see equation 9),

$$dm/d\tau = \gamma m \tag{27}$$
$$m_x = m_0 \, e^{\gamma \tau} = m_0 \, e^{\gamma L/f} \, e^{-\gamma X/f} \tag{27a}$$

Note that when the nutrient front has just reached the exit at L, the population at the top at $X = 0$ is $m_{x=0}$, compare equation (26), and that at $X = L$ the population is m_0, i.e., the indigenous population without growth.

As for the reaction profile, let us consider the usual two limiting cases. By equation (21), and equation (27a)

$$-dS/dX = c_\infty mS/K_{mf} \tag{28}$$

and $$-dS/dX = \frac{c}{f}\infty \frac{S}{K_m} m_0\, e^{\gamma\tau} \tag{29}$$

for *first order* kinetics. $c_\infty = A\gamma_\infty + B_\infty$.

During integration of equations (28) and (29), the delay time τ must be held constant since integration of equation (27a) has already been performed. Here $\tau = (X_T - X)/f$ for a total time of flow $T = X_T/f$ and with the front at $X\tau = L$.

Under zero order conditions, the concentration at $X = L$ is given by

$$S_L = S_0 - LC_\infty\, m_0\, e^{\gamma\tau}/f \tag{30}$$

and the substrate concentration entering the top of the column is S_0.

Although equations (28), (29) and (30) have yet to be tested, they have been presented as a heuristic model to be tested by experiment.

It may be remarked that if we assume that the growth of, for example *Nitrobacter*, is independent of substrate concentration and that the indigenous population in the column is uniform, then the population behind the front depends on the time τ that it has had to grow. Since τ is constant for any volume element of substrate solution as it passes along the column, it is acted upon by the same biomass density (Saunders and Bazin, 1973). It may be noted that according to equations (2) and (29), as L or τ increase, the population appears to continue to multiply without limit.

Obviously growth in a column is limited and logistic growth has been observed (Doner and McLaren, 1976); a population can achieve a maximum and $dm_{max}/dt = 0$. Such a column can show quasi-steady state kinetics for a time (e.g. Ardakani *et al.*, 1975). Under these conditions, integration of equation (25) gives

$$K_m \ln \frac{(NO_2^-)}{(NO_2^-)_0} + (NO_2^-) - (NO_2^-)_0 = -K_f X \tag{25a}$$

which is illustrated in Fig. 4. Of course, if $S_0 < < K_m$ we are left with

$$K_m \ln \frac{S}{S_0} = -K_f X \tag{25b}$$

and if $S_0 > > K_m$ the equation (compare equation 30)

$$(NO_2^-) = (NO_2^-)_0 - K_f X \tag{25c}$$

applies.

Insofar as one is looking at nitrification and denitrification (McCartey and Haug, 1971), the A term in equation (21) is much less than the B term.

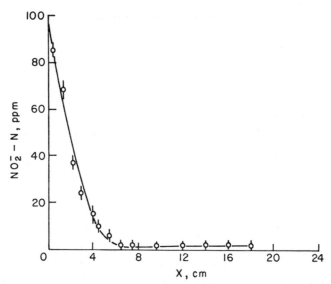

Fig. 4. Plot of equation 25a for oxidation of nitrite to nitrate in a soil column $(NO_2^- - N)_0 = 100$ ppm and $f = 4\cdot2$ cmh^{-1}. Data collected after 28 days of infiltration (Ardakani *et al.*, 1973).

This is not general and the A term is especially important in carbon nutrition and metabolism of soil microorganisms (Wagner, 1975). However, the enzymes of nitrogen oxidation in nitrifiers seem to reside in outer cell membranes and can function in the absence of carbon (as carbon dioxide); in other words, the nitrifier biomass acts much like a soil exoenzyme system in terms of kinetics. The nitrifiers can be treated kinetically within the framework we have developed provided that autocatalytic growth behaviour is included. Similar remarks probably apply to sulphur-oxidizing autotrophs but heterotrophs, with a more complicated nutrition and metabolism, do not seem to have been treated kinetically in soil. Experimental methods for glucose and amino acid metabolism in soil columns are available (Macura, 1966; Hattori, 1973).

3. Consecutive reactions in columns

In a series of reactions such as

$$\text{urea} \xrightarrow{K_u} NH_4^+ \xrightarrow{K_1} NO_2^- \xrightarrow{K_2} NO_3^- \xrightarrow{K_3} N_2O \qquad (31)$$

we need both of the latter two equations (equations 25b and 25c) to approximate the results. Consider first reactions (1) and (2). By equations (1), (4a), (11) and (25b) we have

$$(NH_4^+) = (NH_4^+)_0 \exp\left(-K_1 X/K_{m1}\right) \qquad (25d)$$

$$(NO_2^-) = \frac{K_1'\,(NH_4^+)_0}{K_2' - K_1'} \exp - K_1'X - \exp - K_2'X \qquad (11a)$$

Where $K_1' = K_1/K_{m1}$ and $K_2' = K_2/K_{m2}$ and $K_1 = (B_\infty)_1\,(m_{max})_1/f$ and $K_2 = (B_\infty)_a(m_{max})_2/f$. Bazin and Saunders (1973) have applied these equations to *Nitrosomonas* plus *Nitrobacter* adsorbed to glass beads in a column.

It may be noted that if, for example, by prior enrichment with respect to nitrite oxidizers, $K_2' >> K_1'$ then the nitrite concentration is only a small fraction of the ammonium concentration and it may also be noted that $d(NO_3^-)/dX$ is essentially equal to $-d(NH_4^+)/dX$. This has been observed experimentally by Ardakani et al., 1974.

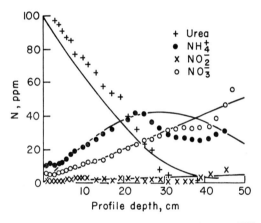

Fig. 5. Hydrolysis and consecutive reactions of urea-nitrogen $[(U)_0 = 100$ ppm] in a soil column. Urea by equation [32]; ammonia by equation [33]; nitrite by equation [34]; nitrate by equation [35]. Ardakani et al., 1975.

If all the steps in equation (31) are first order an equation similar to (11a) may be applied, with an exponential term for each step. In general, however, the K_m for each step must be considered for the concentrations of each chemical species (Table 1). In an experiment with 100 ppm urea–N (U) in a soil column with a uniformly distributed catalyst system (and some extracellular urease) the following equations encompass the data, namely

$$K_m \ln \frac{U}{U_0} + U - U_0 = K_u X \qquad (32)$$

$$NH_4^+ = U_0 - U - K_1 X/f \qquad (33)$$

$$NO_2^- = U_0 \left[1 - \frac{X_a K_2}{U_0 f}\right] \frac{X}{X_a} \qquad (34)$$

$$NO_3^- = U_0 \left[1 - \frac{(NO_2^-)}{U_0} X_a - \frac{X_a K_3}{U_0 f} \right] \frac{X}{X_a} \tag{35}$$

See Fig. 5.

Here X_a is the distance or depth of flow at which the ammonium ions are all utilized by a zero order process; $X_a = U_0 f / K_1$ and X is less than X_a. Equation (32) is identical in form with equation (25a). Equations (33)–(35) represent consecutive zero order conversions. Details may be found elsewhere (McLaren, 1970).

Finally, equations of the form of equation (11a) and (25d) still hold in a column with growing populations if t is replaced by τ, taken as constant in equation (23) before integration (Saunders and Bazin, 1973). This approach to a steady state has yet to be evaluated in practice.

Incidentally, the properties of artificial enzyme matrices are similar to soil columns and the interested reader may consult Zaborsky's monograph (1973) for useful comparison.

References

ARDAKANI M. S., REHBOCK J. T. and MCLAREN A. D. (1973). Oxidation of nitrite to nitrate in a soil column. *Soil Sci. Soc. Am. Proc.* **37**, 53–56.

ARDAKANI M. S., REHBOCK J. T. and MCLAREN A. D. (1974). Oxidation of ammonium to nitrate in a soil column. *Soil Sci. Soc. Am. Proc.* **38**, 96–99.

ARDAKANI M. S., VOLZ M. G. and MCLAREN A. D. (1975). Consecutive steady state reactions of urea, ammonium and nitrite nitrogen in soil. *Can. J. Soil Sci.* **55**, 83–91.

BAZIN M. J. and SAUNDERS P. T. (1973). Dynamics of nitrification in a continuous flow system. *Soil Biol. Biochem.* **5**, 531–544.

BENESI A. and MCLAREN A. D. (1976). A microenvironmental redox shift at a charged surface detected by papain activity. *J. Solid-Phase Biochem.* **1**, 27–32.

BRAMS W. H. and MCLAREN A. D. (1974). Phosphatase reactions in a column of soil. *Soil Biol. Biochem.* **6**, 183–189.

BRAMS W. H., DAY P. R. and MCLAREN A. D. (1975). Effect of hydrodynamic dispersion on phosphatase reactions in a soil column. *Soil Biol. Biochem.* **7**, 223–225.

BREMNER J. M. and ZANTUA M. I. (1975). Enzyme activity in soils at subzero temperatures. *Soil Biol. Biochem.* **7**, 383–387.

BURNS R. G., EL-SAYED M. H. and MCLAREN A. D. (1972). Extraction of an urease-active organo-complex from soil. *Soil Biol. Biochem.* **4**, 107–108.

CAPELLOS C. and BIELSKI H. J. (1972). "Kinetics Systems", John Wiley, New York.

CAWSE P. A. (1975). Microbiology and biochemistry of irradiated soils. *In* "Soil Biochemistry" Vol. 3 (E. A. Paul and A. D. McLaren, Eds), Marcel Dekker, New York.

CERVELLI S., NANNIPIERI P., CECCANTI B. and SEQUI P. (1973). Michaelis constant of soil acid phosphatase. *Soil Biol. Biochem.* **5**, 841–845.

CHO C. M. (1971). Convective transport of ammonium with nitrification in soil. *Can. J. Soil Sci.* **120**, 256–260.

DALAL R. C. (1975). Effect of toluene on the energy barriers in urease activity of soils. *Soil Sci.* **120**, 256–260.

DONER H. and MCLAREN A. D. (1976). *In* "Environmental Biogeochemistry" (J. O. Nriagu, Ed.) pp. 245–258, Ann Arbor Science Pub., Ann Arbor.

ENGASSER J. and HORVATH C. (1975). Electrostatic effects on the kinetics of bound enzymes. *Biochem. J.* **145**, 431–435.

GRAY T. R. G. and WILLIAMS S. T. (1971). Microbial productivity in soil. In Microbes and Biological Productivity. *Symp. Soc. Exp. Biol.* **28**, 255–286.

HATTORI T. (1973). "Microbial Life in the Soil", Marcel Dekker, New York.

IRVING G. C. J. and COSGROVE D. J. (1976). The kinetics of soil acid phosphatase. *Soil Biol. Biochem.* **8**, 335–340.

KIRDA C., STARR J. L., MISRA C., BIGGAR J. W. and NIELSEN D. (1974). Nitrification and denitrification during miscribe displacement in an unsaturated soil. *Soil Sci. Soc. Am. Proc.* **38**, 772–778.

LADD J. and BUTLER J. (1975). *In* "Soil Biochemistry" Vol. 4 (E. A. Paul and A. D. McLaren, Eds), pp. 143–144, Marcel Dekker, New York.

LARSON A. D. and KALLIO R. E. (1954). Purification and properties of bacterial urease. *J. Bacteriol.* **68**, 67–73.

LEES H. and QUASTEL J. H. (1946). Kinetics of, and effects of poisons on, soil nitrification, as studied by soil perfusion. *Biochem. J.* **40**, 803–814.

LLOYD A. B. (1975). Extraction of urease from soil. *Soil Biol. Biochem.* **7**, 357–358.

MACURA J. (1966). *In* "Theoretical and Methodological Basis of Continuous Culture" (I. Matek and Z. Fencl, Eds) pp. 462–492, Academic Press. New York.

MAYAUDON J., BATISTIC J. and SARKAR J. M. (1975). Proprietes des activites proteolytiques extraites des sol frais. *Soil Biol. Biochem.* **7**, 281–286.

MCCARTEY P. L. and HAUG R. T. (1971). *In* "Microbial Aspects of Pollution". (G. Sykes and F. A. Skinner, Eds.) Academic Press, London and New York.

MCLAREN A. D. (1970). Temporal and vectorial reactions of nitrogen in soil. *Can. J. Soil Sci.* **50**, 97–109.

MCLAREN A. D. (1971). Kinetics of nitrification in soil: growth of the nitrifiers. *Soil Sci. Soc. Am. Proc.* **35**, 91–95.

MCLAREN A. D. (1975a). Soil as a system of humus and clay immobilized enzymes. *Chemica Scripta* **8**, 97–99.

MCLAREN A. D. (1975b). Comments on kinetics of nitrification and biomass of nitrifiers in a soil column. *Soil Sci. Soc. Am. Proc.* **39**, 597.

MCLAREN A. D. (1976a). Comments on an autoecological study of microorganisms. *Soil Sci.* **121**, 60–61.

MCLAREN A. D. (1976b). Rate constants for nitrification and denitrification in soil. *Rad. and Environ. Biophys.* **13**, 43–51.

MCLAREN A. D. and BARSHAD I. (1976). Clays, proteins and nonsense. *Soil Sci.* **121**, 188.

MCLAREN A. D. and PACKER L. (1970). Some aspects of enzyme reactions in heterogenous systems. *Adv. Enzymol.* **33**, 245–308.

MCLAREN A. D. and SKUJIŅŠ J. J. (1963). Nitrification by *Nitrobacter* on surfaces and in soil with respect to hydrogen ion concentration. *Can. J. Microbiol.* **9**, 729–731.

MORRILL L. G. and DAWSON J. E. (1962). Growth rates of nitrifying chemoautotrophs in soil. *J. Bact.* **83**, 205–206.

NISHIO M. and FURUSAKA C. (1971). Number of nitrite-oxidizing bacteria in soil percolated with nitrite. *Soil Sci. Pl. Nutr.* **16**, 54–60.

PAULSON D. S. and KURTZ L. T. (1969). Locus of urease activity in soil. *Soil Sci. Soc. Am. Proc.* **33**, 897–901.

POWLSON D. S. (1975). Effect of biocidal treatments on soil organisms. In "Soil Microbiology" (N. Walker, Ed.) pp. 193–224. John Wiley, New York.

RAMIREZ-MARTINEZ J. R. and MCLAREN A. D. (1966). Determination of phosphatase activity in native soils. *Enzymologia* **31**, 23–38.

ROLSTON D. E., RAUSCHKOLB R. S. and HOFFMAN D. L. (1975). Infiltration of organic phosphate compounds in soil. *Soil Sci. Soc. Am. Proc.* **39**, 1089–1094.

ROTINI O. T. and SEQUI P. (1974). Progress realises dans la fertilite du sol et la nutrition des plantes. *Geoderma* **12**, 331–346.

SAUNDERS P. T. and BAZIN M. J. (1973). Non-steady state studies of nitrification in soil. *Soil Biol. Biochem.* **5**, 545–547.

SCHMIDT E. L. (1974). Quantitative autoecological study of microorganisms in soil. *Soil Sci.* **118**, 141–149.

SCHURR J. M. (1964). Enzyme reactions in structurally restricted systems. Ph.D. Thesis, Univ. Calif., Berkeley.

SKUJIŅŠ J. J. (1963). Enzyme and microbial activity in radiation sterilized soil. Ph.D. Thesis, Univ. Calif., Berkeley.

SKUJIŅŠ J. J. and MCLAREN A. D. (1969). Assay of urease activity using ^{14}C-urea in stored, geologically pressured and irradiated soil. *Soil Biol. Biochem.* **1**, 89–99.

SKUJIŅŠ J. J. and MCLAREN A. D. (1971). Urease reaction rates a low water activity. *Space Life Sciences* **3**, 3–11.

TABATABAI M. A. (1973). Michaelis constants of urease in soils and soil fractions. *Soil Sci. Soc. Am. Proc.* **37**, 707–710.

TABATABAI M. A. and BREMNER J. M. (1972). Assay of urease activity in soils. *Soil Biol. Biochem.* **41**, 479–487.

THORNTON J. and MCLAREN A. D. (1975). Enzymatic characterization of soil evidence. *J. Forensic Sci.* **20**, 674–692.

WAGNER G. H. (1975). Microbial growth and carbon turnover. In "Soil Biochemistry", Vol. 3, (E. A. Paul and A. D. McLaren, Eds). pp. 269–304, Marcel Dekker, New York.

ZABORSKY O. (1973). "Immobilized Enzymes", Chemical Rubber Co. Press, Cleveland.

4

Soil Polysaccharidases: Activity and Agricultural Importance

S. KISS, M. DRĂGAN-BULARDA and D. RĂDULESCU

Department of Plant Physiology, Babeş-Bolyai University, Cluj-Napoca, Romania.

I. Introduction

The most abundant organic compounds in nature are polysaccharides. A large part of the plant residues and a smaller part of the animal residues that find their way into soil are polysaccharidic. Consequently, the decomposition of carbohydrate polymers and the subsequent mineralization of the products have a special significance in the biological cycling of carbon and thus the perpetuation of life on our planet. It is therefore logical to attribute a similar significance to the soil polysaccharidases; the enzymes which catalyse the hydrolytic depolymerization of polysaccharides in the soil. In addition, some polysaccharides participate in the aggregation of soil particles and these same polysaccharides can be decomposed through enzymatic hydrolysis. As a result of their role in both the carbon cycle and soil aggregation, polysaccharidases are of great agricultural importance.

Polysaccharidase activity in soil, like that of other enzymes, results from accumulated polysaccharidases as well as the activity of proliferating microorganisms. By definition, accumulated polysaccharidases are regarded as enzymes present and active in a soil independent of immediate microbial proliferation. Sources of this "background" enzyme level are primarily microbial cells although some, no doubt, originate from plant and animal residues.

The accumulated polysaccharidases so far found in soil are the following: α- and β-amylases, cellulase, lichenase, laminarinase, inulase, xylanase, dextrinase, dextranase, polygalacturonase, and levanase (Kiss and Drăgan-Bularda, 1972; Kiss et al., 1975; Skujiņš, 1976).

Under natural soil conditions the polysaccharidases, like other enzymes, are continuously being synthesized and accumulated, inactivated and decomposed. Fluctuations in measured enzyme activity reflect the momentary ratio between these opposite processes. The factors influencing these changes include climatic conditions, soil properties, vegetation and agricultural and silvicultural techniques. All these are interrelated. However, to facilitate a logical review of the literature each factor will be discussed separately. This chapter will concentrate on the influence of crops and agricultural techniques on the activity of polysaccharidases in soil.

II. Influence of Crops on the Polysaccharidase Content of Soil

Crop plants, like other higher plants, contribute to the content of polysaccharidases and other enzymes in soil both directly and indirectly. Their direct contribution is by way of plant polysaccharidases whilst their indirect contribution is connected with the microbial decomposition of plant residues during which microbial polysaccharidases are synthesized. Among plant organs the roots are the most important sources of the soil enzymes. In addition to the endoenzymes contained in root residues, extracellular enzymes secreted by living roots may make a significant contribution to the total activity.

The influence of a given plant on the polysaccharidase content of soil can be assessed by comparing enzyme activity in the rhizosphere and the non-rhizosphere soil; this influence may vary according to the age of the vegetation and the season. Different crop species may influence the soil polysaccharidase content to a different extent. It is also possible that the nature of the previous crops affects the current polysaccharidase content of soil. It is interesting to compare enzyme activities in soil under crop plants to that under a grassland vegetation located in the vicinity. Weed species in a crop are also expected to influence enzyme content in soil. Finally, the correlation, if any, between crop yield and soil polysaccharidase deserves attention.

1. Rhizosphere effect

1.1. Amylase

In a pot experiment on a light grey podzolized soil, Peterson (1961) found much higher amylase activity in the rhizosphere of winter rye, lucerne, clover and carrot than in the non-rhizosphere soil. Ratios between rhizospheric and non-rhizospheric amylase activities were 1·3, 7·2, 1·7 and 5·2, respectively. Similar results were obtained with bean and barley. Chunderova and Zubets (1966a) determined the amylase activity of a sod-podzolic soil cultivated with potatoes and sugar beets. Fallow land provided the non-rhizosphere soil and showed lower activity as compared to the rhizosphere of both vegetables. In pot experiments, also using sod-podzolic soils, the same authors (Chunderova and Zubets, 1966b) demonstrated that amylase activity was higher in the rhizosphere of both maize and oat than in the uncropped soils. In a field experiment on a sod-podzolic soil, Dubovenko and Ulasevich (1968) compared amylase activity in oat rhizosphere with that in the fallow soil; the rhizosphere activity was always higher. Balasubramanian and Patil (1968) and Balasubramanian *et al.* (1970) cultivated finger millet (*Eleusine coracana*) in pots filled with red loamy soil and found that amylase activity was significantly enhanced in the rhizosphere. Narayanaswami and Veerraju (1969) presented data, according to which amylase activity in the rhizosphere of strawberry (*Fragaria vesca*) was greater than four times that in the non-rhizosphere soil. In a pot experiment using a sod-podzolic soil, Sal'nikov *et al.* (1969) and Sal'nikov and Filippova (1973) found that amylase activity in the rhizosphere of germinating spring wheat seeds varied with the temperature and humidity of the soil. The activity increased with increasing temperature (10°C, 20–22°C, 35°C) but decreased at 10°C or 35°C when the humidity of the soil was increased from 60–90% of its water holding capacity. Under field conditions, the rhizosphere soil of horse bean and winter rye showed higher amylase activity than the same sod-podzolic soil in which no plants were grown (Lebedeva and Gomonova, 1972; Lebedeva, 1974). Samples of a grey brown podzolic soil, in a field trial run as a monoculture of cereals, potatoes or sugar beets for 23 years, contained more hydrolases, including amylase, protease and alkaline phosphatase, than the samples taken from the neighbouring fallow (Beck, 1975).

1.2. Cellulase

A nearly 20-fold increase in cellulase activity occurred in the rhizosphere of strawberry plants in comparison with the non-rhizosphere soil (Naraya-naswami and Veerraju, 1969). Smaller but significant increases of cellulase activity were found in the rhizosphere of sugar beets (Bagnyuk and Shche-tinskaya, 1971).

2. Age of vegetation and season

2.1. Amylase

In the West Ukraine, a grey podzolized soil cropped with barley showed higher amylase activity in summer than in either spring or autumn (Peterson and Teterya, 1961). Similar results were obtained in the Leningrad region by Daragan-Sushchova and Katsnel'son (1963) who determined the amylase activity of a weakly podzolized soil under clover, timothy and a mixed clover-timothy stand. Studying the dynamics of enzymatic activity under different soil and climatic conditions of Armenia, Galstyan (1965, 1974) assayed amylase in a leached chernozem and a chestnut soil under winter wheat, and in a brown soil under cotton or lucerne. He concluded that activity was highest in May and lowest in July; a second but smaller peak appeared in October. Amylase activity was nearly twice as high in May than in July in a peaty-bog soil cultivated with hemp in Byelorussia (Kuprevich and Shcherbakova, 1966). The rhizosphere soil of both potatoes and sugar beets was more amylase-active in July than in August 1963 under the conditions of a sod-podzolic soil located in the Leningrad region. But in 1964 the samples collected in July were less active than those taken in August (Chunderova and Zubets, 1966a). Chunderova and Zubets (1966b) also found that amylase activity in the rhizosphere of maize and oat decreased constantly during the 11 June–2 August period but had increased slightly by the end of August. Dubovenko and Ulasevich (1968) registered maximum values of amylase activity in the rhizosphere of young oat plants cultivated on a sod-podzolic soil in the Ukraine. Minimum values were found in the flowering phase.

Amylase activity in the rhizosphere of finger millet, cultivated on a red loamy soil in Bangalore, India, increased with plant age; maximum activity was recorded on the 60th day of plant growth but a decline was observed on the 75th day (Balasubramanian *et al.*, 1970). Ivanov and Baranova (1972) determined amylase activity of three soils (a light grey podzol, a podzol, and a meadow chernozem) located in the Trans-Ural region and culti-vated with spring wheat. The activity increased during the development of the crop, maximum values being found in the second half of summer. The increase was attributed to the growing root system serving as source of amylase. At the same time, Beck (1975) found nearly identical values of hydrolase (including amylase, protease and alkaline phosphatase) activities in samples of an Upper-Bavarian grey brown podzolic soil collected in springtime, summer and late autumn. The crops were wheat, potatoes and sugar beets.

2.2. Cellulase

Studying the cellulase activity of a meadow chernozem in the Ukraine on which were grown pea, horse bean, winter wheat, and sugar beets, Lisoval (1967) and Vlasyuk and Lisoval (1968) established that in the soil under legumes the activity was lowest in May (at the beginning of growth) and highest in July or August (at harvest). At the same time, the soil under winter wheat showed higher activity in May (at tillering time) than in June (at ripening time). In the soil under sugar beets the activity was higher in May than in July. In contrast, under the very continental climate of North Kazakhstan, a dark chestnut soil cultivated with wheat showed lower cellulase activity at tillering time than at ripening time; minimum activity values were registered in the earing period (Karamshuk, 1975).

3. Nature of present crop

3.1. Amylase

Hoffmann (1959) determined amylase activity in the top layers of two plots on the same soil. The amylase activity of each layer was higher in the barley than in the beet plot. Activity in the soil under a hop plantation was low. However, sowing a clover mixture in the same soil resulted, after two months, in almost a 30% increase in amylase activity (Hoffmann and Leibelt, 1961). The weakly podzolized soil studied by Daragan-Sushchova and Katsnel'son (1963) contained a little more amylase under clover or timothy than under a mixed clover-timothy stand.

Spaced single plants of legumes (red clover, white clover) and grasses (cocksfoot, prairie grass) were maintained for at least one year on recent alluvial soils at four sites. The soil collected under the leaf canopy showed that amylase activities were nearly always greater under legumes than under grasses at the same site (Ross, 1965, 1966a, b). Amylase activity was constantly higher in the rhizosphere of potatoes than in that of sugar beets cultivated on the same sod-podzolic soil (Chunderova and Zubets, 1966a). No constant differences were revealed, however, by a similar comparison between maize and oat (Chunderova and Zubets, 1966b). Ambrož (1972) found that the amylase activity of a brown soil was higher under winter wheat than under a clover-grass stand. In a two-year pot experiment, Pronin *et al.* (1972) compared the amylase activity of a fertile and a less fertile sod-podzolic soil under maize and horse bean in pure and mixed stands. Activity under the pure maize stand exceeded that under the pure horse bean stand, in both soils. The activity was highest in the fertile soil under mixed stand and lowest in the less fertile soil under mixed stand.

3.2. Cellulase

According to Romeiko and co-workers (1971), the sod-podzolic soils culti-vated with maize, under the conditions of the Ukrainian forest and steppe zone, contain more cellulase than those cropped with winter wheat, clover or lupin. Cellulase, like the amylase activity of the brown soil studied by Ambrož (1972), was higher under winter wheat than under a clover-grass stand. Cellulase activity was five times as high in the rhizosphere soil of raspberry than in that of currant (Kislitsina, 1972). Depending on the crop, cellulase activity of a sod-podzolic soil decreased in the following order: maize > winter wheat > potatoes > clover (Zubenko, 1973).

3.3. Xylanase

The loam soils of the Field Experimental Station Askov (Denmark) were more xylanase-active under wheat than under beets. But at the Virumgaard Experimental Station soil samples collected from plots situated close to each other and carrying different crops showed the following order of xylanase activity: beets > wheat > potatoes > flax (Sørensen, 1957).

4. Nature of previous crop

4.1. Amylase

Examining soil amylase activity in thirteen hop plantations of different ages (1–30 years) and in the neighbouring field under usual crop rotations, Hoffmann and Leibelt (1961) established that, in general, the soils of younger plantations were more active than the soils on crop rotation whilst those of the older plantations were even less active. Habán (1967a) assayed the carbohydrase (including amylase and invertase) activities of a degraded chernozem under two different five-year crop rotations. The first rotation comprised crops which—in comparison with those of the second rotation— did not contribute significantly to the increase of organic matter content in the soil. The first rotation led to lower soil carbohdrase activity than the second one. Kuprevich and Shcherbakova (1971) measured amylase activity under a sod-podzolic soil cropped continuously or in rotation with lupin, potatoes or rye. Monocultures resulted in slightly lower activity in comparison with crop rotations.

In an investigation of three 23-year monocultures and two crop rotation systems on a grey brown podzol, amylase activity was much higher in soil from cereal monoculture than from potato or sugar beet monoculture. At the same time, the activity under monocultures with cereals and potatoes was distinctly lower than under crop rotations (cereals–root crops–legumes and potatoes–cereals–legumes, respectively). In other words, changes of crops resulted in increases in soil amylase activity. It should be added that even the stubble catch crops were able to cause some increases in soil amylase activity under cereal monoculture (Beck, 1975). Shcherbakova *et al.* (1975) studied the drained plots on a shallow peat soil cultivated with a mixture of clover and timothy or with potatoes for five years. The soil under the legume-grass mixture constantly contained more amylase than that under the monoculture of potatoes.

4.2. Cellulase

Lisoval (1967) compared two crop rotations on plots of the same meadow chernozem (pea–winter wheat–sugar beets and horse bean–winter wheat–sugar beets). Cellulase activity under wheat was nearly three times as high in the second rotation (previous plant: horse bean) than in the first rotation (previous plant: pea) but in respect to cellulase activity under sugar beets, the differences between the two rotations were negligible. Blagoveshchen-skaya and Danchenko (1974) studied a sod-podzolic soil which had been cropped only with maize or with several plants in rotation (maize–maize–maize–sugar beets–mixture of vetch and oats–clover) for 12 years. The maize in monoculture and the maize and sugar beets in crop rotation had been fertilized. Cellulase activity was found to be higher under the mono-culture than under the crop rotation. In the dark chestnut soil studied by Karamshuk (1975), cellulase activity showed a tendency to increase from the first to the third year of wheat cultivation after clean fallow. The accumulation of stubble remains also increased. This meant that the cellu-lase activity, although it increased, was not able to prevent the accumulation of stubble remains. The relatively low rate of cellulolysis and mineralization processes in general can be explained by the continental climate under which this soil is located in North Kazakhstan, especially relevant is its early deep freezing (to 0·5 m or more) and the delayed recovery of its microflora in the spring.

5. Effect of crop plants in comparison with grassland

5.1. Amylase

Hofmann and Hoffmann (1955) found that amylase activity of the 0–10 cm layer of a Bavarian soil was about twice as high under grassland vegetation than in the arable land located in the vicinity. Under the different soil and climatic conditions of Armenia, amylase activity was also higher under grassland than under crop plants. For example, the ratio of amylase activity of a chernozem under natural meadow to the activity of the same soil under winter wheat was 1·75:1 (Galstyan, 1957, 1961a; Galstyan and Markosyan, 1967). In a grey brown podzolic soil studied by Beck (1975), amylase activity under permanent pasture was about 3·5, 4·0 and 7·0 times as high than under the 23-year monocultures of cereals, potatoes and sugar beets, respectively.

6. Effect of weed species

6.1. Amylase

Galstyan (1961b) compared—from the viewpoint of soil enzymology—two cropfields on a dark chestnut soil. The first field was a pure spring wheat stand, without weeds. The other was covered, in nearly equal proportions, by spring wheat and different weeds, among which the predominant species were wild mustard or charlock (*Sinapis arvensis*), chalk plant (*Gypsophila elegans*) and knawel (*Scleranthus annuus*). The weeds diminished the soil amylase activity by 50%.

7. Enzyme activity and crop yield

7.1. Amylase

In three brown semi-desertic stony soils, a direct relationship was found between yields of winter wheat grain (1·58, 2·17 and 2·95 $t\ ha^{-1}$) and soil amylase activity (1·9, 2·8 and 4·1 mg maltose g^{-1} soil 24 h^{-1}, respectively) (Galstyan, 1963). Comparing soil carbohdrase (including amylase and invertase) activities and crop yields in a five-year rotation system on a degraded chernozem, Habán (1967a) drew the conclusion that in most cases higher enzyme activity was associated with higher crop yield and vice versa. However, the correlation between these factors was only significant in some years.

Cultures of the fungus *Trichoderma lignorum*, antagonist of many phytopathogenic microorganisms, were inoculated in sod-podzolic soil. This treatment led, under both glasshouse and field conditions, to a significant increase of crop yields (buck wheat, 34·4%; winter wheat, 12%; potatoes, 11%) and, simultaneously, to an increase in soil amylase activity (Kanivets *et al.*, 1971). Productivity was 46·2% (maize) and 11·3% (horse bean) higher in the fertile than in the less fertile sod-podzolic soils studied by Pronin *et al.* (1972). Amylase activity was also higher in the fertile than in the less fertile soils, but the increase was only 21·2% (under maize) and 5·5% (under horse bean).

In a long-term fertilizer trial on a brown soil, soil amylase activity, measured in the final year, showed no correlation with the crop (maize) yield achieved in that year. For example, the highest activity was registered in the unfertilized soil and the highest crop yield in the plots fertilized with NPK at high dosages; the highest activity was associated with very low crop yield and the highest crop yield with moderate activity (Jäggi, 1974). At the same time, Beck (1975) found a very good relation between biological activity (including amylase activity) in soil and the average values of crop yields in a monoculture and crop rotation experiment run for 23 years. However, he emphasized that an exact comparison was difficult, because the monocultures in particular (to which no manure was applied and no stubble catch crop planted) show strong annual oscillations in crop yields. These oscillations are conditioned by a more or less strong incidence of plant parasites which is itself dependent on climate.

7.2. Cellulase

Both the cellulase activity of a meadow chernozem and the yield of winter wheat were higher when the wheat had been sown after horse bean than after pea (Lisoval, 1967).

III. Influence of Agricultural Techniques on the Polysaccharidase Content of Soil

1. Tillage

1.1. Amylase

In the 0–10 cm layer of a grey podzolized soil, amylase activity was higher in plots submitted to ploughing with mouldboard to 35–37 cm depth than in the plots ploughed without mouldboard to 20 cm depth. In the 15–25 and 30–40 cm layers, amylase was not detectable in any of these plots. The crop was barley. In autumn all plots were ploughed and in spring re-ploughed to 16 cm depth and cropped with table beet (*Beta vulgaris var.*

esculenta). Plots, which in the preceding year had been ploughed to 35–37 cm, again showed higher amylase activity in the 0–10 cm soil layer (Peterson and Teterya, 1961).

Hábán (1966, 1967a, b, 1968a, b, 1971) studied the effects of ploughing depth on the enzyme (including amylase) activities of a degraded cherno-zem, in a five-year field trial on two crop rotations (one with crops which do not contribute and the other with crops which do contribute significantly to the increase of organic matter content in the soil). His results indicated that, in comparison with normal ploughing to 27 cm, ploughing followed by sub-soiling, gradual deepening and amelioration ploughing up to 45 cm, led to significantly higher enzyme activity in the 30–40 cm layer. In contrast, the changes which occurred in the activity of the 5–20 cm layer were small.

In a field experiment described by Romeiko *et al.* (1968) and Romeiko and Dubovenko (1969) a sod-podzolic soil, non-cropped or cropped with oats, was studied for several years. In the non-ploughed soil amylase activity decreased with depth according to the order: humus > podzolic > illuvial layer. Turning of the podzolic and illuvial layers to the surface resulted in increased amylase activity which approached or even exceeded that of the humus layer. Amylase activity also increased in the humus layer after its translocation to the podzolic or illuvial layer. Consequently, for improving biological activity in lower horizons, especially in the podzolic layer, they recommend periodic multi-stage ploughing.

The soil-enzymological effects of 30–35 cm deep rotary tillage and ploughing of a drained shallow peat soil were investigated by Shcherbakova *et al.* (1975). The soil was cropped with a clover–timothy mixture or with potatoes for five years. Both rotary tillage and ploughing led to increased amylase activity in the soil layers examined (0–15, 15–30, 30–45 cm) in the case of both crop rotations, except for the 0–15 cm layer in plots with potatoes. The effect of rotary tillage was uniformly more pronounced than that of ploughing in the plots under clover–timothy. In the case of the potato monoculture, the effect of rotary tillage was stronger in the 30–45 cm layer and that of the ploughing was more evident at the 15–30 cm depth. The investigations resulted in the following recommendations for increas-ing biological activity and fertility of the shallow peat soils: they should be submitted to rotary tillage and cultivated with clover–timothy mixture in the first years after draining, then ploughed to 20–30 cm and cultivated with root crops, especially potatoes.

1.2. Cellulase

Fallow breaking and autumn ploughing of a dark chestnut soil cultivated with wheat led to increased cellulase activity in the 0–10 cm layer (Karam-shuk, 1975).

2. Drainage

2.1. Amylase

Ross (1966b) studied the influence of drainage on a gley soil (a silt loam). Amylase activity was significantly lower in the drained than in the undrained soil. The rate of biochemical conversions occurring in a peaty soil is determined largely by its wetness. This conclusion was made as the result of a seven-year study of enzyme (including amylase) activity in lysimetric test plots with varying ground water levels (Shcherbakova, 1968; Kuprevich and Shcherbakova, 1971). The lysimeter was set up on a cultivated peaty-bog soil of lowland type with a peat layer of 2·5–3·0 m in some places and up to 4 m in others. During the first five years of the experiment enzyme activity was lowest in the plot having a ground water level of 20–70 cm. A decline in the ground water level to 120–150 cm was followed by increased enzymatic activity. With a further drop in the water level to 2 m or more, activity decreased but at all times exceeded that in plots having a high ground water level. During the final two years, the differences in enzymatic activity among the plots became less pronounced. As a general rule, it is possible to demonstrate, in peaty-bog soils of lowland type, a gradual decrease of the enzyme activity in the initial years following drainage.

2.2. Cellulase

Cellulase activity increased with the drop of ground water level in a peaty-bog soil of transition type submitted to drainage for three years (Zagural'-skaya, 1974; Zagural'skaya and Egorova, 1974).

3. Irrigation

3.1. Amylase

Fresh water from a brook and waste water (a mixture, in equal proportion, of waste waters from a dairy farm and domestic sources) were used for spray irrigation of a meadow soil in a two-year field trial. Fresh water brought about an increase of amylase activity in plots treated with mineral fertilizers (NPK) and a decrease in unfertilized plots. Waste water irrigation caused a diminution of amylase activity in both fertilized and unfertilized plots (Röschenthaler and Poschenrieder, 1959). Pastures on a silt loam had been irrigated for several years by flooding, either when 50% of the available moisture in the topsoil had been utilized or at wilting point. Values of soil amylase activity were similar for both irrigation treatments and significantly higher than those found in the non-irrigated soil. Amylase activity was also higher in an irrigated fine sandy loam than in the non-irrigated samples of the same soil (Ross, 1966b). A brown soil without vegetation was irrigated with starch waste water (130 mm) and, in the next vegetation period, cropped with winter wheat or a clover-grass mixture. The analyses carried out during the vegetation period showed that soil amylase activity under both stands increased in irrigated plots in comparison with that in the non-irrigated plots. In a laboratory experiment, in which the soil was irrigated with starch waste water or distilled water, amylase activity was also found to increase under the influence of starch waste water (Ambrož, 1972).

3.2. Cellulase

In the experiments described in the previous section, Ambrož (1972) also established that soil cellulase activity, in contrast to amylase activity, decreased following irrigation with starch waste water. But in a four-year field trial on a common chernozem under different crops, irrigation with fresh water led to increased cellulase activity in the soil (Zakharov et al., 1975).

4. Fertilization and liming

4.1. Amylase

4.1(a) Enzyme substrate

Drobník (1955) amended samples of a brown forest soil and a calcareous brown soil with 1% (dry soil basis) starch, glucose, sucrose or with mixtures of these carbohydrates. After wetting, the samples were incubated at room temperature. Amylase activity determinations carried out on the 21st and 50th days of incubation showed that activity increased in each of the amended samples as compared to the unamended soils. The highest increase occurred in the samples treated with starch, the substrate of amylase. These observations were interpreted as evidence of the induction of amylase synthesis by the soil microorganisms under the influence of starch. In another experiment of Drobník (1957), soil samples were amended with 1% (dry soil basis) starch, glucose, cellulose or gum Arabic and then moistened and incubated at 20°C for 63 days. Again, the increase in amylase activity was highest in the starch-amended samples.

    In a sod-podzolic soil (pH 4·5), the microbial amylase synthesis was more pronounced when the samples, before being incubated at 22–23°C for 105 days, were treated not only with 1·5% (fresh soil basis) starch but also brought to pH 7·0 with CaO (Rybalkina et al., 1964). Samples of a sod-podzolic soil were sterilized by autoclaving. After cooling, they were inoculated with spores of three actinomycete species, treated with sterile starch solution or distilled water, then incubated at 26–27°C. Amylase activity was determined after 10 and 20 days. The results indicated that addition of starch resulted in enhanced production of amylase by the developing actinomycetes (Vladimirova, 1970). Ambrož (1972) attributed the increased amylase activity in the soil irrigated with starch waste water (see section III.3.1) to induction of microbial amylase. Beck (1973) treated samples of three soils (humus sand, fine sandy loam, compost earth) with 0·3% starch plus 0·15% peptone (both dry soil basis). After wetting, the samples were incubated at room temperature for 76 days, during which they were analysed seven times. The data showed that treating soils with starch and peptone led to significantly increased amylase activity. The increase was most pronounced in the humus sand which was initially poor in amylase and less pronounced in the compost earth which was rich in amylase even before treatment. The peak of increase occurred in the first days of incubation. Thereafter, the activity showed a tendency to decrease. However, it exceeded by far the activity of the non-treated soil not only at the end of incubation but even after four months.

All these experiments indicate—in good agreement with the suggestions of others (Ross, 1965a, 1966b; Pancholy and Rice, 1973a, b; Cortez *et al.*, 1975)—that the influence of vegetation on soil amylase activity is largely exerted through the starch content in plant residues which, in turn, induces microbial amylase synthesis.

4.1(b). Organic fertilizers

In a four year trial on a degraded chernozem, the effects of farmyard and green manure were compared. The stubble crop mixture, consisting of maize, sunflower, horse bean and pea, was harvested or ploughed in. Median values of soil amylase activity were a little higher in the manured plots than in the unmanured ones, but nearly identical in the farmyard manure and stubble crop treatments (Habán and Prokopová, 1966).

4.1(c). Mineral fertilizers and lime

In the experiments of Röschenthaler and Poschenrieder (1959) previously mentioned in section III.3.1, NPK fertilization (30:70:140 or 60:100:200 kg ha^{-1}) increased soil amylase activity in fresh water-irrigated plots and decreased it in waste water-irrigated plots in comparison with unfertilized, non-irrigated soil. The increase was more pronounced with the higher NPK dose, while the decrease was stronger with the lower dose. Amylase activity in the rhizosphere of finger millet plants was unchanged after the first foliar application of NPK (as $(NH_4)_2SO_4$, Na_2HPO_4 and KCl) on the 30th day after sowing. However, an increasing trend was observed after subsequent foliar treatments with NPK on the 45th and 60th day as compared to the water-sprayed control plants. Subsequently, activity declined on the 75th day (Balasubramanian and Patil, 1968; Balasubramanian et al., 1970; see also section II.1.1).

Küster and Gardiner (1968) and Küster (1970) carried out a series of pot experiments in which N ($NaNO_3$), P and K (K_2HPO_4) and lime were added singly or in combinations to a milled peat of highmoor type (pH 4·2). The mixtures were then incubated over a period of 140 days. N, PK and NPK additions led, within a few weeks, to a 200–225% increase in amylase activity. Addition of lime (singly or with N, PK or NPK) and the subsequent increase in pH lowered enzyme activity. Lebedeva and Gomonova (1972) and Lebedeva (1974) studied the effect of the long-term use of mineral fertilizers and lime on the enzyme (including amylase) activity of an acid sod-podzolic soil. Continuous use of NPK ($60:60:60$ kg ha^{-1}) in various combinations for 17 years without liming had a negative effect on the soil enzyme content. Thus, under the influence of the physiologically acid fertilizer NH_4NO_3, soil amylase activity declined. Liming of the continuously NH_4NO_3-treated, acid soil only partially restored amylase activity. Lime was more effective when it was applied twice; before continuous fertilizer application and after the 17-year period.

Anhydrous ammonia (at rates of 120 and 280 kg ha^{-1}) was introduced to three soils (sod-podzolic, peaty, sandy loam). After 10 days, amylase activity showed higher values in the treated sod-podzolic soil. In the other two soils ammonia had no obvious effect at a rate of 120 kg ha^{-1} on amylase activity but reduced it when 280 kg ha^{-1} was applied (Zyatchina, 1974).

4.1(d). Organic and mineral fertilizers and lime
Incubation of soil with 2% (dry weight basis) farmyard manure or plant residues (straw, cocksfoot, lucerne) for 144 days resulted in increased amylase activity. The highest increase occurred in the lucerne-treated soil. Under similar conditions, NH_4NO_3 (0·075%) plus K_2HPO_4 (0·055%) with or without farmyard manure (0·2%) (each addition on dry soil basis) did not cause any considerable changes in amylase activity as compared to the untreated soil (Drobník, 1957).

A light grey forest soil was treated with either peat humic fertilizer (0·5 or 1·0 t ha^{-1}) or with 25% aqueous ammonia solution (240 litre ha^{-1}) as a similar ammonia solution had been used for the preparation of peat humic fertilizer. Soil amylase activity increased in both instances. The highest increase occurred in the soil fertilized with ammonia solution, and the lowest in the soils with 0·5 t ha^{-1} peat humic fertilizer (Miroshnichenko, 1962). Yarchuk (1965) studied the changes which occur in the enzyme activities of peat during preparation of peat–mineral–ammoniacal fertilizers. Peat of lowland type was treated with 20% aqueous ammonia solution (20 litres t^{-1} peat), kainit (15 kg t^{-1}) and Thomas slag (15 kg t^{-1}) in several combinations and allowed to ferment for 3–5 months. Peat treated with all these substances became more amylase-active, less toxic and more valuable as a fertilizer in comparison with the non-treated peat and the peat treated only with kainit or kainit and Thomas slag.

A sod-podzolic soil cropped with potatoes was fertilized with NPK (60:60:60 kg ha^{-1}) in form of ammonium nitrate, superphosphate and potassium chloride, with farmyard manure (30 t ha^{-1}) or with both NPK and manure. Rhizospheric amylase activity was highest in the unfertilized soil. NPK reduced the activity stronger than did the manure. The lowest diminution occurred in the NPK + manure variant (Chunderova and Zubets, 1966a). In another experiment (Chunderova and Zubets, 1966b), the effects of peat and peat humic fertilizer on the amylase activity of sod-podzolic soils under maize or oat were studied. Following fertilizer application, amylase activity in the rhizosphere decreased in the first period of vegetation, but later—simultaneous with the uptake of nutrients by the developing plants—the enzyme activity increased and even exceeded that of the unfertilized soil.

In most of the experiments already mentioned in section III.1, Habán (1967b, 1968b, c), in addition to the effects of ploughing depth, studied the influence of fertilization (NPK + farmyard manure) on soil enzyme (including amylase) activity. His results show that, in the topsoil (5–20 cm layer), differences in the enzymatic activity between the fertilized and unfertilized plots were statistically insignificant. In the subsoil layer (30–40 cm), however, enzymatic activity was significantly higher in the fertilized than in the unfertilized plots. Romeiko et al. (1968) (see section III. 1.1) also studied the influence of fertilizer and lime on soil amylase activity. NPK (60:60:60 kg ha^{-1}), farmyard manure (80 t ha^{-1}) and lime (3 t ha^{-1}) were applied to a sod-podzolic soil. Mixing them with the soil resulted in increased activity, not only in the humus layer but also in the podzolic horizon. In a long-term fertilizer trial on an initially acid sod-podzolic soil, Semenov et al. (1974) found the highest amylase activity in the plots which had been limed and fertilized with farmyard manure and NPK. NPK plus manure or lime brought about lower increases than the unlimed and unfertilized controls.

The effects of long-term fertilization of an irrigated red loamy soil on enzyme (including amylase) activity were studied by Balasubramanian et al. (1974). Ammonium sulphate, potassium sulphate and superphosphate (at the rate of 14·3, 125·6 and 376·6 kg ha^{-1}, respectively) in various combinations, as well as farmyard manure (12·4 t h^{-1} each year or every alternate year) were applied. Maximum amylase activity was observed in the plot which received farmyard manure continuously. In the long-term fertilizer trial already referred to in section II.7.1, Jäggi (1974) found that farmyard manure, dried sewage sludge, industrial compost, pure mineral fertilizers (ammonium nitrate, Thomas slag, potassium chloride) applied singly or in combinations caused the reduction of amylase activity. The reduction was lowest in the farmyard-manure-treated plot and highest in the plot moderately fertilized with NPK. Periodic farmyard-manuring of a grey brown podzolic soil under a 23-year potato monoculture increased hydrolase (including amylase) activities in the soil. Foliar fertilization of a monoculture of sugar beets also resulted in increased soil amylase activity (Beck, 1975).

The soil-enzymological consequences of using coal as fertilizer will also be mentioned in this section. In the investigations of Kermen (1967) a Polish brown coal was used. It contained 73·7% organic matter in which the humic acids represent 45·2% and the fulvic acids 4·3%. Samples of a loamy sand (4 kg) were mixed with sand (3 kg), treated with NPK (ammonium nitrate, Thomas slag, potassium chloride) or with NPK plus 0·5 or 5 g powdered coal. During the four-month experimental period no obvious differences occurred between the NPK and NPK + coal treatments with regard to soil amylase activity. At the same time, the oxidized coal (20 t ha^{-1}) used by Miroshnichenko et al. (1969) for the amelioration of a bog-like saline soil, considerably increased soil amylase activity.

This section will also contain a short description of investigations in which soil enzyme activities were connected with the hygienic problems of incorporating sewage sludge into soil. Ukhtomskaya (1952) determined soil enzyme activity in experimental plots used for final disposal of sewage sludge. Sludge in the amounts of 1000, 2000 or 5000 t ha^{-1} was worked into an acid loamy sand at 20 cm depth, in May or in October. Enzyme activity was periodically analysed, together with other parameters of sanitary importance. The results indicated that soil amylase activity greatly increased immediately after the incorporation of sludge. Later, the activity decreased and within 1–1·5 years returned to the level it had shown before sludge incorporation. The only exception was the soil treated with the highest amount of sludge. Amylase activity may thus be regarded as an index of the self-purification capacity of soil.

Since application of sewage sludge to soil is likely to become an important agricultural technique in the future, Suess et al. (1975) initiated detailed investigations on the effects of untreated, heat-treated and irradiated sewage sludge on both plants and soil. Sludge, in quantities of 130, 400 and 800 m^3 ha^{-1}, was applied and worked into the soil at four experimental plots. After 14 days, the soil was analysed and it was found that amylase activity increased with increasing quantities of sewage sludge. The increase was between ≈ 150–300% when untreated or irradiated sludge was applied but only between ≈ 100–200% when heat-treated sludge was used. In greenhouse experiments, sludge was incorporated to soil in 100–500 m^3 ha^{-1} amounts. The observations, made over a period of 3–4 weeks, indicated that amylase activity rose in a manner which depended on the soil and the quantity of untreated and irradiated sludge; with increased applications of heat-treated sludge, activity remained either stationary or decreased noticeably.

Finally, in this section, the soil-enzymological aspects of seed inoculation will be briefly described. Filippova and Kolotova (1971) inoculated horse bean seeds with Nitragin (a specific *Rhizobium* culture). Some seeds were soaked in an 0·03 % ammonium molybdate solution, and others were treated with both Nitragin and molybdenum. The seeds were then sown in a sod-podzolic soil. Enzyme activity in the rhizosphere of growing plants was assayed. It was found, however, that none of the treatments had any effect on the rhizospheric amylase activity.

4.2. Cellulase

4.2(a). Enzyme substrate

Cellulase activity of a chernozem-like meadow soil was higher under spear-grass (*Stipa*) containing 45·0% cellulose in root residues than under tansy (*Tanacetum*), the root residues of which contained only 16·8% cellulose (Kozlov et al., 1968). Buffer and water extracts, from soils incubated for 4 weeks with cellulose powder, exhibited cellulase activity when carboxy-methylcellulose was used as substrate. However, the two extracts from non-treated soils were generally without cellulase activity (Drozdowicz, 1971). Cellulose (with or without $(NH_4)_2SO_4$ or K_2HPO_4) was added to three organic soils, and induced an increase of the cellulase content during incu-bation (10 days at 28°C). This increase was shown to fluctuate according to the soil type and the N or P enrichment. In the acid peat soil, the increase of cellulase (C_1) activity using cellulose as substrate was N dependent, while increase of C_x activity (carboxymethylcellulose as substrate) was both N and P dependent. In the calcic peat, the increase of both C_1 and C_x was strictly P dependent. In the mor of the podzol, C_1 increase was independent of N and P whilst increases in C_x required both N and P. In the light of these findings, it is evident that cellulolytic activity under natural conditions (no cellulose, N and P treatments) is limited by different factors in each of these organic soils. Cellulolytic activity is limited by P or N deficiency in the acid peat and by P deficiency in the calcic peat. In the mor of the podzol, neither N nor P deficiencies are entirely responsible for the weak cellulolysis (Kong and Dommergues, 1972).

Ambroz (1973) treated fresh samples of a rendzina with 0·5% cellulose, pectin or glucose. During the ten-day incubation, cellulase activity greatly increased in the cellulose-treated soil. The increase was slight in the pectin-treated samples, while glucose actually brought about a slight decrease of cellulase activity. Drăgan-Bularda and Kiss (1973) added cellulose powder, carboxymethylcellulose, cellobiose or glucose in 0·5 or 1% amounts (dry soil basis) to a brown forest soil and a leached chernozem. After wetting, the soils were incubated at room temperature for 30 to 60 days, then analysed to determine their cellulase activity. The analytical data and their statistical evaluation showed that cellulose induced cellulase production by the microorganisms in both soils but carboxymethylcellulose, cellobiose and glucose had no such effect. The induction was more pronounced with 1% than with 0·5% cellulose powder. The microorganisms of the brown forest soil produced more cellulase than those of the leached chernozem.

The importance of the cellulose content of plant residues for soil cellulase activity was also emphasized by Pancholy and Rice (1973a, b).

4.2(b). Mineral fertilizers and lime

Cellulase activity in a sod-podzolic soil under sown meadow fertilized with NPK (80–90:60–70:80–100 kg ha^{-1}) annually for 18 years and limed every fifth year was considerably higher than in the soil of a natural, unfertilized meadow situated in the same territory (Lopatin and Kuz'mina, 1974). The dark chestnut soils studied by Andreyuk and Dul'gerov (1975) were treated with $(NH_4)_2SO_4$ (0·025–2·0 g 100 g^{-1} soil) and incubated at 30°C for 20 days. Increasing the fertilizer amount from 0·025 g to 0·25 g was associated with increasing cellulase activity, but at higher dosages, the activity diminished.

4.2(c). Organic and mineral fertilizers and lime

In the experiments of Yarchuk (1965) concerning enzymological aspects of the preparation of peat-mineral-ammoniacal fertilizers (see section III. 4.1(d)), treating peat with ammonia, kainit and Thomas slag and the subsequent 3–5-month fermentation led, besides an increased amylase activity, also to an increased cellulase activity. A meadow chernozem cropped with pea, winter wheat or horse bean showed higher cellulase activity under the influence of fertilization. The effect of NPK + farmyard manure was stronger than that of NPK, but in the plots with sugar beets, only NPK had a constant increasing effect on cellulase activity (Lisoval, 1967; Vlasyuk and Lisoval, 1968).

In long-term fertilizer trials, Rawald (1970a, b) found that annual median values of cellulase were slightly higher or lower in fertilized (NPK + farmyard manure) than in unfertilized plots, depending on the soil type. Farmyard manure—as compared to peat—enhanced the cellulase activity of soil. According to the data of Zubenko (1973), the cellulase activity of a sod-podzolic soil cropped with maize, winter wheat, potatoes or clover was constantly higher in the plots fertilized with NPK and farmyard manure than in the unfertilized plots. The sod-podzolic soil studied by Blagovesh-chenskaya and Danchenko (1974) was annually fertilized for 12 years and limed every fifth year. Rates of application were N, 45 or 90 kg, P, 60 kg, K, 90 kg, farmyard manure 30 t and lime 3 t ha^{-1}. The plots were under maize monoculture or crop rotation. Soil cellulase activity was slightly higher in the variant with $N_{90}P_{60}K_{90}$ plus farmyard manure than in the $N_{90}P_{60}K_{90}$ variant, on both maize monoculture and crop rotation. Mineral N at the 90 kg rate—as compared to the 45 kg rate—brought about some decrease in cellulase activity which was more evident in the case of crop rotation.

In field trials in which a sod-podzolic soil was treated with NPK, farmyard manure and straw, Vizla and Vinkalne (1976) established that the different treatments increased soil cellulase activity (and crop yield) according to the following order: $N_{70}P_{76}K_{140}$ + farmyard manure (30 t ha^{-1}) > $N_{55}P_{76}K_{140}$ + straw (3 t ha^{-1}) + N_{15} as NH$_4$OH > $N_{55}P_{76}K_{140}$ + straw (3 t ha^{-1}) + N_{15} as NH$_4$NO$_3$ > $N_{70}P_{76}K_{140}$.

4.3. Laminarinase

4.3(a). Enzyme substrate
Fungal cell walls containing or lacking β-1, 3-glucan were incorporated in a kaolinite paste moulded into aggregates that were subsequently incubated on soil. The developing microorganisms produced β-1, 3-glucanase in the aggregates prepared from cell walls rich in β-1, 3-glucan, i.e., the enzyme substrate induced the microbial synthesis of enzyme. Glucanase activity was measured by using laminarin as substrate (Jones and Webley, 1968; Webley and Jones, 1969).

4.4. Xylanase

4.4(a). Enzyme substrate
Portions of a loam soil with low xylanase activity were amended with 0·5% (dry soil basis) xylan, mannan, cellulose, starch, xylose, arabinose or glucose, or with powdered wheat straw in amounts from 0–2% (dry soil basis), then wetted and incubated at 25°C for 16 days. By the end of incubation, xylanase activity increased six times in the xylan-treated samples, but only 1·3 times with the other carbohydrates. Straw caused a 2–6-fold increase, depending on quantity used. Similar results were obtained with two other soils (Sørensen, 1955, 1957). In another experiment, which lasted up to 700 days, Sørensen (1969) used a sandy soil amended with xylan or glucose. Xylanase activity in the xylan-amended samples was much higher than that in the glucose-treated soil. Although it decreased with prolonged incubation, even on the 90th day it exceeded (by about 11 times) the activity of the glucose-treated soil. All these data of Sørensen indicate that the xylanase content of the soil appears to be primarily a function of the amount of

xylan added which gives rise to an accelerated excretion of the adaptive enzyme xylanase, besides an increase of the microflora. The small rise observed on addition of non-xylan carbon sources may be caused by a general increase of the microflora.

4.4(b). Organic and mineral fertilizers

Sørensen (1955, 1957) determined the xylanase activity in samples of loam and sand soils, collected from unfertilized, farmyard-manured and mineral fertilizer-treated plots. The largest activity was found in the manured plots and the smallest in the unfertilized plots.

4.5. Dextranase

4.5(a). Enzyme substrate

Two soils (a brown forest and a leached chernozem) were amended with 0·5 or 1% (dry soil basis) dextran or glucose, wetted and incubated at room temperature. The analysis, performed after 21 days, indicated that dextran induced the production of dextranase by proliferating soil microorganisms, while glucose did not. Initial dextranase activity was higher in the brown forest soil than in the leached chernozem, but the effect of dextran on dextranase formation was more pronounced in the leached chernozem (Drăgan-Bularda and Kiss, 1972).

In a study on the effects of molasses on microbial enzyme production and water stable aggregation, samples of three soils (solonchak, alluvial and leached chernozem) were treated with molasses in amounts from 0·1–4% (dry soil basis). After wetting and 30-day incubation, their dextranase activity was determined. The results showed that the sucrose present in molasses induced sequentially the microbial production of dextransucrase and dextranase in each soil, i.e., the sucrose-induced dextransucrase cata-lysed the synthesis of dextran which induced the synthesis of dextranase. The inducing effect was proportionate to the amount of molasses. Dextran synthesis (dextransucrase activity) exceeded dextran hydrolysis (dextranase activity). Enzyme activities correlated significantly with water stable aggregation. The correlation supports the idea that in molasses-treated soils the enzymatically synthesized dextran contributes to the water stable aggregation of soil particles; amelioration of molasses-treated soil is partly due to soil enzyme activities (Drăgan-Bularda and Kiss, 1977).

4.6. Polygalacturonase

4.6(a). Enzyme Substrate

Activity of pectinolytic enzymes (polygalacturonase and pectin lyase) in a garden soil treated with 2% pectin suddenly increased on the second day of incubation, reached a peak on the 7th day, then gradually decreased and returned to the initial level after 45 days. The increased activity proved that incorporation of pectin in soil induced the synthesis and excretion of pectinolytic enzymes by microorganisms (Kaiser and Monzon de Asconegui, 1971; Monzon de Asconegui and Kaiser, 1972).

4.7. Levanase

4.7(a). Enzyme substrate

Incubating samples of seven soils with 2·5% (dry soil basis) levan led to increased levanase activity due to the induction of the microbial levanase (Kiss et al., 1965). In the experiment of Drăgan-Bularda and Kiss (1977)—see section III.4.5(a)—the sucrose present in molasses also sequentially induced the microbial production of levansucrase and levanase in each of the soils studied (the sucrose-induced levansucrase catalysed the synthesis of levan which induced the synthesis of levanase). Levansucrase activity was higher than levanase activity. Activities of these enzymes also correlated significantly with water stable aggregation of soil particles. In molasses-treated soils the enzymatically synthesized levan, like dextran, contributes to the water stable aggregation.

5. Pesticide application

5.1. Amylase

5.1(a). Herbicides

Zubets (1967) studied the effects of many pre-emergent herbicides in sugar beet plots on a sod-podzolic soil. In 1963, alipur (16·5% cycluron (3-cyclooctyl-1, 1-dimethylurea) + 11·5% chlorbufam (1-methylprop-2-ynyl N-(3-chlorophenyl) carbamate)), dalapon (2, 2-dichloropropionic acid), murbetol (14·0% endothal (7-oxabicyclo (2, 2, 1) heptane-2, 3-dicarboxylic acid) + 8·4% propham (isopropyl carbanilate)) and endothal (at rates of 1, 15, 10 and 2 kg ha^{-1}, respectively) considerably decreased soil amylase activity; the greatest decrease occurred in the endothal-treated soils. In 1964, the effect of these herbicides was slight. In other plots cultivated with maize, simazine (2-chloro-4,6-bis(ethylamino)-s-triazine), chlorazine (2-chloro-4, 6-bis (diethylamino)-s-triazine), 2, 4-D(2, 4-dichlorophenoxyacetic acid) butyl ester (at 1, 5 and 1 kg ha^{-1} respectively) applied pre-emergently, slightly lowered soil amylase activity but in 1964, treatments with simazine (2 kg ha^{-1}) and simazine plus 2, 4-D butyl ester (2 + 0·4 kg ha^{-1}) slightly increased the activity. The results obtained in 1965 were similar to those found in 1964 (Chunderova and Zubets, 1969). DNOC

(4,6-dinitro-o-cresol), nitrophen (2,4-dichlorophenyl-p-nitrophenyl ether), 2,4-D pyramin (4-amino-2-methyl-5-pyridine methanol) were applied pre-emergently to potato plots on a peaty soil of transition type. Rates of application were: 7·5, 15, 1 and 6 kg ha^{-1}, respectively. Amylase activity diminished, especially in the pyramin-treated plots in which a 20% decrease occurred (Zubets, 1968).

In the pot experiments performed by Mereshko (1969) chernozemic soils were surface-treated with atrazine (2-chloro-4-(ethylamino)-6-(isopropy-lamino)-s-triazine), simazine or 2, 4-D at 3 and 30 kg ha^{-1} rates. The soils were cropped with maize or left uncropped. Enzyme analyses were made on the 30th day after herbicide application. He concluded that none of the herbicides studied had a significant effect on the amylase activity of either cropped or uncropped soils. Similar results were reported by Bliev (1973a) who applied mixtures of dalapon and 2, 4-D at rates of 20 + 2 or 60 + 6 kg ha^{-1} to three sod-podzolic soils and a leached chernozem under labora-tory conditions. Enzyme analyses were carried out one, two and three months after herbicide application. In another laboratory experiment (Bliev, 1973b), simazine (2 or 6 kg ha^{-1}) or mixtures of TCA (trichloro-acetic acid) + 2, 4-D (60 + 2 or 180 + 6 kg ha^{-1}) were added to a sod-podzolic soil. No significant changes occurred in amylase activity during the first two months. In the third month, however, amylase activity increased in the soil treated with simazine (6 kg ha^{-1}) or with TCA + 2,4-D, in comparison with the activity measured in the untreated soil. In the experiments of Beck (1973), a humus sand was treated with aresin (mono-linuron-3-(p-chlorophenyl)-1-methoxy-1-methylurea), tribunil (N'N-dimethyl-N'-(2-benzthiazolyl)urea) or simazine at rates of 3 or 200 μg g^{-1} soil under laboratory conditions and 3 μg g^{-1} soil under field conditions. One and 3·5 months after herbicide application, amylase activity remained practically unchanged in the aresin- and tribunil-treated soils. At the same time, the simazine treatment led, in the first month, to a considerable decrease in amylase activity ($\approx 20\%$ in the laboratory experiment and $\approx 35\%$ in the field experiment) but by the end of the experimental period amylase activity approached the initial level (laboratory experiment) or even exceeded it (field experiment).

Under laboratory conditions, application of simazine at the normal and ten-fold rates (2·47 and 24·7 kg ha^{-1}) to a red sandy loam soil reduced amylase activity only slightly, during the ten-day observation period (Balasubramanian et al., 1973; Balasubramanian and Siddaramappa, 1974).

Long-term field trials and laboratory experiments of Goguadze (1975) indicated that amylase activity of herbicide-treated krasnozems under a tea plantation remained at the level of the activity in the untreated soil. Sima-zine, atrazine, monuron (3-(p-chlorophenyl)-1, 1-dimethylurea), dalapon (6–8 kg ha^{-1}) and their mixtures were applied at different rates.

The influence of potassium azide (KN_3) on soil amylase activity was studied by Kelley and Rodriguez-Kabana (1975). Azide was applied at the rates of 56, 112, 168 and 224 kg ha^{-1} to plots of loamy sand. Half of each plot was covered with a polyethylene sheet and the other half was water-sealed. The plots were planted with loblolly pine (*Pinus taeda*) seed. Soil samplings 2, 5 and 16 weeks after the KN_3-treatment revealed that the overall response of amylase activity to the azide-treatment was a decline, still evident at the final sampling. This response was better defined under plastic-sealed plots where a more proportionate response to increasing azide rates was observed.

Maize had been grown annually for nine years with or without yearly application of 3·4 kg ha^{-1} atrazine in plots on a silty clay loam with a small amount of silt loam soil. Analyses indicated that long-term atrazine applications did not decrease soil amylase level (Cole, 1976).

5.1(b). Fungicides

Finger millet plants, grown in pots filled with red loamy soil, were sprayed with blastin (pentachlorobenzyl alcohol) (diluted to 0·1 % with water) when the plants were 30 days old and the same treatment was repeated on the 45th and 60th day. Although amylase activity in the rhizosphere was reduced by the first spray treatment, a subsequent increase in activity was recorded. This increase appeared to be retained in the rhizosphere even 15 days after the final spraying (Balasubramanian and Patil, 1968; Balasubramanian *et al.*, 1970).

A humus sandy soil and a compost soil were treated with benomyl (methyl-1(carbamoyl)-2-benzimidazol carbamate) at rates of 30 and 300 μg g soil^{-1} and incubated at room temperature for up to four months. It had been found that benomyl did not affect amylase activity of soil (Hofer *et al.*, 1971). Dithane M-45 (maneb) (manganese ethylene bisdithiocarbamate) applied at the normal rate (748 litres of 0·2% solution ha^{-1}) and at a 10-times normal rate to a red sandy loam soil had no marked effect on soil amylase activity (Balasubramanian *et al.*, 1973).

5.1(c). Insecticides

In the experiments of Balasubramanian and Patil (1968) and Balasubramanian *et al.* (1970) the influence of dimecron (2-chloro-2-diethyl carbamoyl-1-methyl vinyl-dimethyl phosphate) (diluted to 0·02% with water) on amylase activity in the rhizosphere of finger millet plants was studied. Amylase activity was consistently higher in the rhizosphere of dimecron-sprayed plants than in the rhizosphere of untreated plants, throughout the experimental period (75 days).

5.2. Cellulase

5.2(a). Herbicides

According to Mereshko (1969), the herbicides tested (see section III. 5.1(a)). did not cause any essential changes in soil cellulase activity. Long-term atrazine application (Cole, 1976; see section III.5.1(a)) led to only a slight decrease in cellulase activity. Cellulase in soil was also resistant to a single application of 2, 4-D used as an arboricide (Zagural'skaya and Klein, 1976).

5.3. Laminarinase

Lethbridge *et al.* (1976) mention that Suffix (2-(N-benzoyl-3, 4- dichloro-anilino) propionic acid) and diallate (S-2, 3-dichloroallyl N, N-diisopropyl-thiolcarbamate) have an inhibitory effect on β-1, 3 glucanase.

IV. Conclusions

In this chapter we have emphasized the significance of soil polysacchari-dases to the carbon cycle, soil aggregation and consequently agriculture in general. We have reviewed literature concerned with the influence of crop plants and agricultural techniques on amylase, cellulase, laminarinase, xylanase, dextranase, polygalacturonase and levanase. Many aspects of the influence of crops on soil polysaccharidase activities have been described (e.g., rhizosphere, nature of current crop, nature of previous crop, crop in comparison with grassland vegetation, weed species in crops, crop yield). In addition, the important influence of agricultural techniques on poly-saccharidase content of soil has been discussed, including tillage, drainage, irrigation, fertilization, liming and pesticide application.

References

AMBROŽ Z. (1972). Biological processes in soils irrigated with starch waste waters. *Acta Univ. Agr., Brno, Fac. Agron.* **20**, 575–580.

AMBROŽ Z. (1973). Study of the cellulase complex in soils. *Rostl. Výroba* **19**, 207–212.

ANDREYUK E. I. and DUL'GEROV A. N. (1975). Effect of some ecological factors on the enzymatic activity in soil. *Tr. IV S'ezda Mikrobiol. Ukr., Kiev* **1975**, 58–60.

BAGNYUK V. M. and SHCHETINSKAYA L. I. (1971). Method for the determination of cellulase activity in soil and bottom sediments. *Mikrobiol. Biokhim. Issled. Pochv, Mater. Nauch. Konf., Kiev* **1969**, 118–120.

BALASUBRAMANIAN A. and PATIL R. B. (1968). Soil enzymes in relation to soil fertility and plant growth. *Proc. 1st All-India Symp. Agr. Microbiol., Hebbal, Bangalore* **1968**, 52–56.

BALASUBRAMANIAN, A. and SIDDARAMAPPA R. (1974). Effect of simazine application on certain microbiological and chemical properties of a red sandy loam soil. *Mysore J. Agr. Sci.* **8**, 214–219.

BALASUBRAMANIAN A., BAGYARAJ D. J. and RANGASWAMI G. (1970). Studies on the influence of foliar application of chemicals on the microflora and certain enzyme activities in the rhizosphere of *Eleusine coracana* Gaertn. *Pl. Soil* **32**, 198–206.

BALASUBRAMANIAN A., SIDDARAMAPPA R. and OBLISAMI G. (1973). Studies on the effect of biocides on microbiological and chemical properties of soil. I. Effect of simazine and dithane M-45 on soil microflora and certain soil enzymes. *Pesticides* **7**, 13.

BALASUBRAMANIAN A., SHANTARAM M. V., EMMIMATH V. S. and SIDDARAMAPPA R. (1974). Effect of a permanent manurial and cropping schedule on microbial populations and enzyme activities in the new permanent manurial plots at Coimbatore, Tamilnadu. *Madras Agr. J.* **61**, 183–186.

BECK T. (1973). Über die Eignung von Modellversuchen bei der Messung der biologischen Aktivität von Böden. *Bayer. Landw. Jahrb.* **50**, 270–288.

BECK T. (1975). Der Einfluss langjähriger Monokultur auf die Bodenbelebung im Vergleich zur Fruchtfolge. *Landw. Forsch., Sonderheft* 31/2, 268–276.

BLAGOVESHCHENSKAYA Z. K. and DANCHENKO N. A. (1974). Activity of soil enzymes during long-term application of fertilizers under continuous maize and crops of a farm crop rotation. *Pochvovedenie* (10), 124–130.

BLIEV YU. K. (1973a). Effect of herbicides on the biological activity of soils. *Pochvovedenie* (7), 61–68.

BLIEV YU. K. (1973b). Activity of enzymes following composting of soil with herbicides. *In* "Khimicheskii Ukhod za Lesom" (V. P. Bel'kov, Ed.) pp. 157–159. Lenizdat, Pskov. Otd., Pskov.

CHUNDEROVA A. I. and ZUBETS T. P. (1966a). Rhizospheric microflora of potatoes and characteristics of its changes under the influence of organic and mineral fertilizers. *Nauch. Tr., Sev.-Zapad. Nauch. Issled. Inst. Sel. Khoz.* (10), 219–227.

CHUNDEROVA A. I. and ZUBETS T. P. (1966b). Effect of organic fertilizers on the activity of enzymes in podzolic soil. *In* "Nauka–Sel'skokhozyaistvennomu Proizvodstvu" pp. 195–201. Lenizdat, Leningrad.

CHUNDEROVA A. I. and ZUBETS T. P. (1969). Effect of herbicides on the biological processes in sod-podzolic soil cropped with maize without inter-row tillage. *Byull. Vses. Nauch.-Issled. Inst. Sel'skokhoz. Mikrobiol.* (14/2), 52–59.

COLE M. A. (1976). Effect of long-term atrazine application on soil microbial activity. *Weed Sci.* **24**, 473–476.

CORTEZ J., BILLÈS G. and LOSSAINT P. (1975). Étude comparative de l'activité biologique des sols sous peuplement arbustifs et herbacés de la garrigue méditerranéenne. II. Activités enzymatiques. *Rev. Écol. Biol. Sol* **12**, 141–156.

DARAGAN-SUSHCHOVA A. YU. and KAT'SNELSON R. S. (1963). Effect of meadow grasses on the enzymatic activity of soils. *Tr. Bot. Inst. Akad. Nauk SSSR, Ser. III* (14), 160–171.

DRĂGAN-BULARDA M. and KISS S. (1972). Dextranase activity in soil. *Soil Biol. Biochem.* **4**, 413–416.

DRĂGAN-BULARDA M. and KISS S. (1973). Influence of enzyme substrate on cellulase production by soil microorganisms. *Stud. Univ. Babes-Bolyai, Ser. Biol.* (1), 137–144.

DRĂGAN-BULARDA M. and KISS S. (1977). Effects of molasses on microbial enzyme production and water stable aggregation in soils. *In* "Soil Biology and Conservation of the Biosphere" (J. Szegi, Ed.) pp. 397–403. Akad. Kiadó, Budapest.

DROBNIK J. (1955). Degradation of starch by the enzyme complex of soils. *Folia Biol., Prague* 1, 29–40.

DROBNIK J. (1957). Biological transformations of organic substances in the soil. *Pochvovedenie* (12), 62–71.

DROZDOWICZ A. (1971). The behaviour of cellulase in soil. *Rev. Microbiol., Brazil* 2, 17–23.

DUBOVENKO E. K. and ULASEVICH E. I. (1968). Study of the activity of some enzymes in the rhizosphere of agricultural plants. *Sb. Dokl. Simp. Ferment. Pochvy, Minsk* 1967, 320–327.

FILIPPOVA K. F. and KOLOTOVA S. S. (1971). Effect of molybdenum and Nitragin on the biochemical processes of microorganisms in the horse bean plants and the rhizosphere. *Mikrobiol. Biokhim. Issled. Pochv, Mater. Nauch. Konf., Kiev* 1969, 188–193.

GALSTYAN A. SH. (1957). Study of the amylolytic activity in soil. *Byull. Nauch.-Tech. Inform., Arm. Nauch.-Issled. Inst. Zemledel.* (2), 4–6.

GALSTYAN A. SH. (1961a). Carbohydrase activity in soil. *Dokl. Akad. Nauk Arm. SSR* 32, 101–104.

GALSTYAN A. SH. (1961b). Weed infestation of field lowers biological activity of soil. *Izv. Akad. Nauk Arm. SSR, Biol. Nauki* 14, 69–74.

GALSTYAN A. SH. (1963). Evaluation of the fertility status of soil with enzymatic reactions. *In* "Mikroorganizmy v Sel'skom Khozyastve" pp. 327–336. Izd. Mosk. Univ., Moscow.

GALSTYAN A. SH. (1965). Dynamics of the enzymatic processes of soils. *Dokl. Akad. Nauk Arm. SSR* 40, 39–42.

GALSTYAN A. SH. (1974). "Fermentativnaya aktivnost' pochv Armenii". Izd. Ayastan, Erevan.

GALSTYAN A. SH. and MARKOSYAN L. V. (1967). Study of the β-fructofuranosidase activity in soil. *Tr. Inst. Pochvoved. Agrokhim., Erevan* (3), 311–325.

GOGUADZE V. D. (1975). Effect of herbicides on the biological activity of krasnozems under tea plantation. *V S'ezd Vses. Mikrobiol. Obshch., Sekts. Sel'skokhoz. Mikrobiol., Erevan* 1975, 73–74.

HABÁN L. (1966). Effect of subsoiling on the activity of microbial processes in the soil. *Ved. Pr. Výsk. Úst. Rastl. Výroby Pieštanoch* (4), 139–152.

HABÁN L. (1967a). Effect of ploughing depth and crop plants on the microflora and enzymatic activity of soil. *Ved. Pr. Výsk. Úst. Rastl. Výroby Pieštanoch* (5), 157–169.

HABÁN L. (1967b). Effect of deepening of the arable layer on some biological properties of degraded chernozem. *Ved. Pr. Výsk. Úst. Rastl. Výroby Pieštanoch* (5), 203–214.

HABÁN L. (1968a). Effect of soil deepening on some biological properties of degraded chernozem. *Sb. Ref. Mezin. Věd. Symp., Brno* 1966, 393–396.

HABÁN L. (1968b). Effect of amelioration ploughing on the microbial activity of degraded chernozem. *Rostl. Výroba* **14**, 953–962.

HABÁN L. (1968c). Effect of fertilization on the soil microflora and enzymatic activity. *Ved. Pr. Výsk. Úst. Rastl. Výroby Piešťanoch* (6), 67–79.

HABÁN L. (1971). Contribution to the problem of the influence of deep loosening of arable land on some biological properties of soil. *Poľnohospodárstvo* **17**, 85–92.

HABÁN L. and PROKOPOVÁ A. (1966). Green manuring and its effect on soil microflora. *Ved. Pr. Výsk. Úst. Rastl. Výroby Piešťanoch* **4**, 127–138.

HOFER I., BECK T. and WALLNÖFER P. (1971). Der Einfluss des Fungizids Benomyl auf die Bodenmikroflora. *Z. Pflanzenkr. Pflanzenschutz* **78**, 398–405.

HOFFMANN G. (1959). Verteilung und Herkunft einiger Enzyme im Boden. *Z. Pflanzenernähr., Düng., Bodenkd.* **85**, 97–104.

HOFFMANN G. and LEIBELT W. (1961). Der Enzymgehalt in Böden aus Hopfenanlagen der Anbaugebiete Hallertau und Jura. *Bayer. Landw. Jahrb.* **38**, 780–791.

HOFMANN E. and HOFFMANN G. (1955). Über das Enzymsystem unserer Kulturböden. VI. Amylase. *Z. Pflanzenernähr., Düng., Bodenkd.* **70**, 97–104.

IVANOV N. A. and BARANOVA R. P. (1972). Dynamics of enzymatic and biochemical processes in some soils of the forest steppe Trans-Ural region. *Tr. Sverdlovsk. Sel'skokhoz. Inst.* **26**, 24–39.

JÄGGI W. (1974). Bodenmikrobiologische Untersuchungen in einem Düngungsversuch. *Schweiz. Landw. Forsch.* **13**, 531–547.

JONES D. and WEBLEY D. M. (1968). A new enrichment technique for studying lysis of fungal cell walls in soil. *Pl. Soil* **28**, 147–157.

KAISER P. and MONZON DE ASCONEGUI M. A. (1971). Mesure de l'activité des enzymes pectinolytiques dans le sol. *Biol. Sol* **14**, 16–19.

KANIVETS I. I., KRIVICH N. Ya. and KLUGEN Z. A. (1971). The fungus *Trichoderma lignorum* and its role in improving the soil fertility and increasing the yields and quality of agricultural crops. *Vestn. Sel'skokhoz. Nauki* (2), 40–45.

KARAMSHUK Z. P. (1975). Cellulase activity of dark chestnut soil on fallow-cereal rotation. *Izv. Sib. Otd. Akad. Nauk SSSR* (5/1), 11–16.

KELLEY W. D. and RODRIGUEZ-KABANA R. (1975). Effects of potassium azide on soil microbial populations and soil enzymatic activities. *Can. J. Microbiol.* **21**, 565–570.

KERMEN J. (1967). Effect of brown coal on the biological activity of soil. *Rocz. Nauk Roln., Ser. A* **93**, 91–124.

KISLITSINA V. P. (1972). Transformation of carbon-containing substances in soils of Central Siberia as characterized by the cellulase activity. *Symp. Biol. Hung., Proc. Symp. Soil Microbiol., Budapest, 1970* **11**, 143–145.

KISS S. and DRĂGAN-BULARDA M. (1972). Polysaccharidases in soil. *Contrib. Bot., Cluj* pp. 377–384.

KISS S., BOARU M. and CONSTANTINESCU L. (1965). Levanase activity in soils. *Symp. Methods Soil Biol., Bucharest* **1965**, 129–136.

KISS S., DRĂGAN-BULARDA M. and RĂDULESCU D. (1975). Biological significance of enzymes accumulated in soil. *Advan. Agron.* **27**, 25–87.

KONG K. T. and DOMMERGUES Y. (1972). Limitation de la cellulolyse dans les sols organiques. II. Étude des enzymes du sol. *Rev. Écol. Biol. Sol* **9**, 629–640.

144   S. KISS, M. DRĂGAN-BULARDA AND D. RĂDULESCU

KOZLOV K. A., KISLITSINA V. P., MARKOVA YU. A. and MIKHAILOVA E. N. (1968). Some problems of evidencing enzymatic activity in soils. *Sb. Dokl. Simp. Ferment. Pochvy, Minsk* **1967**, 66–89.

KUPREVICH V. F. and SHCHERBAKOVA T. A. (1966). "Pochvennaya Enzimologiya." Izd. Nauka i Tekhnika, Minsk.

KUPREVICH V. F. and SHCHERBAKOVA T. A. (1971). Comparative enzymatic activity in diverse types of soil. *In* "Soil Biochemistry" (A. D. McLaren and J. J. Skujiņš, Eds) pp. 167–201. Marcel Dekker, New York.

KÜSTER E. (1970). Der Einfluss von Ca- und N-Gaben auf mikrobielle Aktivitäten in Moorböden und Torf. *Landw. Forsch., Sonderheft* **25/2**, 115–124.

KÜSTER E. and GARDINER J. J. (1968). Influence of fertilizers on microbial activities in peatland. *Proc. 3rd Int. Peat Congr., Quebec* **1968**, 314–317.

LEBEDEVA L. A. (1974). Effect of lime on the yield of plants and the properties of the soil as a function of its application times during the prolonged use of mineral fertilizers. *Khim. Sel. Khoz.* (10), 3–7.

LEBEDEVA L. A. and GOMONOVA N. F. (1972). Effect of the long-term use of mineral fertilizers and lime on the properties of sod-podzolic soil and the yield of plants. *Khim. Sel. Khoz.* (9), 2–8.

LETHBRIDGE G., BULL A. T. and BURNS R. G. (1976). Soil enzymes as monitors of agrochemical pollution. *Proc. Soc. Gen. Microbiol.* **4**, 26.

LISOVAL A. P. (1967). Cellulase activity of soil under various conditions of plant growing. *Dobr. Polyakh Ukr., Tez. Dop. Nauk. Konf., Kiev* **1967**, 127–132.

LOPATIN V. D. and KUZ'MINA T. S. (1974). Biological activity of soil under sown meadows. *In* "Pochvennye Issledovaniya v Karelii" pp. 197–202. Karel. Filial Akad. Nauk SSSR, Petrozavodsk.

MERESHKO M. YA. (1969). Effect of herbicides on the biological activity of microflora in chernozemic soils. *Mikrobiol. Zh., Kiev* **31**, 525–529.

MIROSHNICHENKO L. A. (1962). Microbiological basis of the application of peat humic fertilizers. *In* "Guminovye Udobreniya. Teoriya i Praktika Ikh Primeneniya", Part 2 pp. 215–231. Izd. Sel'skokhoz. Lit., Kiev.

MIROSHNICHENKO L. A., ZABRODINA L. V., SIDOROVA I. N. and KOROBUSHKINA E. D. (1969). Microflora of soils during their desalinization. *Izv. Biol.-Geogr. Nauch.-Issled. Inst., Irkutsk. Univ.* **21**, 191–205.

MONZON DE ASCONEGUI M. A. and KAISER P. (1972). L'utilisation des produits de décomposition de la pectine par "*Azotobacter chroococcum*". *Ann. Inst. Pasteur, Paris* **122**, 1009–1028.

NARAYANASWAMI R. and VEERRAJU V. (1969). Enzymatic studies in the rhizosphere of strawberry, *Fragaria vesca* L. *Indian J. Exp. Biol.* **7**, 126–127.

PANCHOLY S. K. and RICE E. L. (1973a). Soil enzymes in relation to old field succession: Amylase, cellulase, invertase, dehydrogenase, and urease. *Soil Sci. Soc. Am. Proc.* **37**, 47–50.

PANCHOLY S. K. and RICE E. L. (1973b). Carbohydrases in soil as affected by successional stages of revegetation. *Soil Sci. Soc. Am. Proc.* **37**, 227–229.

PETERSON N. V. (1961). Sources of soil-enzyme enrichment. *Mikrobiol. Zh., Kiev* **23**, 5–11.

PETERSON N. V. and TETERYA G. M. (1961). Biological activity of some soils in dependence of the techniques of their tillage. *Mikrobiol. Zh., Kiev* **23**, 19–24.

PRONIN V. A., VORONKOVA F. V. and YAKOVLEV A. A. (1972). Interactions between plants in mixed stands in connection with the peculiarities of microbiological and biochemical processes in soil. *In* "Fiziologo-Biokhimicheskie Osnovy Vzaimodeistviya Rastenii v Fitotsenozakh" (A. M. Grodzinskii, Ed.) pp. 116–121. Izd. Naukova Dumka, Kiev.

RAWALD W. (1970a). Über die Beeinflussung bodenbiologischer Aktivitäten durch organische Düngung. *Microbiol., Lucr. Conf. Nat. Microbiol. Gen. Apl., Bucharest* **1968**, 467–476.

RAWALD W. (1970b). Über die bodenenzymatische Aktivität als Komponente der bodenbiologischen Aktivität, insbesondere im Hinblick auf die Beurteilung des Fruchtbarkeitszustandes des Bodens, sowie Aspekte der Aufgaben bodenenzymatischer Forschung. *Zentralbl. Bakteriol., Parasitenkd., Infektionskr. Hyg., Abt.* II **125**, 363–384.

ROMEIKO I. N. and DUBOVENKO E. K. (1969). Biological activity of the soil as an index of its fertility. *In* "Puti Povysheniya Plodorodiya Pochv" pp. 67–72. Izd. Urozhai, Kiev.

ROMEIKO I. N., DUBOVENKO E. K. and ULASEVICH E. I. (1968). Biochemical activity of microorganisms in sod-podzolic soil as influenced by different techniques of land cultivation. *Sb. Dokl. Simp. Ferment. Pochvy, Minsk* **1967**, 309–320.

ROMEIKO I. N., DUBOVENKO E. K., ZAKHAROVA V. I., DUDCHENKO V. G., MALINS'KA S. M., ULYASHOVA R. M., KUCHERENKO V. I. and CHECHEL'NITS'KA L. M. (1971). Biological activity of soils and symbiosis of root nodule bacteria with leguminous plants as influenced by agrotechnical procedures. *Mikrobiol. Zh., Kiev* **33**, 724–726.

RÖSCHENTHALER R. and POSCHENRIEDER H. (1959). Zur Kenntnis der Wirkung von Abwasser auf den Mikroorganismengehalt und die biologische Aktivität eines Wiesenbodens aus dem Verregnungsgebiet Triesdorf. *Mitt. Landkultur, Moor-Torfwirt.* **7**, 87–94.

ROSS D. J. (1965). A seasonal study of oxygen uptake of some pasture soils and activities of enzymes hydrolysing sucrose and starch. *J. Soil Sci.* **16**, 73–85.

ROSS D. J. (1966a). Bodenenzyme und Fruchtbarkeit. *Umschau Wiss. Tech.* **66**, 443.

ROSS D. J. (1966b). A survey of activities of enzymes hydrolysing sucrose and starch in soils under pasture. *J. Soil Sci.* **17**, 1–15.

RYBALKINA A. V., KONONENKO E. V. and VASILENKO E. S. (1964). Microflore active et son rôle dans les processus du sol. *Trans. 8th Int. Congr. Soil Sci., Bucharest* **3**, 753–759.

SAL'NIKOV A. I. and FILIPPOVA K. F. (1973). Effect of soil temperature, moisture and microflora on spring wheat seed germination. *Uch. Zap. Perm. Univ.* (263), 60–68.

SAL'NIKOV A. I., VAKULENKO L. V. and FILIPPOVA K. F. (1969). Effect of temperature on microbiological processes in the rhizosphere of germinating spring wheat seeds. *Uch. Zap. Perm. Univ.* (219), 165–169.

SEMENOV V. A., LEVINA V. I., VESELKINA R. V., CHUNDEROVA A. I., CHUBAROV A. P. and BEREZOVSKII V. A. (1974). Changes in properties of sod-podzolic soils under the influence of various systems of fertilization in crop rotation. *Nauch. Tr., Sev.-Zap. Nauch.-Issled. Inst. Sel. Khoz.* (29), 38–59.

SHCHERBAKOVA T. A. (1968). Effect of ground water level on the enzymatic activity of peaty-bog soil. *Sb. Dokl. Simp. Ferment. Pochvy, Minsk* **1967**, 289–297.

146   S. KISS, M. DRĂGAN-BULARDA AND D. RĂDULESCU

SHCHERBAKOVA T. A., KOROBOVA G. Ya., VOLKOV A. E., BOROD'KO S. N., SHIMKO N. A. and VOLODINA L. A. (1975). Biological activity of shallow peat soils and its changes under the influence of melioration and reclamation. *Probl. Poles'ya* (4), 228–247.

SKUJIŅŠ J. (1976). Extracellular enzymes in soil. CRC *Crit. Rev. Microbiol.* **4**, 383–421.

SØRENSEN H. (1955). Xylanase in the soil and the rumen. *Nature* **176**, 74.

SØRENSEN H. (1957). Microbial decomposition of xylan. *Acta Agric. Scand.* Suppl. 1, 1–86.

SØRENSEN L. H. (1969). Fixation of enzyme protein in soil by the clay mineral montmorillonite. *Experientia* **25**, 20–21.

SUESS A., ROSOPULO A., BORCHERT H., BECK T., BAUCHHENSS J. and SCHURMANN G. (1975). Experience with a pilot plant for the irradiation of sewage sludge. Results on the effect of differently treated sewage sludge on plants and soil. *In* "Radiation for a Clean Environment" pp. 503–533. IAEA, Vienna.

UKHTOMSKAYA F. I. (1952). Role of enzymes in self-purification of soil. *Gig. Sanit.* (11), 46–49.

VIZLA R. R. and VINKALNE M. O. (1976). Role of straw as organic fertilizer in the fertility of soils. *In* "Problemy Nakopleniya i Ispol'zovaniya Organicheskikh Udobrenii" (S. G. Skoropanov, Ed.) pp. 79–83. Beloruss. Nauch.-Issled. Inst. Pochvoved. Agrokhim., Minsk.

VLADIMIROVA O. V. (1970). Hydrolytic activity of actinomycetes in soil. *Mikrobiol. Zh., Kiev* **32**, 301–305.

VLASYUK P. A. and LISOVAL A. P. (1968). Effect of plants and fertilizers on the activity of some enzymes in soil. *Sb. Dokl. Simp. Ferment. Pochvy, Minsk* **1967**, 10–23.

WEBLEY D. M. and JONES D. (1969). Techniques for the study of localized microbial activity in soil. *Trans. 9th Int. Congr. Soil Sci., Adelaide,* 1968 **3**, 657–664.

YARCHUK I. I. (1965). Effect of fermentation on the agrochemical and biochemical properties of peat-mineral-ammoniacal fertilizers. *Torf. Promysh.* (2), 33–35.

ZAGURAL'SKAYA L. M. (1974). Role of microorganisms in the decomposition of organic matter of bog soils. *In* "Pochvennye Issledovaniya v Karelii" pp. 179–189. Karel. Filial Akad. Nauk SSSR, Petrozavodsk.

ZAGURAL'SKAYA L. M. and EGOROVA R. A. (1974). Cellulase and proteolytic activity of bog soils. *In* "Puti Izucheniya i Osvoeniya Bolot Severo-Zapada Evropeiskoi Chasti SSSR" (N. I. P'yavchenko, Ed.) pp. 87–93. Izd. Nauka, Leningrad. Otd., Leningrad.

ZAGURAL'SKAYA L. M. and KLEIN L. A. (1976). Effect of the arboricide 2, 4-D on the enzymatic activity of soil. *In* "Vozdeistvie 2, 4-D na Biogeotsenozy Listvenno-Sosnovykh Molodnyakov" pp. 132–137. Karel. Filial Akad. Nauk SSSR, Petrozavodsk.

ZAKHAROV I. S., TARAN N. P. and TOLOCHKINA S. A. (1975). Biological activity and formation of humic substances in soil in dependence of irrigation of a common chernozem in Moldavia. *V S'ezd Vses. Mikrobiol. Obshch., Sekts. Sel'skokhoz. Mikrobiol., Erevan* **1975**, 110.

ZUBENKO V. F. (1973). The organic-matter balance in different rotations on sod-podzolic soils. *Agrokhimiya* **4**, 61–68.

ZUBETS T. P. (1967). Effect of herbicides on the enzymatic activity in sod-podzolic soil. *In* "Issledovaniya Molodykh Uchenykh" pp. 105–113. Lenizdat, Leningrad.

ZUBETS T. P. (1968). Effect of herbicides on the processes of mineralization of organic matter in peaty soil. *Nauch. Tr., Sev.-Zapad. Nauch.-Issled. Inst. Sel. Khoz.* (12), 46–50.

ZYATCHINA G. P. (1974). Effect of anhydrous ammonia on the biochemical activity of soils. *Mater. Konf. Molodykh Uchenykh Kormoproizv., Moscow* **1974,** 119–122.

5

Urease Activity in Soils

J. M. BREMNER and R. L. MULVANEY

Department of Agronomy, Iowa State University, Ames, Iowa, USA

I. Introduction

Urease (urea amidohydrolase, EC 3.5.1.5) is the enzyme that catalyses the hydrolysis of urea to carbon dioxide and ammonia. It occurs in a large number of higher plants and microorganisms (particularly bacteria) and has been detected in the gastric mucosa of man and several animals. Its presence in soil was first indicated by work by Rotini (1935). Subsequent pioneer work by Conrad (1940a, b; 1942a, b; 1943) provided convincing evidence that soils contain urease and indicated that this enzyme is responsible for the conversion of urea nitrogen to ammonium nitrogen in soils treated with urea.

Urease is unique among soil enzymes in that it greatly affects the fate and performance of an important fertilizer (urea). For this reason, it has been studied more intensively than other soil enzymes and has a very extensive literature. Some aspects of this literature are discussed in articles by Briggs and Spedding (1963), Durand (1965), Voets and Dedeken (1966), Skujiņš (1967, 1976), Kuprevich and Shcherbakova (1971a, b), Sequi (1974) and

Kiss *et al.* (1975) but no comprehensive review is available. Attempts to review and evaluate information concerning soil urease have undoubtedly been discouraged by the numerous contradictions in the literature on this enzyme. Some of these contradictions can be attributed to defects in experimental techniques, but others are difficult to explain and there is a clear need for further research to account for the divergent findings in studies of soil urease activity and to obtain reliable information concerning the nature and properties of soil urease.

The origin of the urease in soils is still obscure. Most workers subscribe to the view that soil urease is largely, if not entirely, of microbial origin, but there is very little basis for this conclusion or for speculation by other workers that soil urease is largely derived from plants. The state of the urease in soils is also obscure. Most workers believe that soil urease is a free enzyme accumulated through release of urease from living and disintegrated microbial cells and is essentially a microbial extracellular enzyme. It is evident, however, that this enzyme must be associated with, and protected by, soil constituents because, if truly free, it would be rapidly decomposed or inactivated in soils. Several workers have suggested that the urease in soils is protected by humus or clay colloids (Conrad, 1940b; Pinck and Allison, 1961; McLaren, 1963, 1975), and some have speculated that this protection arises through immobilization of the enzyme within organic colloids during humus formation (Burns *et al.*, 1972a, b).

II. Properties of Ureases from Various Sources

Urease is the trivial name for enzymes with the systematic name of urea amidohydrolase and refers to hydrolases which act on C–N bonds (non-peptide) in linear amides. The term "ureases" would be more appropriate because it is now known that urease activity is exhibited by several protein species from many different sources (Reithel, 1971).

Most of our knowledge concerning the properties of urease is derived from studies of the urea-hydrolysing enzyme in jack beans (*Canavalia ensiformis*). This was the first enzyme to be crystallized, and it has been studied very extensively (for reviews, see Sumner, 1951; Sumner and Somers, 1953; Varner, 1960; Reithel, 1971). It is rather surprising, therefore, that several cherished beliefs concerning jack bean urease have recently been shown to be erroneous. For example, although it has long been believed that jack bean urease is absolutely specific for urea and completely devoid of metals, recent work has shown that this enzyme will act upon hydroxyurea, dihydroxyurea and semicarbazide (Fishbein *et al.*,

1965; Fishbein, 1969; Gazzola et al., 1973) and that it contains nickel and is probably a nickel metalloenzyme (Dixon et al., 1975).

The molecular weight of jack bean urease is about 480 000. Despite extensive research, the subunit structure of the oligomer is still obscure (Gorin et al., 1962; Reithel et al., 1964; Gorin and Chin, 1965). The species of molecular weight 480 000 contains about forty-seven sulphydryl groups, and it has been estimated that 4–8 of these groups are essential for activity (Gorin and Chin, 1965).

Although noted for its high specificity, jack bean urease is probably more remarkable for its enormous efficiency as a hydrolase. The mechanism by which it hydrolyses urea has been the subject of considerable controversy, but work by Blakeley et al. (1969) leaves very little doubt that hydrolysis occurs via carbamate according to the following equation:

$$NH_2CONH_2 \xrightarrow[H_2O]{urease} NH_2COOH + NH_3 \xrightarrow{H_2O} CO_2 + 2NH_3 \xrightarrow{H_2O} H_2CO_3 + 2NH_3$$

The kinetics of jack bean urease have been studied very extensively, and a vast amount of data has been reported. Some of this data is summarized in Tables 1 and 2, which show Michaelis constants (K_m values) and activation energy (Ea) values for ureases from various sources. The wide divergence in the kinetic data reported for jack bean urease can be at least partly attributed to the wide variety of conditions adopted in different kinetic studies (see Tables 1 and 2).

It is well established that jack bean urease hydrolyses urea according to zero-order kinetics when the substrate (urea) concentration is not limiting, but according to first order kinetics when the substrate concentration is limiting (Laidler and Hoare, 1949; Sumner, 1951; Sumner and Somers, 1953). It is also well established that this enzyme is most active at about 65°C (Sumner, 1951) and is inactivated at temperatures above 70°C (Van Slyke and Cullen, 1914). The optimum pH for jack bean urease appears to depend on the buffer and urea concentration used to study the effect of pH (Howell and Sumner, 1934). It usually lies between 6·0 and 7·0, but may be as high as 8·0. This is illustrated in Table 3, which shows pH optima reported for ureases from various sources.

Although the properties of jack bean urease are often assumed to hold for ureases from other sources, it is evident from work summarized in Tables 1–3 and from other investigations that urease preparations from different sources exhibit different properties. For example, Andersen et al. (1969) found that the molecular weight of urease isolated from Proteus mirabilis (151 000) was much smaller than that of jack bean urease (480 000), and Valmikinathan et al. (1968) found that urease isolated from the seeds of

TABLE 1

K_m values for ureases from various sources

Source of urease[a]	K_m $(10^{-3}\mathrm{M})$	Conditions			Reference
		Buffer	pH	Temp. (°C)	
Soybean	25·0	None	—	20	Van Slyke and Cullen (1914)
Soybean	20·0	None	—	25	Talsky and Klunker (1967)
Soybean	19·0	Phosphate	7·0	25	Talsky and Klunker (1967)
Soybean	55·5	Thiosulphate	7·0	25	Talsky and Klunker (1967)
Soybean	476·0	Sulphite	7·0	25	Talsky and Klunker (1967)
Jack bean (5)[b]	9·8–11·6	Phosphate	7·0	25	Peterson et al. (1948)
Jack bean	3·0	Maleic acid	7·0	25	Harmon and Niemann (1949)
Jack bean	4·0	Tris–H_2SO_4	7·1 or 8·0	21	Wall and Laidler (1953)
Jack bean	4·0	Maleic acid	5·0	21	Lynn (1967)
Jack bean	6·4	Tris–H_2SO_4	7·5	21	Lynn (1967)
Jack bean	3·3	Phosphate	7·0	38	Blakeley et al. (1969)
Jack bean	4·0	Tris–HCl	7·4	25	Fishbein (1969)

Jack bean	3·1	Phosphate	5·0	38	Gazzola et al. (1973)
Jack bean	5·5	Tris–HCl	7·0	20	Pettit et al. (1976)
Jack bean	19·0	Phosphate	7·0	20	Pettit et al. (1976)
BP	130·0	Phosphate	6·7	20	Larson and Kallio (1954)
BP	40·0	Phosphate	7·7	20	Larson and Kallio (1954)
CR	30·0	Phosphate	7·0	37	Lister (1956)
CI	9·1	Tris–acetic acid	7·0	30	Malhotra and Rani (1969)
CI	3·3	Tris–H_2SO_4	7·0	30	Malhotra and Rani (1969)
Soil	213·0	None	—	25	Paulson and Kurtz (1970)
Soils (7)[b]	1·3–7·0	Tris–H_2SO_4	9·0	37	Tabatabai (1973)
Soil	2·1	None	—	†	Ardakani et al. (1975)
Soil	52·3	Tris–HCl	7·0	20	Pettit et al. (1976)
Soil	62·5	Phosphate	7·0	20	Pettit et al. (1976)

[a] BP, *Bacillus pasteurii*; CR, *Corynebacterium renale*; CI, *Cajanus indicus*.
[b] Values in parentheses indicate number of samples used.
† Not reported.

TABLE 2

Activation energy (Ea) values for ureases from various sources

Source of urease[a]	Ea (Kcal. mole^{-1})	Urea concentration (10^{-3} M)	Conditions			Reference
			Buffer	pH	Temp. range (°C)	
Soybean	8·7–11·7	250	Phosphate	7·0	0·2–50	Sizer (1940)
Soybean	8·7	250	Phosphate	7·0	22–40	Talsky and Klunker (1967)
Soybean	16·3	883	Phosphate	7·0	3–9	Talsky and Klunker (1967)
Jack bean	11·7	250	Phosphate	7·0	0–23	Sizer (1939)
Jack bean	8·7	250	Phosphate	7·0	23–40	Sizer (1939)
Jack bean	8·8	250	Phosphate	7·0	5–20	Kistiakowsky and Lumry (1949)
Jack bean	12·5	5	Phosphate	6·6	20–50·5	Laidler and Hoare (1950)
Jack bean	5·9	1496	Phosphate	6·2	20–50·5	Laidler and Hoare (1950)
Jack bean	8·8	33	Maleic acid	7·0	10–25	Kistiakowsky and Rosenberg (1952)
Jack bean	6·8	5	Tris–H$_2$SO$_4$	7·1	12–30	Wall and Laidler (1953)
Jack bean	9·7	250	Tris–H$_2$SO$_4$	7·1	12–30	Wall and Laidler (1953)
Jack bean	8·5	5	Tris–H$_2$SO$_4$	8·0	13–30	Wall and Laidler (1953)

Jack bean	11·1	250	Tris–H$_2$SO$_4$	8·0	13–30	Wall and Laidler (1953)
Jack bean	11·0	50	Maleic acid	6·5	2–43	Lynn and Yankwich (1962)
Jack bean	4·0	4	Maleic acid	6·5	1–29	Lynn (1967)
Jack bean	0	4	Tris–H$_2$SO$_4$	7·5	4·5–26	Lynn (1967)
Jack bean	8·0	70	Tris–H$_2$SO$_4$	8·0	13–25	Miller et al. (1968)
Jack bean	7·2	50	Phosphate	7·0	20–38	Blakeley et al. (1969)
Jack bean	6·0	5	Tris–maleic acid	8·0	12–36·5	Ramachandran and Perlmutter (1976)
Jack bean	11·1	250	Tris–maleic acid	8·0	12–36·5	Ramachandran and Perlmutter (1976)
PV	8·7–14·4	250	Phosphate	7·0	0·2–50	Sizer (1941)
BP	9·9	500	Phosphate	6·7	10–24·5	Larson and Kallio (1954)
BP	4·4	500	Phosphate	6·7	24·5–50	Larson and Kallio (1954)
CR	7·8	500	Phosphate	7·0	15–45	Lister (1956)
Soil	22·6	100[a]	None	—	15–33	Rachinskiy and Pel'tser (1967)
Soil	9·8	429[a]	None	—	2–45	Gould et al. (1973)
Soils (15)[b]	3·9–6·0	6000 or 24 000[a]	None	—	22–42	Dalal (1975b)
Soils (15)[b]	18·5–24·5[c]	6000 or 24 000[a]	None	—	22–47	Dalal (1975b)

[a] PV, *Proteus vulgaris*; BP, *Bacillus pasteurii*; CR, *Corynebacterium renale*.
[b] Values in parentheses indicate number of samples used.
[c] Values obtained when soils were treated with toluene (0·2 ml toluene g^{-1} soil).
[a] Expressed as μg urea g^{-1} soil.

TABLE 3

Optimum pH for ureases from various sources

Source of urease[a]	Optimum pH	Conditions		Reference
		Urea concentration (10^{-3} M)	Buffer	
Jack bean	6·4	417	Acetate	Howell and Sumner (1934)
Jack bean	6·5	417	Citrate	Howell and Sumner (1934)
Jack bean	6·9	417	Phosphate	Howell and Sumner (1934)
Jack bean	6·7	16·7	Acetate	Howell and Sumner (1934)
Jack bean	6·7	16·7	Citrate	Howell and Sumner (1934)
Jack bean	7·6	16·7	Phosphate	Howell and Sumner (1934)
Jack bean	8·0	250	Tris–H_2SO_4	Wall and Laidler (1953)
Jack bean	6·0–6·5	0·8	Tris–citrate	Fishbein et al. (1965)
Jack bean	6·0–7·0	70	Tris–H_2SO_4	Miller et al. (1968)
Jack bean	6·5–7·0	4	Tris–maleic acid	Fishbein (1969)
CR	7·5	500	Phosphate	Lister (1956)
Soils (4)[b]	6·5–7·0	100[c]	Phosphate	Hofmann and Schmidt (1953)
Soil	9·0	2·4[c]	Tris–H_2SO_4	Tabatabai and Bremner (1972)
Soil	8·8	2·1[c]	Phosphate	May and Douglas (1976)
Soil	6·5	360[c]	Phosphate	Pettit et al. (1976)

[a] CR, *Corynebacterium renale*.
[b] Values in parentheses indicate number of samples used.
[c] Expressed as mg urea g^{-1} soil.

Glyciridia maculata was active at temperatures up to 90°C but exhibited no activity at 30°C or below! The properties of soil urease differ significantly from those of ureases from other sources. For example, several investigators have found that the Michaelis constant, activation energy and optimum pH values for soil urease are significantly higher than corresponding values for jack bean urease (see Tables 1–3). It should be noted, however, that it is much more difficult to obtain reliable kinetic data for enzymes in heterogeneous environments such as soil than for enzymes in homogeneous solutions (McLaren and Packer, 1970; Cervelli *et al.*, 1973; Irving and Cosgrove, 1976; Skujiņš, 1976).

III. Urease Activity in Soils

1. Assay

Numerous methods have been used for assay of urease activity in soils (e.g. Conrad, 1940b; Hofmann and Schmidt, 1953; McLaren *et al.*, 1957; Stojanovic, 1959; Hoffman and Teicher, 1961; Anderson, 1962; Vasilenko, 1962; Hofmann, 1963; Niekerk, 1964; Porter, 1965; Skujiņš, 1965; McGarity and Myers, 1967; Simpson, 1968; Paulson and Kurtz, 1969a; Skujiņš and McLaren, 1969; Thente, 1970; Douglas and Bremner, 1971; Kozlovskaya *et al.*, 1972; Tabatabai and Bremner, 1972; Norstadt *et al.*, 1973; Zantua and Bremner, 1975a; May and Douglas, 1976; Searle and Speir, 1976). Most of these methods involve estimation of the ammonium released on incubation of toluene-treated soil with buffered urea solution (e.g. Hofmann and Schmidt, 1953; Stojanovic, 1959; Hoffman and Teicher, 1961; McGarity and Myers, 1967; Tabatabai and Bremner, 1972). Others involve estimation of the urea decomposed or the carbon dioxide released on incubation of soil with urea (e.g. Conrad, 1940b; Porter, 1965; Skujiņš, 1965; Simpson, 1968; Skujiņš and McLaren, 1969; Douglas and Bremner, 1971; Norstadt *et al.*, 1973; Zantua and Bremner, 1975a). Several methods adopted have not involved use of a buffer to control pH (e.g. Conrad, 1940b; Porter, 1965; Simpson, 1968; Douglas and Bremner, 1971; Zantua and Bremner, 1975a) or addition of toluene to inhibit microbial activity (e.g. McLaren *et al.*, 1957; Simpson, 1968; Paulson and Kurtz, 1969a; Skujiņš and McLaren, 1969; Zantua and Bremner, 1975a). Most of the methods proposed must be considered empirical because no studies to evaluate them have been reported. For example, the possibility that methods involving estimation of the ammonium released on incubation of urea-treated soil may be vitiated by fixation or volatilization of this ammonium during the assay procedure has received little attention and it has not been demonstrated that methods involving long incubation times are not invalidated by microbial activity during incubation. Few attempts have been made to determine the optimal conditions for assay of soil urease activity by buffer methods, and the various buffers employed have differed greatly in composition, pH and concentration.

It is unfortunate that a detailed discussion of the various methods proposed for assay of soil urease activity is beyond the scope of this review because many of the divergent findings in research on soil urease can be attributed to defects in methods adopted for assay of urease activity, and there is an urgent need for critical evaluation of these methods. The methods which have been most thoroughly evaluated are the non-buffer method proposed by Zantua and Bremner (1975a) and the buffer method proposed by Tabatabai and Bremner (1972). The non-buffer method cited is essentially a scaled-down version of the method proposed by Douglas and Bremner (1971), the only significant difference being that toluene is omitted. It involves determination of the amount of urea hydrolysed on incubation of the soil sample with urea at 37°C for 5 h, urea hydrolysis being estimated by colorimetric determination of urea in the extract obtained by shaking the incubated soil sample with 2M KCl containing a urease inhibitor (phenylmercuric acetate) and filtering the resulting suspension. The buffer method of Tabatabai and Bremner (1972) involves determination of the ammonium released on incubation of the soil sample with THAM (tris-H_2SO_4) buffer (pH 9·0), urea and toluene at 37°C for 2 h, ammonium release being determined by shaking the incubated soil sample with 2·5 M KCl containing a urease inhibitor (silver sulphate) and by steam-distilling an aliquot of the resulting soil suspension with magnesium oxide. Both methods give precise results but the buffer method gives markedly higher values than the non-buffer method and detects urease activity that does not occur when soils are treated with urea in the absence of buffer (Table 4). The non-buffer method provides a very good index of the ability of soils to hydrolyse urea under natural conditions (Zantua and Bremner, 1975a), and its results are not affected by inclusion of toluene (Table 4).

It is important to emphasize that the choice of method for assay of urease activity in soils should depend upon the purpose of the assay because this has frequently been overlooked in research on soil urease. If, as in most investigations, the purpose is to obtain an index of the ability of the urease in the soil sample under study to hydrolyse urea under natural conditions, the non-buffer method described is obviously superior to the buffer method. If, on the other hand, the purpose is to detect urease in soils or soil fractions, the buffer method should be preferred because it detects urease activity not detected by the non-buffer method.

TABLE 4

Comparison of methods of assaying urease activity in soils (Zantua and
Bremner, 1975a)

Soil[a]	pH	Organic matter content (%)	Urease activity[b] Buffer method	Non-buffer method[c]
Buckner sa	6·1	0·54	10·7	0 (0)
Storden 1	8·0	0·54	39·6	14·2 (14·1)
Dickinson sa	6·5	0·99	19·9	4·7 (4·7)
Ida sil	7·9	1·58	72·9	18·9 (18·9)
Canyon sacl	7·9	1·60	98·7	23·6
Belinda sil	6·3	2·16	46·2	34·3 (33·9)
Lindley 1	5·0	3·02	31·3	18·9 (19·0)
Regina c	7·5	3·76	133·0	42·5 (42·8)
Muscatine sicl	6·0	4·05	53·6	28·3
Fargo sic	7·5	4·84	233·8	84·9 (85·3)
Webster cl	6·9	5·27	90·6	37·6 (37·5)
Paaloa sic	3·6	5·51	6·4	0 (0)
Hayden sal	6·9	5·78	90·8	61·3 (61·2)
Harps cl	7·9	5·78	165·2	70·8
Primghar sicl	7·0	6·14	128·7	70·8
Marcus sicl	6·6	6·86	81·5	37·8 (38·0)

[a] Sa, sand; l, loam; sil, silt loam; sacl, sandy clay loam; c, clay; sicl, silty clay
loam; sic, silty clay; cl, clay loam; sal, sandy loam.

[b] Expressed as μg urea hydrolysed at $37°C$ g^{-1} soil h^{-1}.

[c] Values in parentheses are those obtained by non-buffer method when toluene
was added ($0·1$ ml toluene g^{-1} soil).

Although toluene is commonly used to inhibit microbial growth and
assimilation of enzymatic products in assay of enzyme activity in soils, the
extensive literature concerning the effects of toluene indicates that this
reagent causes more problems than it solves (for discussions of this literature, see Kiss and Boaru, 1965; Skujiņš, 1967, 1976; Kiss et al., 1975).
Many workers have found that toluene can significantly affect the results
obtained by buffer or non-buffer methods of assaying soil urease activity
(e.g. Rotini, 1935; Conrad, 1942a; Galstyan, 1965a; Kiss and Boaru,
1965; McGarity and Myers, 1967; Roberge, 1968; Thente, 1970; Tabatabai and Bremner, 1972; Dalal, 1975a). McGarity and Myers (1967) found
that the values obtained with Australian soils by a buffer method of assaying
urease activity were substantially reduced by addition of toluene, Dalal
(1975a) found that addition of this reagent greatly reduced the values
obtained with Trinidad soils by a non-buffer method of assaying urease
activity. McGarity and Myers (1967) concluded that the urease activity
measured in the presence of toluene is derived from extracellular urease

adsorbed on soil colloids, whereas the activity measured in the absence of toluene includes activity derived from metabolizing ureolytic micro-organisms. Galstyan (1965a) also found that addition of toluene led to a decrease in the values obtained in assay of soil urease activity by a buffer method but he concluded that toluene has an inhibitory effect on urease. Support for this conclusion was provided by Thente (1970), who found that toluene reduced the activity of jack bean urease. Tabatabai and Bremner (1972) found that the results obtained by their buffer method of assaying urease activity in soils were increased by the addition of toluene. Other workers (e.g. Rotini, 1935; Conrad, 1942a) have also noted that toluene treatment of soil can lead to an increase in urease activity, several have suggested that toluene (a plasmolytic agent) releases urease from microorganisms (e.g. Kiss and Boaru, 1965; Skujiņš, 1967; Thente, 1970).

The possibility that radiation sterilization may be useful for assay of urease activity in soils and research on soil urease was suggested by the observation that a Californian soil exhibited urease activity after steriliza-tion by irradiation with an electron beam (McLaren et al., 1957). Many studies of the effects of high energy radiation on soil urease activity have been reported (e.g. Vela and Wyss, 1962; Niekerk, 1964; Roberge and Knowles, 1968a, b; Skujiņš and McLaren, 1969; Roberge, 1970; Thente, 1970; Pettit et al., 1976), but these studies have given contradictory results and are difficult to interpret (for discussions, see McLaren, 1969; Cawse, 1975; Skujiņš, 1976). Briefly, they show that irradiation of soil can lead to a decrease or increase in urease activity and that the effect of high energy radiation depends upon the soil and the radiation dose. For example, Skujiņš and McLaren (1969) found that urease activity in some soils was increased by exposure to 4 Mrad, whereas Roberge and Knowles (1968a) found that urease in a black spruce humus was almost completely inacti-vated by 4–5 Mrad. Thente (1970) concluded that neither radiation sterilization nor toluene should be used for assay of urease activity in soils and suggested that, if the incubation time is short enough, errors due to microbial activity during the assay will be insignificant. Support for this conclusion has been provided by recent work showing that the results obtained by the 5 h incubation method of assaying soil urease activity pro-posed by Zantua and Bremner (1975a) are not affected by addition of toluene (Table 4) or of organic substances known to promote production of urease by soil microorganisms (Zantua and Bremner, 1976).

Since several workers have assayed soil urease activity or studied urea decomposition in soils by estimating the $^{14}CO_2$ released through hydrolysis of ^{14}C-labelled urea by soil urease (Simpson and Melsted, 1963; Rachinskiy and Pel'tser, 1965, 1967; Skujiņš, 1965; Skujiņš and McLaren, 1968, 1969; Pel'tser, 1972a, b; Norstadt et al., 1973), attention should be drawn to problems in the use of ^{14}C-urea. One problem overlooked in several investigations is that the $^{14}CO_2$ produced by hydrolysis of ^{14}C-labelled urea in soils is not evolved quantitatively. This problem was recognized by Skujiņš and McLaren (1968, 1969), who proposed use of an acidic (pH 5·5) buffer for assay of urease activity in soils by determination of the $^{14}CO_2$ released from ^{14}C-labelled urea (see also Skujiņš, 1965). Unfortunately this buffer method has the obvious disadvantage of allowing assay of soil urease activity only under acidic conditions known to retard urease activity. Another problem overlooked in use of ^{14}C-labelled urea is the possibility of isotope effects. This possibility cannot be ignored because Rabinowitz et al. (1956) found that jack bean urease hydrolysed ^{12}C-urea about 10% faster than ^{14}C-urea at 37°C.

Most assays of soil enzyme activity have been performed on air-dried soils and this holds for assays of soil urease activity. The practice of air-drying soils before assays of enzyme activity is open to criticism because it is well established that air-drying of soils can significantly affect the activities of soil enzymes and that the effect depends on the enzyme. For example, Ross (1965) found that air-drying of soils led to a reduction in invertase and amylase activity, whereas Tabatabai and Bremner (1970) found that it increased arylsulphatase activity. Skujiņš (1967) reviewed literature pertaining to preservation of soil samples for assay of enzyme activity and concluded that the best method of preservation depends upon the enzyme under assay.

Zantua and Bremner (1975b) recently evaluated different methods of preserving soil samples for assay of urease activity and found that the following drying or storage treatments of field-moist Iowa soils had no effect on the values obtained by their non-buffer method of assaying soil urease activity: freeze-drying ($-60°C$) for 60 h; air-drying (22°C) for 48 h; oven-drying (55°C) for 24 h; storage at 20, 5, -10 or $-20°C$ for times ranging from one day to three months. They also found that no change in urease activity occurred when air-dried soils were stored at 21–23°C for times ranging from one week to one year. In subsequent work concerning the stability of soil urease, they found that no loss of urease activity could be detected when field-moist soils were air-dried and stored at 21–23°C for two years and that the following treatments of field-moist soils had no effect on urease activity: drying for 24 h at temperatures ranging from 30 to 60°C; storage for six months at temperatures ranging from -20 to 40°C; incubation under aerobic or waterlogged conditions at 30 or 40°C for six months (Zantua and Bremner, 1977a). Also, Pancholy and Rice (1972) found that no loss in urease activity occurred when field-moist soils were stored in closed containers at 5 or 21°C for 15 or 30 days. However, there are reports that air-drying and air-dry storage of soils can lead to an increase or decrease in urease activity (e.g. Vasilenko, 1962; McGarity and

Myers, 1967; Cerna, 1968; Skujiņš and McLaren, 1969; Thente, 1970; Gould *et al.*, 1973; Speir and Ross, 1975) and that storage of field-moist soils at 5–20°C or at sub-zero temperatures can also lead to an increase or decrease in urease activity (e.g. Tagliabue, 1958; McGarity and Myers, 1967; Speir and Ross, 1975; Pettit *et al.*, 1976).

It is difficult to account for the divergent findings in studies of the effects of drying and storing soils on soil urease activity. It seems likely, however, that they are at least partly due to differences in the techniques used to study these effects (see Zantua and Bremner, 1975b). Another possible explanation is that the effects of drying and storage depend upon the amount and type of plant residue in the soil samples studied. Most of the samples studied by Zantua and Bremner (1975b, 1977a) were from cultivated soils used for corn and soybean production, whereas other workers (e.g. McGarity and Myers, 1967; Speir and Ross, 1975) have studied samples from soils under pasture, which normally contain much greater amounts of unhumified plant material than do cultivated soils.

2. Levels of activity

Although numerous assays of urease activity have been performed on soils, the wide divergence in the methods adopted for these assays makes it very difficult to review current information concerning the levels of urease activity in soils. Most assays have been performed by methods involving use of buffer and toluene, but the buffer, pH, incubation temperature and time of incubation have varied greatly, and the results have been expressed in a variety of units. To facilitate comparison of data reported, we have re-calculated urease activity values obtained by buffer and non-buffer methods involving incubation at 37°C so that all values are expressed in terms of μg urea hydrolysed at 37°C g^{-1} soil h^{-1}. The results (Table 5) illustrate the wide range in urease activity values reported for surface soils in different parts of the world. It is difficult to account for the exceptionally high values reported by Dalal (1975a) for Trinidad soils, particularly since these values were obtained by non-buffer methods, which usually give lower values than buffer methods. The lowest value obtained with these soils by a non-buffer method not involving addition of toluene is much higher than the highest values obtained with Malaysian and Iowa soils by an essentially identical method.

Studies relating to the effects of soil properties on the level of urease activity have indicated that urease activity tends to increase with organic matter content and that sandy or calcareous soils tend to have a lower activity than heavy-textured or non-calcareous soils (Koepf, 1954; Galstyan, 1958; Briggs and Spedding, 1963; McGarity and Myers, 1967; Skujiņš, 1967; Myers and McGarity, 1968; Skujiņš and McLaren, 1969).

TABLE 5

Comparison of urease activity values reported for surface soils

Soils		Organic C (%)		pH		Urease activity (μg urea hydrolysed at 37°C g^{-1} soil h^{-1})			Ref.[b]
Location	No. of samples	Range	Mean	Range	Mean	Method of assay[a]	Range	Mean	
Ceylon	216	†	1·70	†	5·1	B (pH 6·7), +T	15–191	112	A
Australia	100	0·16–5·88	2·41	4·8–6·7	5·6	B (pH 6·7), +T	22–416	122	B
U.S.A. (Iowa)	21	0·30–6·73	2·60	4·6–8·0	6·4	NB, −T	0–113	39	C
Malaysia	20	0·64–3·44	1·15	4·2–4·9	4·6	B (pH 9·0), +T	11–189	82	D
						NB, −T	0·5–30	11	
						B (pH 9·0), +T	5–41	20	
Trinidad	15	1·20–19·1	4·17	4·0–6·8	5·1	NB, −T	707–2964	1285	E
						NB, +T	70–1828	732	
India	6	0·24–0·98	0·50	4·2–6·7	5·2	B (pH 6·7), +T	20–162	56	F
						B (pH 6·7), −T	22–206	68	
Egypt	6	0·72–0·86	0·77	7·2–8·1	7·7	B (pH 6·8), +T	10–51	25	G
Australia	5	1·45–3·52	2·45	4·5–8·9	6·6	B (pH 8·8), +T	20–130	66	H

[a] B, buffer method; NB, non-buffer method; T, toluene.

[b] Reference. A, Silva and Perera, 1971; B, McGarity and Myers, 1967; C, Zantua et al., 1977; D, Tan and Bremner, 1977; E, Dalal, 1975a; F, Ananthanarayana and Mithyantha, 1970; G, El-Essawi et al., 1973; H, May and Douglas, 1976.

† Not reported.

These studies have also suggested that soils under dense vegetation tend to have a high urease activity and that saline and gleyed soils tend to have a low urease activity (Gibson, 1930a; Hofmann et al., 1953; Galstyan, 1959a, b, 1960; Myers and McGarity, 1968). The effect of soil type is illustrated by Table 6, which shows some of the data obtained by McGarity and Myers (1967) in an extensive survey of urease activity in surface (0–3 in.) samples of soils under pasture in northern New South Wales (see also Silva and Perera, 1971).

Many workers have demonstrated that urease activity in soil profiles decreases markedly with depth (e.g. Hofmann and Schmidt, 1953; Koepf, 1954; Galstyan, 1958, 1960; Hofmann, 1963; Myers and McGarity, 1968; Musa and Mukhtar, 1969; Bhavanandan and Fernando, 1970; Gould et al., 1973; Tabatabai, 1977). Myers and McGarity (1968) and Gould et al. (1973) found that urease activity in soil profile samples was significantly correlated with organic carbon content. However, Myers and McGarity (1968) concluded from a detailed study of profiles of five great soil groups in Australia that, although the level and distribution of urease activity in these profiles were closely related to organic matter content, this relationship was modified within particular horizons by other factors (e.g. pH, texture and gleying).

Statistical studies of the relationships between urease activity and other soil properties have been reported by Koepf (1954), McGarity and Myers (1967), Myers and McGarity (1968), Silva and Perera (1971), Gould et al. (1973), Pancholy and Rice (1973), Dalal (1975a), Tabatabai (1977) and Zantua et al. (1977).

Koepf (1954) found that urease activity in thirty German soils was positively correlated with clay and organic matter content. McGarity and Myers (1967) found that urease activity in 100 Australian surface soils was highly correlated with organic carbon and weakly correlated with pH. They also observed a highly significant relationship between urease activity and organic carbon in profile samples of five great soil groups, but did not detect a significant correlation between urease activity and pH (Myers and McGarity, 1968). Silva and Perera (1971) found that urease activity in rubber soils of Ceylon (Sri Lanka) was significantly related to pH, but was not significantly correlated with organic carbon in two of six soil series studied. Gould et al. (1973) observed a significant relationship between urease activity and organic carbon in profile samples of an Alberta soil, and Tabatabai (1977) found that urease activity in surface and profile samples of Iowa soils was significantly correlated with organic carbon. However, Pancholy and Rice (1973) found that urease activity in nine Oklahoma surface soils was not significantly correlated with organic carbon or pH and concluded that urease activity in these soils was determined by the type of vegetation.

TABLE 6

Analyses of surface soils in northern New South Wales (McGarity and Myers, 1967)

Soils		Organic carbon (%)		pH		Urease activity[a]	
Great soil group	No. of samples	Range	Mean	Range	Mean	Range	Mean
Krasnozem	20	1·72–5·40	3·98	4·85–5·50	5·19	27·1–116·6	62·8
Chocolate	20	1·14–5·88	3·20	5·05–6·55	5·74	17·3–85·5	44·5
Yellow podzolic	20	0·48–4·12	1·84	5·00–6·30	5·51	6·1–49·3	14·7
Gley podzolic	20	0·55–5·03	1·76	4·84–6·00	5·27	8·2–37·7	16·7
Red-brown earth	20	0·16–2·99	1·27	5·23–6·70	6·04	9·2–53·8	32·3

[a] Expressed as mg ammonium-N released 100 g^{-1} soil 6 h^{-1} (37°C).

The most detailed studies of the relationships between urease activity and other soil properties have been by Dalal (1975a) and Zantua et al. (1977). Results of these studies are summarized in Table 7. Dalal (1975a) found that urease activity in fifteen Trinidad surface soils was correlated highly significantly with organic carbon and cation-exchange capacity. He also found that urease activity was significantly correlated with oxalate-extractable (amorphous) iron or aluminium, but not with pH or clay. Zantua et al. (1977) found that urease activity in twenty-one diverse Iowa surface soils was correlated highly significantly with organic carbon, total nitrogen and cation-exchange capacity. Urease activity also was significantly correlated with clay, sand and surface area, but not with pH, silt or calcium carbonate equivalent. Multiple regression analyses of the data obtained by these workers indicated that most of the variation in urease activity observed in the soils studied could be accounted for by organic matter content. This is in harmony with the conclusion from other studies that organic soil constituents contribute substantially to protection of native soil urease (see Section III. 6).

TABLE 7

Correlations between urease activity and other soil properties

| | Correlation coefficient (r) | | |
| | Trinidad soils[a] | | Iowa |
Soil property	A	B	soils[b]
Organic carbon	0·79***	0·96***	0·72***
Total nitrogen	†	†	0·71***
Cation-exchange capacity	0·89***	0·91***	0·67***
Surface area	†	†	0·45*
Clay	0·71**	0·48	0·53*
Sand	†	†	−0·47*
Silt	†	†	0·39
pH	0·46	0·21	−0·01
CaCO₃ equivalent	†	†	−0·11
Oxalate-extractable Fe	0·72**	0·74**	†
Oxalate-extractable A1	0·67**	0·69**	†

* Significant at 5 % level.
** Significant at 1 % level.
*** Significant at 0·1 % level.
[a] Data for 15 surface soils (Dalal, 1975a). Urease activity was assayed by non-buffer method (A, toluene present; B, toluene absent).
[b] Data for 21 surface soils (Zantua et al., 1977). Urease activity was assayed by non-buffer method (toluene absent). Results by this method were not affected by addition of toluene.
† Not determined.

3. Factors affecting activity

3.1. Substrate concentration

Studies of the effect of varying the substrate (urea) concentration in assay of urease activity in soils by buffer and non-buffer methods have shown that the rate of hydrolysis of urea by soil urease increases with increase in urea concentration until the amount of urea added is sufficient to saturate the enzyme with substrate (Douglas and Bremner, 1971; Tabatabai and Bremner, 1972; Dalal, 1975a; Zantua and Bremner, 1977b). This is illustrated by Fig. 1, which shows data obtained by Zantua and Bremner

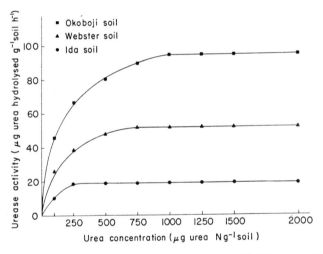

Fig. 1. Effect of urea concentration on urease activity in soils (Zantua and Bremner, 1977b).

(1977b) in a study of the effect of varying the urea concentration in assay of soil urease activity by the non-buffer method of Zantua and Bremner (1975a). It is evident from Fig. 1 that urease activity in soil is affected by urea concentration only when the amount of urea added is a limiting factor in the assay procedure and that, under such conditions, soil urease activity will be increased by adding larger amounts of urea. This accounts for the finding by several workers that the rate of urea hydrolysis in soils treated with small amounts of urea is much slower than that observed with large amounts of urea (e.g. Conrad, 1942b, 1943; Broadbent et al., 1958; Fisher and Parks, 1958; Simpson and Melsted, 1963; Overrein and Moe, 1967; Pel'tser, 1972b; Gould et al., 1973).

3.2. Water level

Several workers have found that urease activity in soil is not significantly affected by the water level (Overrein, 1963; Skujiņš and McLaren, 1967, 1969; Delaune and Patrick, 1970; Gould et al., 1973; Zantua and Bremner, 1977b). This is illustrated by Table 8, which shows data obtained by Zantua and Bremner (1977b) in a study of the effect of varying the water level in assay of urease activity in Iowa soils by the non-buffer method of Zantua and Bremner (1975a). Skujiņš and McLaren (1967, 1969) detected urease activity in air-dried soils equilibrated at 80% relative humidity and found that activity in an air-dried soil equilibrated at 100% relative humidity was close to that observed when this soil contained 50% water (Skujiņš and McLaren, 1969). There are reports, however, that urease activity in soil is increased (Rachinskiy and Pel'tser, 1965) or decreased (Jones, 1932; Simpson and Melsted, 1963; Wang et al., 1966; Pel'tser, 1972a; Dalal, 1975a; Sankhayan and Shukla, 1976) by an increase in water level.

TABLE 8

Effect of water level on urease activity in soils (Zantua and Bremner, 1977b)[a]

Water level (ml g^{-1} soil)	Urease activity (μg urea hydrolysed at $37°C$ g^{-1} soil h^{-1})	
	Webster soil	Hayden soil
0·2	51·6	56·8
0·3	51·6	56·4
0·4	51·7	56·7
0·6	51·6	56·4
0·8	52·0	56·7
1·0	51·9	56·6

[a] Soil (5 g) was incubated for 5 h at 37°C after treatment with 1–5 ml of water containing 5 mg of N as urea.

It is difficult to account for the divergent findings concerning the effect of water level on urease activity in soils. It is obvious from Fig. 1, however, that this effect will depend upon the urea concentration.

3.3. Temperature

Numerous studies have shown that urease activity in soils increases with increase in temperature from 10 to 40°C (e.g. Gibson, 1930b; Conrad, 1942b; Broadbent et al., 1958; Fisher and Parks, 1958; Kalinkevich, 1961;

Chin and Kroontje, 1963; Overrein, 1963; Simpson and Melsted, 1963; Wang *et al.*, 1966; Overrein and Moe, 1967; Rachinskiy and Pel'tser, 1967; Tanabe and Ishizawa, 1969; Gould *et al.*, 1973; Dalal, 1975a, b; Zantua and Bremner, 1977b). Zantua and Bremner (1977b) found that urease activity in Iowa soils increased very markedly with increase in temperature from 40 to 70°C, but decreased rapidly with increase in temperature from 70 to 80°C. This is illustrated in Fig. 2, which shows data obtained by these workers in a study of the effect of temperature on soil urease activity as assayed by the non-buffer method of Zantua and Bremner (1975a). Although inactivation of soil urease has been detected at 65–70°C (see Galstyan, 1965b; Pettit *et al.*, 1976), urease activity is not completely destroyed when soils are heated at 75°C for 24 h (Zantua and Bremner, 1977a) or at 80–90°C for 48 h (Conrad, 1940a, b). There is evidence, however, that soil urease is completely inactivated when soils are heated at 105°C for 24 h (Zantua and Bremner, 1977a) or at 115°C for 15 h (Rotini, 1935). Studies discussed in Section III.1. have shown that soil urease is not destroyed when soils are stored at sub-zero temperatures (-10 to -33°C).

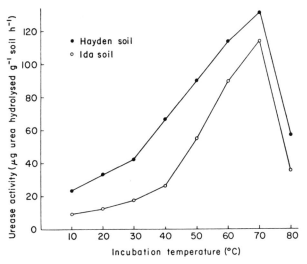

Fig. 2. Effect of temperature on urease activity in soils (Zantua and Bremner, 1977b).

Bremner and Zantua (1975) recently investigated the possibility that enzyme activity can occur in soils at sub-zero temperatures. They found that urease activity, phosphatase activity and arylsulphatase activity could be detected in soils at -10 or -20°C, but not in soils at -30°C. They concluded that the enzyme activity detected at -10 and -20°C resulted from enzyme-substrate interaction in unfrozen water at the surfaces of soil particles. Support for this explanation was obtained from experiments showing that hydrolysis of urea by jack bean urease occurs at −10 or −20°C in the presence, but not in the absence, of clay minerals or autoclaved soils.

3.4. pH

Although several studies of the effects of liming and acidifying materials on urease activity in soils have been reported (e.g. Conrad, 1943; Vasilenko, 1962; Wang et al., 1966; Delaune and Patrick, 1970; Pel'tser, 1972a; Zantua and Bremner, 1977b), these studies have very limited value in regard to assessment of the effect of pH on soil urease activity.

Studies involving use of buffers to determine the effect of pH on soil urease activity have been reported by Hofmann and Schmidt (1953), Tabatabai and Bremner (1972), May and Douglas (1976) and Pettit et al. (1976). Hofmann and Schmidt (1953) and Pettit et al. (1976) found that the optimum pH for soil urease was 6·5–7·0, whereas Tabatabai and Bremner (1972) and May and Douglas (1976) found that it was 8·8–9·0. This divergence may be related to differences in the buffers and urea concentrations adopted in these investigations (see Table 3). The optimum pH for soil urease reported by Tabatabai and Bremner (1972) and May and Douglas (1976) is much higher than that reported for jack bean urease (see Table 3).

3.5. Oxygen

Overrein (1963) found that oxygen had a significant effect on the rate of hydrolysis of urea added to an Indiana soil, whereas Delaune and Patrick (1970) found that oxygen had no effect on the rate of hydrolysis of urea added to a Crowley silt loam. It is difficult to account for Overrein's (1963) finding because there is good reason to believe that urea added to soils is hydrolysed largely, if not entirely, by native soil urease (see Kiss et al., 1975), and there is no apparent reason why the activity of this enzyme should be affected by oxygen. Zantua and Bremner (1977b) found that oxygen had no effect on the results obtained in assay of urease activity in soils by the non-buffer method of Zantua and Bremner (1975a).

3.6. Drying and re-wetting

Zantua and Bremner (1977a) found that, although air-drying of field-moist Iowa soils had no effect on urease activity, a significant decrease in urease activity occurred when air-dried soils were rewetted and incubated under aerobic or waterlogged conditions. Fig. 3 shows data obtained by these

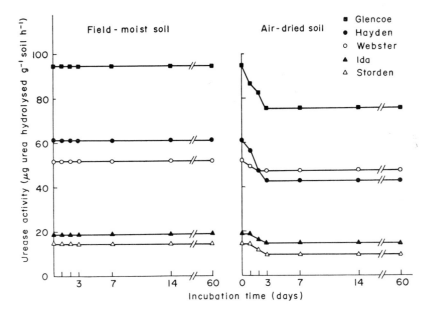

Fig. 3. Effect on urease activity of incubating (30°C, 60% WHC) field-moist and air-dried soils under aerobic conditions for various times (Zantua and Bremner, 1977a).

workers in a study of the effect of incubating field-moist and air-dried soils under aerobic conditions for various times. It can be seen that, whereas incubation of field-moist soils had no effect on urease activity, incubation of re-wetted air-dried soils led to an appreciable reduction in urease activity. Similar results were obtained when incubation was performed under waterlogged conditions, and repeated drying-incubation treatments of field-moist soils were found to have much the same effect as a single treatment. These findings suggest that air-drying of field-moist soils leads to release of urease from protected sites and that the urease thus released is rapidly decomposed when air-dried soils are re-wetted and incubated under aerobic or waterlogged conditions.

3.7. Other factors

Studies concerning the effects on soil urease activity of drying and storing field-moist soils at different temperatures are discussed in Section III.1. Work relating to the effects of soil properties on soil urease activity is discussed in Section III.2, and studies concerning the effects of various soil amendments and urease inhibitors on soil urease activity are discussed in Sections III.4 and III.7, respectively.

4. Effects of various soil amendments on urease activity

4.1. Organic materials

Since soils contain urease-producing microorganisms, it is not surprising that many investigators have found that urease activity in soils can be increased by addition of glucose or other materials that promote microbial activity (e.g. Gibson, 1930b; Conrad, 1942a; Vasilenko, 1962; Chin and Kroontje, 1963; Wang et al., 1966; Moe, 1967; Musa and Mukhtar, 1969; Tanabe and Ishizawa, 1969; Delaune and Patrick, 1970; Balasubramanian et al., 1972; Laugesen, 1972; Zantua and Bremner, 1976, 1977b). Evidence that this increase depends upon the amount and type of organic material added is presented in Table 9, which shows results obtained by Zantua and Bremner (1976, 1977b) in studies of the effects of different amounts of glucose, starch, cellulose, animal manures, plant materials and sewage sludges on urease activity in a Webster soil. The most significant finding in these studies is that, although some of the urease activity produced on treatment of soil with organic materials persists for several weeks, the urease activity of soil amended with organic materials eventually becomes identical to that of the unamended soil. This indicates that the urease activity of unamended soils reflects their capacity for protection of urease and that urease in excess of this capacity is decomposed or inactivated (see Section III.6).

4.2. Fertilizers

Zantua and Bremner (1977b) recently studied the effects of fourteen fertilizer materials on urease activity in Iowa soils. Their results showed that the following N, P, K or S fertilizers had no effect on soil urease activity when applied at rates equivalent to 500 ppm (soil basis) of N, P, K or S: ammonium sulphate, ammonium nitrate, ammonium carbonate, sodium nitrate, potassium nitrate, monocalcium phosphate, dicalcium phosphate, monoammonium phosphate, diammonium phosphate, ground rock phosphate, potassium chloride, potassium sulphate, calcium sulphate and magnesium sulphate.

Studies concerning the effect of urea on urease activity in soils have been reported by Vasilenko (1962), Paulson and Kurtz (1969a), Balasubramanian et al. (1972), Lloyd and Sheaffe (1973) and Zantua and Bremner (1976). These studies clearly merit special attention because, besides being the substrate for urease, urea is a very important nitrogen fertilizer. Data reported by Vasilenko (1962), Balasubramanian et al. (1972) and Lloyd and Sheaffe (1973) indicated that treatment of soils with urea can lead to very marked fluctuations in urease activity, but the methods adopted by these

TABLE 9

Effects of organic materials on urease activity in a Webster soil (Zantua and Bremner, 1976, 1977b)[a]

Material added[b]	Urease activity (μg urea hydrolysed at 37°C g⁻¹ soil h⁻¹)[c]							
	Incubation time (days)							
	0	1	3	5	10	20	30	60
None	51·9	52·0	51·9	52·0	51·8	51·9	51·8	51·9
Glucose (0·2 mg)	51·9	52·0	66·0	63·7	61·3	51·9	51·9	51·8
Glucose (5·0 mg)	51·9	198·2	184·0	155·7	108·5	99·0	80·2	52·0
Starch (5·0 mg)	51·9	63·7	75·5	84·9	75·5	75·5	66·0	52·0
Cellulose (5·0 mg)	51·9	54·3	75·5	141·6	108·5	103·8	70·8	52·0
Beef cattle manure (5·0 mg)	84·9	94·4	108·5	108·5	99·1	70·8	52·0	51·9
Dairy cattle manure (5·0 mg)	80·2	94·4	103·5	109·4	108·3	80·2	56·6	52·0
Sheep manure (5·0 mg)	66·0	70·8	84·9	108·5	108·3	80·2	51·9	51·8
Corn (5·0 mg)	61·3	63·7	108·0	120·3	109·7	84·9	54·3	51·9
Corn (10·0 mg)	70·8	66·0	136·9	155·7	118·0	84·9	54·3	51·9
Orchard grass (5·0 mg)	54·3	54·3	89·7	103·8	82·2	80·2	51·9	51·8
Alfalfa (5·0 mg)	52·8	63·7	89·7	94·4	70·8	66·0	51·8	51·8
Sewage sludge (20·0 mg)	51·9	52·3	54·7	53·1	52·3	51·8	51·9	51·8
Sewage sludge (30·0 mg)	51·9	53·6	59·0	55·4	52·8	51·9	51·8	51·9

[a] Soil (5 g) was incubated (30°C, 60% WHC) for various times after treatment specified.
[b] Amount of material added per g of soil is indicated in parentheses.
[c] Assayed by method of Zantua and Bremner (1975a).

workers to study these effects are open to criticism (see Zantua and Bremner, 1976). Paulson and Kurtz (1969a) found that the increase in urease activity resulting from incubation of a Drummer silty clay loam with urea (13 μg urea N g^{-1} soil) and glucose (1 mg g^{-1} soil) was greater than the increase resulting from incubation with ammonium sulphate (13 μg ammonium N g^{-1} soil) and glucose (1 mg g^{-1} soil) and concluded that treatment of this soil with a small amount of urea promoted production of urease by soil microorganisms. However, they did not study the effect of incubating this soil with urea or ammonium sulphate in the absence of glucose or with glucose in the absence of urea or ammonium sulphate, and it is evident from work reported by Zantua and Bremner (1976) that the experiments performed by Paulson and Kurtz (1969a) do not permit valid conclusions concerning the effect of urea on urease activity in soils. Zantua and Bremner (1976) found that treatment of soils with urea had no effect on urease activity.

4.3. Liming materials

Moe (1967) and Pel'tser (1972a) found that urease activity in soil was decreased by addition of liming materials, which provided support for Galstyan's (1958) conclusion that soils rich in calcium carbonate tend to have a low urease activity because Ca^{2+} has a detrimental effect on urease-producing microorganisms. However, other workers have found that urease activity in soils can be increased by addition of liming materials (Conrad,

TABLE 10

Effects of liming materials on urease activity in soils (Zantua and Bremner, 1977b)

Soil	Treatment	Urease activity[a] (μg urea hydrolysed at 37°C g^{-1} soil h^{-1})
Webster	None	51·9 (6·7)
	CaO (2 mg Ca g^{-1} soil)	103·8 (9·1)
	CaO (4 mg Ca g^{-1} soil)	240·7 (10·2)
	CaCO$_3$ (4 mg Ca g^{-1} soil)	51·9 (7·1)
	CaCO$_3$ (10 mg Ca g^{-1} soil)	51·9 (7·2)
Hayden	None	56·6 (6·7)
	CaO (2 mg Ca g^{-1} soil)	99·1 (9·7)
	CaO (4 mg Ca g^{-1} soil)	188·8 (10·5)
	CaCO$_3$ (4 mg Ca g^{-1} soil)	56·6 (7·0)
	CaCO$_3$ (10 mg Ca g^{-1} soil)	56·6 (7·1)

[a] Assayed by method of Zantua and Bremner (1975a). Values in parentheses indicate pH of soil suspension (1 ml water g^{-1} soil).

1943; Vasilenko, 1962; Wang *et al.*, 1966; Delaune and Patrick, 1970; Zantua and Bremner, 1977b). Table 10 shows some of the data obtained by Zantua and Bremner (1977b) in a study of the effects of different amounts of liming materials on urease activity in Iowa soils. The data reported show that urease activity was increased by addition of calcium oxide, which markedly increased soil pH, but not by addition of calcium carbonate, which had very little effect on soil pH.

Pel'tser's (1972a) finding that urease activity in soil was decreased by addition of liming materials can be attributed to defects in his experimental technique, but it is difficult to account for Moe's (1967) finding that urease activity in a Plainfield sand was greatly decreased by addition of $Ca(OH)_2$.

4.4. Nitrification inhibitors

Bremner and Douglas (1971a) found that 2-chloro-6-(trichloromethyl)-pyridine (N-Serve), 2-amino-4-chloro-6-methyl-pyrimidine (AM), 2, 4-diamino-6-trichloromethyl-*s*-triazine (CL 1580), sulphathiazole (ST) and thirteen other compounds patented as inhibitors of nitrification in soils had very little, if any, effect on soil urease activity when applied at the rate of 50 ppm of soil. Also, Bundy and Bremner (1974a) found that 4-amino-1, 2, 4-triazole, which has been patented as a soil nitrification inhibitor under the trade name of ATC, did not significantly affect urea hydrolysis in soils when applied at the rate of 50 ppm of soil.

4.5. Pesticides

Numerous studies of the effects of pesticides on urease activity in soils have been reported, the most recent being those of Cervelli *et al.* (1976), Gauthier *et al.* (1976) and Lethbridge and Burns (1976). Most of these studies are cited in a recent article by Kiss *et al.* (1975).

The extensive literature on the effects of pesticides on soil urease activity includes studies of the effects of insecticides, fungicides and nematocides, but largely relates to the effects of herbicides. This literature can not be discussed adequately here but is included in Chapter 7 of this volume. It can be stated, however, that the effects reported differ markedly for different pesticides and depend upon many factors, including the soil type, the pesticide dose and the conditions of the study.

4.6. Heavy metals

The recent trend towards use of agricultural land for disposal of sewage sludges and effluents containing substantial amounts of heavy metals has generated interest in the effects of heavy metals on urease activity in soils because many investigations have shown that urease isolated from plants or microorganisms is inactivated by very small amounts of Ag^+, Hg^{2+}, Cu^{2+} and other heavy metal ions (e.g. Shaw, 1954; Shaw and Raval, 1961; Bahadur and Chandra, 1962; Toren and Burger, 1968; Hughes *et al.*, 1969). Bremner and Douglas (1971a) found that the amounts of Ag^+, Hg^{2+}, Cu^{2+} and other heavy metal ions needed to effect substantial inhibition of soil urease were much greater than the amounts needed for equivalent inhibition of jack bean or soybean urease. They also found that the inhibitory effects of 50 ppm (soil basis) of different metallic cations on soil urease activity decreased in the order: $Ag^+ > Hg^{2+} > Au^{3+} > Cu^{2+}$, $Cu^+ > Co^{2+}$, Pb^{2+}, As^{3+}, Pb^+, Cr^{3+}, $Ni^{2+} >$ others (Table 11). Similar results were obtained in recent work by Tabatabai (1977). It is important to note that these studies were performed with water-soluble metallic salts and do not permit valid conclusions concerning the effects on soil urease activity of heavy applications of sewage sludges containing water-insoluble forms of heavy metals. Recent work by Zantua and Bremner (1977b) has shown that the results obtained in assay of urease activity in soils are not affected by addition of large amounts of heavy metals in the form of sewage sludges or by addition of as much as 2000 ppm (soil basis) of Cu in water-insoluble forms (e.g. as Cu, Cu_2O or $CuCO_3$).

TABLE 11

Effects of metallic compounds on urease activity in soils (Bremner and Douglas, 1971a)

Compound	Metallic cation[a]	% Inhibition of urease activity	
		Fayette soil	Webster soil
Silver nitrate	Ag^+	65	60
Silver sulphate	Ag^+	63	61
Cuprous chloride	Cu^+	16	14
Lead nitrate	Pb^+	3	2
Mercuric chloride	Hg^{2+}	42	38
Mercuric sulphate	Hg^{2+}	40	36
Cupric chloride	Cu^{2+}	16	13
Cupric sulphate	Cu^{2+}	14	15
Lead chloride	Pb^{2+}	4	4
Cobaltous chloride	Co^{2+}	4	6
Nickelous chloride	Ni^{2+}	1	2
Gold chloride ($HAuCl_4$)	Au^{3+}	18	20
Arsenic chloride	As^{3+}	4	2
Chromium chloride	Cr^{3+}	3	2
Others[b]	†	0	0

[a] Rate of addition of each cation was equivalent to 50 ppm of soil.
[b] NaCl, Na_2SO_4, KCl, $CaCl_2$, $BaCl_2$, $ZnCl_2$, $MnCl_2$, $AlCl_3$, $FeCl_3$.
† Na^+, K^+, Ca^{2+}, Ba^{2+}, Zn^{2+}, Mn^{2+}, Al^{3+}, Fe^{3+}.

5. Extraction of urease from soils

Studies relating to extraction of urease from soils have been reported by Conrad (1940b), Haig (1955), Briggs and Segal (1963), Tanabe and Ishizawa (1969), Burns et al. (1972a, b), McLaren and Pukite (1972), Nannipieri et al. (1974, 1975), Lloyd (1975), McLaren et al. (1975) and Pettit et al. (1976). Conrad (1940b) and Haig (1955) were unsuccessful in attempts to extract urease from Californian soils, but the other workers cited have reported detection of urease activity in soil extracts or isolation of soil fractions exhibiting urease activity.

Briggs and Segal (1963) reported isolation of "free" urease from a New Zealand forest soil by a very simple procedure involving extraction of 25 kg of soil with Sorensen's phosphate buffer (pH 6·0) at 5°C and separation of the crystalline material precipitated by treatment of the filtered and centrifuged extract with cold acetone. By re-crystallizing this material, they obtained 12 mg of a product which contained 8·78% nitrogen and showed a urease activity of 75 Sumner units mg^{-1}. A detailed examination of this material indicated that it was a mixture of proteins exhibiting urease activity and that the molecular weights of these proteins ranged from 42 000 to 217 000.

Considering the problems in isolation of a small amount of a labile enzyme such as urease from a material as complex as soil, it is difficult to account for the remarkable purity and activity of the urease preparation isolated by Briggs and Segal (1963). Their work has been widely cited as evidence for the existence of "free" (extracellular) enzymes in soils, but it has been questioned by Lloyd (1975) on the grounds that "it is often difficult to obtain a measurable amount of urease from some soils, let alone sufficient for a crystalline preparation", and it has not been mentioned in recent articles relating to extraction of urease from soils.

Recent attempts to extract urease from soils and isolate urease-active fractions have involved a wide variety of extraction and fractionation procedures. These procedures are much more complicated than the procedure described by Briggs and Segal (1963) and cannot be discussed adequately here (for discussions, see Pettit et al., 1976; Skujiņš, 1976). Burns et al. (1972a), McLaren et al. (1975) and Pettit et al. (1976) have described methods of isolating soil fractions that exhibit urease activity resistant to the action of proteolytic enzymes. Recent work by Nannipieri et al. (1974, 1975) has indicated that neutral 0·1 M sodium pyrophosphate (the reagent proposed by Bremner and Lees (1949) for extraction of organic matter from soils under mild conditions) is an effective extractant of soil urease and extracts extracellular urease without damaging ureolytic microorganisms. Nannipieri et al. (1974) found that 30–40% of the urease in a podzol was extracted by treatment with 0·1 M sodium pyrophosphate (pH 7·1) at 37°C for 18 h and that extraction of urease-active organic matter by pyrophosphate preceded extraction of inactive organic material by this reagent. Further evidence that soil urease is not uniformly associated with soil organic matter was obtained in work reported by McLaren *et al.* (1975).

6. Stability of urease in soils

Urease isolated from plants or microorganisms is rapidly decomposed by microorganisms and by proteolytic enzymes (proteases), and several investigations have shown that jack bean urease is decomposed or inactivated very rapidly when added to soils (e.g. Stojanovic, 1959; Roberge, 1970; Pettit *et al.*, 1976; Zantua and Bremner, 1977a). Moreover, it is well established that soils exhibit substantial proteolytic activity (see Ladd and Butler, 1972). It is difficult, therefore, to explain why practically all soils exhibit appreciable urease activity or to account for the detection of urease activity in stored and geologically-preserved soils (Skujiņš and McLaren, 1968, 1969).

The first evidence that native soil urease is more stable than urease added to soils was obtained by Conrad (1940b), who hypothesized that organic soil constituents protect urease against microbial degradation and other processes leading to decomposition or inactivation of enzymes. Support for this hypothesis has been provided by studies showing that urease activity in soils is correlated very highly significantly with organic matter content (see Section III.2) and by the isolation of clay-free organic fractions of soils exhibiting urease activity that is not destroyed by proteolytic enzymes (Burns *et al.*, 1972a, b).

Although the literature since 1940 supports Conrad's (1940b) conclusion that native soil urease is more persistent than urease added to soils, very few studies relating to the stability of native soil urease have been reported. However, recent work has shown that the native urease in Iowa soils is remarkably stable and has provided evidence that different soils have different levels of urease activity determined by the ability of their constituents to protect urease against microbial degradation and other processes leading to inactivation of this enzyme (Zantua and Bremner, 1975b, 1976, 1977a). Some of the evidence for these conclusions is summarized in Table 12, which shows the effects of various treatments of field-moist soils. Three findings merit attention. The first is that, although jack bean urease was inactivated very rapidly when added to the soils studied, prolonged incubation of field-moist samples of these soils at 30 or 40°C under aerobic or waterlogged conditions had no effect on their urease activity (Zantua and Bremner, 1977a). The second is that treatment of field-moist soils with proteolytic enzymes which cause rapid destruction of jack bean urease (trypsin and pronase) did not lead to a decrease in urease activity (Zantua and Bremner, 1977a). The third is that, although a marked increase in urease activity was observed in soils treated with glucose and other organic materials that promote production of urease by soil microorganisms, this increase was temporary and could not be detected after a few weeks. With each soil studied, urease activity increased after addition of jack bean urease or of organic materials promoting microbial activity, but subsequently decreased and stabilized at the level observed before addition of these substances (Zantua and Bremner, 1976, 1977a). The only apparent explanation of these findings is that the urease activity of unamended soils reflects their capacity for protection of urease and that urease in excess of this capacity is decomposed or inactivated.

TABLE 12

Effects of various treatments of field-moist Iowa soils on urease activity (Zantua and Bremner, 1975b, 1976, 1977a)

Treatment of soil	Effect on urease activity
Dried at 30, 40, 50 or 60°C for 24 h	None
Dried at 75°C for 24 h	Partial loss of activity
Dried at 105°C for 24 h	Complete loss of activity
Autoclaved (120°C) for 2 h	Complete loss of activity
Leached with water	None
Stored at −20, −10, 5, 10, 20, 30 or 40°C for 6 months	None
Incubated at 30 or 40°C under aerobic or waterlogged conditions for 6 months	None
Air-dried and stored at 21–23°C for 2 years	None
Incubated at 30°C after addition of jack bean urease	Increase followed by decrease to original activity
Incubated at 30°C after treatment with organic materials	Increase followed by decrease to original activity
Incubated at 30°C after treatment with proteolytic enzymes (pronase or trypsin)	None

It should be noted that the work reported by Zantua and Bremner (1975b, 1976, 1977a) was confined to Iowa soils used for corn and soybean production and that their findings may not, therefore, apply to other types of soils. Previous work had indicated that incubation of soils under aerobic conditions can lead to marked changes in urease activity (e.g. Vasilenko, 1962; McGarity and Myers, 1967; Balasubramanian et al., 1972; Lloyd and Sheaffe, 1973). Also, Myers and McGarity (1968) found that subsurface samples of some Australian podzolic and gley soils exhibited no urease activity and deduced from this observation that waterlogging may cause rapid destruction of urease or lead to production of substances that inhibit urease activity. The finding by Zantua and Bremner (1977a) that treatment of soils with trypsin or pronase had no effect on urease activity is also divergent from findings in previous work, because Conrad (1940b) reported that treatment of a Nord loam with trypsin for two days led to a very marked (c. 40%) reduction in urease activity, and work reported by Burns et al. (1972a) indicated that treatment of a Dublin clay loam with pronase for 24 h led to an appreciable (14%) reduction in urease activity (see also Pettit et al., 1976). Possible explanations of the divergence between the results discussed have been suggested (Zantua and Bremner, 1976).

It is very difficult to account for the remarkable stability of soil urease. Pettit et al. (1976) recently suggested that this urease is:

> immobilized within the organic matter of the organo–mineral complex during humus formation and that the organic matter has pores large enough to allow the passage of substrate (urea and water) and product (ammonia and carbon dioxide) molecules but not of a sufficient size to allow the entry of proteolytic enzymes or the escape of the urease itself.

Previous theories to account for the stability of soil urease have been discussed by McLaren (1963), Burns et al. (1972b) and Pettit et al. (1976).

7. Inhibition of urease activity in soils

The rapidly increasing importance of urea as a nitrogen fertilizer in world agriculture (see Cooke, 1969; Tomlinson, 1970; Harre et al., 1971; Engelstad and Hauck, 1974; Harre and Douglas, 1976) has emphasized the need for research to find methods of reducing the problems encountered in use of this fertilizer. These problems result largely from the rapid hydrolysis of urea to ammonium carbonate through soil urease activity and the concomitant rise in pH and liberation of ammonia. They include damage to germinating seedlings and young plants, nitrite and (or) ammonia toxicity and gaseous loss of urea nitrogen as ammonia (see Gasser, 1964; Tomlinson, 1970; Engelstad and Hauck, 1974).

One approach to reducing the problems associated with the use of urea as a fertilizer is to find compounds that will retard urea hydrolysis when applied to soils in conjunction with fertilizer urea. This approach has received considerable attention (e.g. Volk, 1961; Moe, 1967; Waid and Pugh, 1967; Paulson and Kurtz, 1969b; Pugh and Waid, 1969a, b; Bremner and Douglas, 1971a, b; Douglas and Bremner, 1971; Bremner and Douglas, 1973; Bundy and Bremner, 1973; Fernando and Roberts, 1976), and many compounds have been patented as inhibitors of urea hydrolysis in soils (see Table 13). Most of these inhibitors are inorganic or organic compounds previously shown to be potent inhibitors of urease isolated from plants or

TABLE 13
Compounds patented as inhibitors of urea hydrolysis in soils

Compounds	Patent
Dithiocarbamates	Hyson (1963), Tomlinson (1967), Geissler et al. (1970)
Boron-containing compounds (borax, boric acid, etc.)	Sor et al. (1966), Sor (1968), Sor et al. (1968), Geissler et al. (1970), Sor et al. (1971), Besekau et al. (1974)
Urea derivatives[a]	Sor et al. (1966), Geissler et al. (1970)
Formaldehyde	Sor (1968)
Salts of heavy metals having atomic weights > 50	Sor (1968, 1969)
Fluorine-containing compounds (sodium fluoride, potassium fluoride, etc.)	Sor (1968), Geissler et al. (1970)
Quinones and polyhydric phenols	Anderson (1969, 1970)
Antimetabolites[b]	Peterson and Walter (1970)
Miscellaneous[c]	Geissler et al. (1970)

[a] Methylurea, dimethylurea, thiourea, phenylurea, t-butylurea, n-butylurea.

[b] Pyridine-3-sulphonic acid, desthiobiotin, oxythiamine chloride, γ-benzene hexachloride, o-chloro-p-aminobenzoic acid.

[c] Heavy metal ions, halogens, cyanides, addition compounds of copper formate or copper acetate with urea, urea complexes with amines, urea cupric cyanide, hydroxylamine, coordination complexes of urea and boron trifluoride, copper tetrafluoroborate, sulphuric acid esters and quinones, aldehydes, copper chelates.

microorganisms (e.g. salts of heavy metals, dithiocarbamates, dihydric phenols, quinones). Peterson and Walter (1970), however, have claimed that urea hydrolysis in soils can also be retarded through addition of compounds that function as antimetabolites towards urease-producing microorganisms and have patented five of these compounds (see Table 13). According to their patent, these antimetabolites retard urea hydrolysis in soils, not by inhibiting the activity of urease, but by controlling the production of this enzyme by soil microorganisms. However, recent work to evaluate three of these compounds (pyridine-3-sulphonic acid, desthiobiotin and oxythiamine chloride) showed that they did not retard urea hydrolysis in soils or reduce gaseous loss of urea N as ammonia even when applied at rates far exceeding those recommended for inhibition of urea hydrolysis (Mulvaney and Bremner, 1977). This is not surprising in view of evidence that urea added to soil is hydrolysed by the urease present before addition of urea (see Section III.1 and Kiss et al., 1975) and that treatment of soils with urea does not promote production of urease by soil microorganisms (Zantua and Bremner, 1976).

In most studies to evaluate different compounds as soil urease inhibitors, evaluation has been performed by determining the effect of the test compound on ammonia evolution from soils treated with urea. This is an indirect and unreliable method of assessing the effectiveness of urease inhibitors (see Pugh and Waid, 1969b; Douglas and Bremner, 1971) and it is too time-consuming for evaluation of a large number of compounds. This problem was overcome by the development of a rapid method of evaluation involving determination of the effect of the test compound on the amount of urea hydrolysed on incubation of soil with urea at 37°C for 5 h (Douglas and Bremner, 1971). Bremner and Douglas (1971a) used this procedure to evaluate more than 100 compounds as soil urease inhibitors and found that dihydric phenols and quinones were the most effective organic compounds tested and that silver and mercury salts were the most effective inorganic compounds. This is illustrated in Table 14, which shows data obtained in a comparison of the organic and inorganic compounds found to be most effective as soil urease inhibitors in the work reported by Bremner and

TABLE 14

Comparison of organic and inorganic compounds as inhibitors of urease activity in soils (Bremner and Douglas, 1971a)

	% Inhibition of urease activity[b]	
Compound[a]	Range	Average
Organic:		
Catechol	71–77	74
Phenylmercuric acetate	64–71	67
Hydroquinone	60–69	64
p-Benzoquinone	56–68	62
2, 5-Dichloro-p-benzoquinone	58–68	62
2, 6-Dichloro-p-benzoquinone	52–63	58
1, 2-Naphthoquinone	42–48	44
Phenol	41–43	42
Sodium-p-chloromercuribenzoate	32–38	35
4-Chlorophenol	30–37	35
2, 5-Dimethyl-p-benzoquinone	29–35	32
N-Ethylmaleimide	23–38	25
Acetohydroxamic acid	13–16	14
Inorganic:		
Silver sulphate	45–52	48
Mercuric chloride	35–39	37
Gold chloride	17–19	18
Cupric sulphate	13–14	14

[a] Rate of addition of each compound was equivalent to 50 ppm of soil.
[b] Values for three soils.

Douglas (1971a). The data in Table 14 show that six of the organic compounds studied were more effective than the most effective inorganic compound and that several compounds found to be potent inhibitors of bacterial and plant ureases (e.g. N-ethylmaleimide, acetohydroxamic acid, cupric sulphate) were not highly effective as inhibitors of soil urease activity. The results obtained in this investigation with different metallic salts are discussed in Section III.4.6 (see Table 11).

Bundy and Bremner (1973) studied the effects of thirty-four substituted p-benzoquinones on urease activity in soils. They found that the effectiveness of these compounds as soil urease inhibitors depends largely upon their substituent groups. Methyl-, chloro-, bromo- and fluoro-substituted p-benzoquinones had a very marked inhibitory effect on soil urease activity, whereas phenyl-, t-butyl- and hydroxy-substituted p-benzoquinones had little, if any, effect. Table 15 shows results obtained by Bundy and Bremner (1973) in a study of the effects of 50 ppm (soil basis) of selected p-benzoquinones on the recovery of urea N as urea, exchangeable ammonium and ammonia after incubation of a urea-treated sandy soil at 20°C for fourteen days. The data reported show that, whereas methyl- and chloro-substituted p-benzoquinones markedly retarded urea hydrolysis and greatly reduced

TABLE 15

Effects of various quinones on recovery of urea N as urea, exchangeable ammonium and ammonia after incubation of urea-treated soil for 14 days (Bundy and Bremner, 1973)[a]

| | Recovery of urea N (%) | | |
| | As urea | As exchangeable ammonium | As ammonia |
Quinone			
None	0	32·8	62·8
p-Benzoquinone	75·5	17·9	2·4
Methyl-p-benzoquinone	87·0	9·5	0·5
2, 3-Dimethyl-p-benzoquinone	90·5	6·5	0·1
2, 5-Dimethyl-p-benzoquinone	90·5	6·6	0·1
2, 6-Dimethyl-p-benzoquinone	89·5	7·5	0·1
2, 5-Dichloro-p-benzoquinone	77·8	15·0	2·4
2, 6-Dichloro-p-benzoquinone	76·0	17·8	1·9
2, 5-Dihydroxy-p-benzoquinone	0	32·8	62·0
Tetrahydroxy-p-benzoquinone	0	33·5	62·6
2, 5-Diphenyl-p-benzoquinone	0	36·5	59·4
2, 5-Dibromo-3, 6-dihydroxy-p-benzoquinone	0	31·9	63·3

[a] 10-g samples of soil (Thurman sand) were incubated (20°C, 50% WHC) for 14 days after treatment with 1000 ppm of N as urea and 50 ppm of quinone specified.

184 J. M. BREMNER AND R. L. MULVANEY

volatilization of urea N as ammonia, hydroxy- and phenyl-substituted
p-benzoquinones had little, if any, effect on these processes. Considered
with the findings of Bremner and Douglas (1971a), the results of this work
indicate that 2, 3-dimethyl-, 2, 5-dimethyl- and 2, 6-dimethyl-p-benzo-
quinone are the most effective of the compounds thus far proposed for
retardation of urea decomposition in soils. However, the discovery (see
Table 15) that these methyl-substituted p-benzoquinones are not much
more effective than their parent compound (p-benzoquinone), which is
relatively inexpensive, led Bundy and Bremner (1973) to conclude that p-
benzoquinone is more promising than other quinones proposed for reduction of the problems associated with the use of urea as a fertilizer.

Recent unpublished work by the authors has shown that the effects of
hydroquinone on urea transformations in soils are identical to those of p-
benzoquinone. This suggests that hydroquinone is the most promising
of the compounds so far suggested for reduction of the problems associated with the use of fertilizer urea because it is cheaper than-p benzoquinone and considerably less expensive than other compounds proposed
as soil urease inhibitors.

Tomlinson (1970) found that, when applied at the rate of 4 ppm of soil,
hydroquinone and 2, 5-dimethyl-p-benzoquinone increased the response of
ryegrass to urea in pot experiments but that, when applied at the rate of
2 ppm of soil, neither of these compounds significantly increased the
response of ryegrass or winter wheat to urea in field experiments.

Bremner and Douglas (1973) studied the effects of 50 ppm (soil basis) of
eight urease inhibitors (catechol, hydroquinone, p-benzoquinone, 2, 5-
dimethyl-p-benzoquinone, phenylmercuric acetate, N-ethylmaleimide,
sodium p-chloromercuribenzoate and acetohydroxamic acid) on transformations of urea N in urea-treated soils incubated at 20°C and found that
2, 5-dimethyl-p-benzoquinone had the greatest ability, and acetohydroxamic acid the least ability, to retard urea hydrolysis and reduce gaseous loss
of urea N as ammonia. They also found that the effects of these urease
inhibitors differed markedly with different soils and were greatest with
sands and sandy loams and least with a clay loam. Pugh and Waid (1969b)
similarly found that the ability of acetohydroxamic acid to retard gaseous
loss of urea N as ammonia from soils treated with urea was greatest with
sands and sandy loams. This suggests that the potential value of urease
inhibitors may be greatest with light-textured soils.

Bundy and Bremner (1974b) studied the effects of ten urease inhibitors
on nitrification in soils treated with ammonium sulphate. The inhibitors
used (catechol, hydroquinone, p-benzoquinone, 2, 3-dimethyl-p-benzo-
quinone, 2, 5-dimethyl-p-benzoquinone, 2, 6-dimethyl-p-benzoquinone,
2, 5-dichloro-p-benzoquinone, 2, 6-dichloro-p-benzoquinone, sodium
p-chloromercuribenzoate and phenylmercuric acetate) were those found

most effective in previous work to evaluate more than 130 compounds as soil urease inhibitors (Bremner and Douglas, 1971a; Bundy and Bremner, 1973). Their effects on nitrification were compared with those of three compounds patented as soil nitrification inhibitors (N-Serve, AM and ST). Most of the urease inhibitors studied had little effect on nitrification when applied at the rate of 10 ppm of soil, but had marked inhibitory effects when applied at the rate of 50 ppm. None inhibited nitrification as effectively as N-Serve but phenylmercuric acetate inhibited more effectively than did AM or ST when applied at the rate of 10 ppm.

May and Douglas (1975) studied the effects of seven urease inhibitors (catechol, p-benzoquinone, 2, 5-dimethyl-p-benzoquinone, hydroquinone, 1, 2-naphthoquinone, benzohydroxamic acid and phenylmercuric acetate) on germination of wheat and alfalfa seeds. They found that none of these compounds affected germination when applied at the rate of 10 ppm of soil and that only 2, 5-dimethyl-p-benzoquinone caused severe reduction in germination when applied at the rate of 50 or 100 ppm of soil.

To summarize, current information suggests that simple quinones and phenols are the most promising of the compounds thus far proposed for reduction of the problems associated with the use of urea as a fertilizer. Other compounds proposed are either considerably less effective or much more expensive, and several may be phytotoxic or have other adverse effects (e.g. metal accumulation problems could result from the use of compounds containing mercury).

IV. Conclusions

Since all indications are that urea will become the most important fertilizer in world agriculture within the next twenty-five years, there seems very little doubt that research on soil urease will increase greatly during this period and that a substantial effort will be made to reduce the practical problems resulting from the rapid hydrolysis of fertilizer urea by soil urease.

The need for further research concerning the nature and properties of soil urease is obvious from the numerous contradictions in the literature. There is also a clear need for research concerning urease in the roots and leaves of different plants. There is very little doubt that, in the absence of growing plants, the fate of fertilizer nitrogen added to soils as urea is controlled largely by soil urease. This may not be true, however, in the presence of growing plants because urease in plant roots may contribute significantly to conversion of fertilizer urea to ammonium, and research on urease in plant roots is clearly needed to assess this possibility. Also, research on urease in plant leaves is needed to assess the possibility that, when urea is applied as a foliar fertilizer, hydrolysis of this urea by urease in plant leaves may lead to significant gaseous loss of urea nitrogen as ammonia.

The need for evaluation of methods used to assay urease activity in soils deserves emphasis because many of the divergent findings in research on soil urease can be attributed to defects in these methods. The problems associated with methods based on determination of the ammonium released on incubation of soil with urea should receive greater attention, and the limitations of both buffer and non-buffer methods of assaying soil urease activity should be recognized. A considerable amount of research on soil urease has been vitiated by failure to recognize that, when a buffer method is adopted for assay of soil urease activity, the value obtained merely indicates the activity that occurs in the buffer selected and cannot be taken as an indication of the activity that occurs in the absence of this buffer.

Since it seems likely that a substantial effort will be made to find urease inhibitors that will retard urea hydrolysis when applied to soil in conjunction with fertilizer urea, attention should be drawn to problems in this type of research. One problem is that, to have potential practical value, any compound proposed for inhibition of urease activity in soil must be cheap, effective and safe to use, and all indications are that, to be effective when applied to urea fertilizer granules, the inhibitor must dissolve and move with the fertilizer urea when these granules are applied to soils. It may prove very difficult to meet these requirements. Another problem is that very little is known about the ability of different plants to utilize urea in comparison with their ability to utilize ammonium or nitrate. A further problem is that we lack the type of information needed for assessment of the potential practical benefits of controlling urease activity in soils under different conditions.

There is an obvious need for research to account for the remarkable stability of native soil urease. Elucidation of the mechanism by which this enzyme is protected could contribute significantly to development of immobilized enzymes for commercial purposes.

In conclusion, it should be noted that, besides being applied to soil as a nitrogen fertilizer, urea is added to many soils in the form of animal urine. It has been estimated that approximately 45 g of urea are voided daily by sheep and that approximately 140 g are voided daily by mature cattle (Gasser, 1964). Where animals are confined to feedlots, the resulting heavy applications of urea as urine to small areas of land lead to release of substantial amounts of ammonia to the atmosphere via urease activity, and this can cause an air-pollution problem. Moreover, studies by Hutchinson and Viets (1969) have indicated that ammonia evolved from cattle feedlots is absorbed by, and may promote eutrophication of, lakes and streams in the vicinity. The possibility that these problems can be reduced by application of urease inhibitors deserves attention.

References

ANANTHANARAYANA R. and MITHYANTHA M. S. (1970). Urease activity of the soils of M.R.S., Hebbal. *Mysore J. Agric. Sci.* **4**, 109–111.

ANDERSEN J. A., KOPKO F., SIEDLER A. J. and NOHLE E. G. (1969). Partial purification and properties of urease from *Proteus mirabilis. Federation Proc.* **28**, 764.

ANDERSON J. R. (1962). Urease activity, ammonia volatilization and related microbiological aspects in some South African soils. *Proc. 36th Congr. of the S. Afr. Eng. Technol. Ass.* 97–105.

ANDERSON J. R. (1969). Fertilizing process and composition, British Patent No. 1,142,245.

ANDERSON J. R. (1970). Fertiliser compositions, US Patent Office 3,515,532.

ARDAKANI M. S., VOLZ M. G. and MCLAREN A. D. (1975). Consecutive steady state reactions of urea, ammonium and nitrite nitrogen in soil. *Can. J. Soil Sci.* **55**, 83–91.

BAHADUR K. and CHANDRA V. (1962). A study of the inhibitory influence of Ag⁺, Au³⁺, and Cu²⁺ ions on the decomposition of urea by soyabean (*Glycine soja*) urease. *Proc. Nat. Acad. Sci. India, Sect.* **A32**, 72–82.

BALASUBRAMANIAN A., SIDDARAMAPPA R. and RANGASWAMI G. (1972). Effect of organic manuring on the activities of the enzymes hydrolysing sucrose and urea and on soil aggregation. *Pl. Soil* **37**, 319–328.

BESEKAU E., HARZFELD G., KUEMMEL R., MATZEL W., MUEHLENBRUCH D. and THOMAS G. (1974). Nitrogen-containing fertilizer, East German Patent No. 108,730 (*Chem. Abstr.* **83**, 26886g).

BHAVANANDAN V. P. and FERNANDO V. (1970). Studies on the use of urea as a fertilizer for tea in Ceylon. II. Urease activity in tea soils. *Tea Quart.* **41**, 94–106.

BLAKELEY R. L., WEBB E. C. and ZERNER B. (1969). Jack bean urease (EC 3.5.1.5). A new purification and reliable rate assay. *Biochemistry* **8**, 1984–1990.

BREMNER J. M. and DOUGLAS L. A. (1971a). Inhibition of urease activity in soils. *Soil Biol. Biochem.* **3**, 297–307.

BREMNER J. M. and DOUGLAS L. A. (1971b). Decomposition of urea phosphate in soils. *Soil Sci. Soc. Am. Proc.* **35**, 575–578.

BREMNER J. M. and DOUGLAS L. A. (1973). Effects of some urease inhibitors on urea hydrolysis in soils. *Soil Sci. Soc. Am. Proc.* **37**, 225–226.

BREMNER J. M. and LEES H. (1949). Studies on soil organic matter. II. The extraction of organic matter from soil by neutral reagents. *J. Agric. Sci.* **39**, 274–279.

BREMNER J. M. and ZANTUA M. I. (1975). Enzyme activity in soils at subzero temperatures. *Soil Biol. Biochem.* **7**, 383–387.

BRIGGS M. H. and SEGAL L. (1963). Preparation and properties of a free soil enzyme. *Life Sci.* **1**, 69–72.

BRIGGS M. H. and SPEDDING D. J. (1963). Soil enzymes. *Sci. Prog., London* **51**, 217–225.

BROADBENT F. E., HILL G. N. and TYLER K. B. (1958). Transformations and movement of urea in soils. *Soil Sci. Soc. Am. Proc.* **22**, 303–307.

BUNDY L. G. and BREMNER J. M. (1973). Effects of substituted *p*-benzoquinones on urease activity in soils. *Soil Biol. Biochem.* **5**, 847–853.

BUNDY L. G. and BREMNER J. M. (1974a). Effects of nitrification inhibitors on transformations of urea nitrogen in soils. *Soil Biol. Biochem.* **6**, 369–376.

BUNDY L. G. and BREMNER J. M. (1974b). Effects of urease inhibitors on nitrification in soils. *Soil Biol. Biochem.* **6**, 27–30.

BURNS R. G., EL-SAYED M. H. and MCLAREN A. D. (1972a). Extraction of an urease-active organo-complex from soil. *Soil Biol. Biochem.* **4**, 107–108.

BURNS R. G., PUKITE A. H. and MCLAREN A. D. (1972b). Concerning the location and persistence of soil urease. *Soil Sci. Soc. Am. Proc.* **36**, 308–311.

CAWSE P. A. (1975). Microbiology and biochemistry of irradiated soils. *In* "Soil Biochemistry", Vol. 3 (E. A. Paul and A. D. McLaren, Eds) pp. 213–267. Marcel Dekker, New York.

CERNA S. (1968). The influence of desiccation of the structural soil upon the intensity of biochemical reactions. *Acta Univ. Carol. Biol.* **4**, 285–288.

CERVELLI S., NANNIPIERI P., CECCANI B. and SEQUI P. (1973). Michaelis constant of soil acid phosphatase. *Soil Biol. Biochem.* **5**, 841–845.

CERVELLI S., NANNIPIERI P., GIOVANNINI G. and PERNA A. (1976). Relationships between substituted urea herbicides and soil urease activity. *Weed Res.* **16**, 365–368.

CHIN W.-T. and KROONTJE W. (1963). Urea hydrolysis and subsequent loss of ammonia. *Soil Sci. Soc. Am. Proc.* **27**, 316–318.

CONRAD J. P. (1940a). Hydrolysis of urea in soils by thermolabile catalysis. *Soil Sci.* **49**, 253–263.

CONRAD J. P. (1940b). The nature of the catalyst causing the hydrolysis of urea in soils. *Soil Sci.* **50**, 119–134.

CONRAD J. P. (1942a). The occurrence and origin of ureaselike activities in soils. *Soil Sci.* **54**, 367–380.

CONRAD J. P. (1942b). Enzymatic vs. microbial concepts of urea hydrolysis in soils. *Agron. J.* **34**, 1102–1113.

CONRAD J. P. (1943). Some effects of developing alkalinities and other factors upon ureaselike activities in soils. *Soil Sci. Soc. Am. Proc.* **5**, 171–174.

COOKE G. W. (1969). Fertilisers in 2000 A.D. *Phosphorus in Agriculture, Bull. Doc.* No. 53, pp. 1–13. International Superphosphate and Compound Manufacturers' Association, Paris.

DALAL R. C. (1975a). Urease activity in some Trinidad soils. *Soil Biol. Biochem.* **7**, 5–8.

DALAL R. C. (1975b). Effect of toluene on the energy barriers in urease activity of soils. *Soil Sci.* **120**, 256–260.

DELAUNE R. D. and PATRICK W. H. (1970). Urea conversion to ammonia in water-logged soils. *Soil Sci. Soc. Am. Proc.* **34**, 603–607.

DIXON N. E., GAZZOLA C., BLAKELEY R. L. and ZERNER B. (1975). Jack bean urease (EC 3.5.1.5). A metalloenzyme. A simple biological role for nickel? *J. Am. chem. Soc.* **97**, 4131–4133.

DOUGLAS L. A. and BREMNER J. M. (1971). A rapid method of evaluating different compounds as inhibitors of urease activity in soils. *Soil Biol. Biochem.* **3**, 309–315.

DURAND G. (1965). Les enzymes dans le sol. *Rev. Ecol. Biol. Sol* **2**, 141–205.

EL-ESSAWI T. M., KISHK F. M. and ABDEL-GHAFFAR A. S. (1973). Preliminary studies on some soil enzymes. *Alexandria J. Agr. Res.* **21**, 125–131.

ENGELSTAD O. P. and HAUCK R. D. (1974). Urea—will it become the most popular nitrogen carrier? *Crops Soils* **26**, 11–14.

FERNANDO V. and ROBERTS G. R. (1976). The partial inhibition of soil urease by naturally occurring polyphenols. *Pl. Soil* **44**, 81–86.

FISHBEIN W. N. (1969). Urease catalysis. III. Stoichiometry, kinetics, and inhibitory properties of a third substrate: dihydroxyurea. *J. Biol. Chem.* **244**, 1188–1193.

FISHBEIN W. N., WINTER T. S. and DAVIDSON J. D. (1965). Urease catalysis. I. Stoichiometry, specificity, and kinetics of a second substrate: hydroxyurea. *J. Biol. Chem.* **240**, 2402–2406.

FISHER W. B. Jr. and PARKS W. L. (1958). Influence of soil temperature on urea hydrolysis and subsequent nitrification. *Soil Sci. Soc. Am. Proc.* **22**, 247–248.

GALSTYAN A. SH. (1958). Enzymatic activity in some Armenian soils. IV. Urease activity in soil. *Dokl. Akad. Nauk Arm. SSR* **26**, 29–39.

GALSTYAN A. SH. (1959a). Soil enzymes. *Soobshch. Lab. Agrokhim. Akad. Nauk Arm. SSR* **2**, 19–25.

GALSTYAN A. SH. (1959b). Activity of enzymes and respiratory intensity in soils. *Dokl. Akad. Nauk SSSR* **127**, 1099–1102.

GALSTYAN A. SH. (1960). Enzyme activity in saline soils. *Dokl. Akad. Nauk Arm. SSR* **30**, 61–64.

GALSTYAN A. SH. (1965a). A method of determining the activity of hydrolytic enzymes in soil. *Soviet Soil Sci.* **2**, 170–175.

GALSTYAN A. SH. (1965b). Effect of temperature on activity of soil enzymes. *Dokl. Akad. Nauk Arm. SSR* **40**, 177–181.

GASSER J. K. R. (1964). Urea as a fertilizer. *Soils Fertil.* **27**, 175–180.

GAUTHIER S. M., ASHTAKALA S. S. and LENOIR J. A. (1976). Inhibition of soil urease activity and nematocidal action of 3-amino-1, 2, 4-triazole. *Hort. Science* **11**, 481–482.

GAZZOLA C., BLAKELEY R. L. and ZERNER B. (1973). On the substrate specificity of jack bean urease (urea amidohydrolase, EC 3.5.1.5). *Can. J. Biochem.* **51**. 1325–1330.

GEISSLER P. R., SOR K. and ROSENBLATT T. M. (1970). Urease inhibitors, US Patent Office 3,523,018.

GIBSON T. (1930a). The decomposition of urea in soils. *J. Agric. Sci.* **20**, 549–558.

GIBSON T. (1930b). Factors influencing the decomposition of urea in soils. *Zentralbl. Bakteriol. Abt.* 2 **81**, 45–60.

GORIN G. and CHIN C. C. (1965). Urease. IV. Its reaction with N-ethylmaleimide and with silver ion. *Biochim. Biophys. Acta* **99**, 418–426.

GORIN G., FUCHS E., BUTLER L. G., CHOPRA S. L. and HERSH R. T. (1962). Some properties of urease. *Biochemistry* **1**, 911–916.

GOULD W. D., COOK F. D. and WEBSTER G. R. (1973). Factors affecting urea hydrolysis in several Alberta soils. *Pl. Soil* **38**, 393–401.

HAIG A. D. (1955). Some characteristics of esterase- and urease-like activity in the soil. Ph.D. Thesis, University of California, Davis.

HARMON K. M. and NIEMANN C. (1949). The competitive inhibition of the urease-catalyzed hydrolysis of urea by phosphate. *J. Biol. Chem.* 177, 601–605.

HARRE E. A. and DOUGLAS J. R. (1976). World crop needs for nitrogen fertilizers—will the supply be adequate? *1st International Symposium on Nitrogen Fixation Proceedings, Pullman, Washington* 2, 674–692.

HARRE E. A., GARMAN W. H. and WHITE W. C. (1971). The world fertilizer market. *In* "Fertilizer Technology and Use", 2nd edition (R. A. Olson, T. J. Army, J. J. Hanway and V. J. Kilmer, Eds) pp. 27–55. Soil Science Society of America, Madison, Wisconsin.

HOFFMAN G. and TEICHER K. (1961). Ein kolorimetrisches Verfahren zur Bestimmung der Ureaseaktivität in Böden. *Z. Pflanzenernähr. Düng. Bodenkd.* 95, 55–63.

HOFMANN E. (1963). Urease. *In* "Methods of Enzymatic Analyses" (H. U. Bergmeyer, Ed.) pp. 913–916. Academic Press, New York.

HOFMANN E. and SCHMIDT W. (1953). Über das Enzym-system unserer Kulturböden —II. Urease. *Biochem. Z.* 324, 125–127.

HOFMANN E., WOLF E. and SCHMIDT W. (1953). Vergleich zwischen Enzymgehalt und anderen Eigenschaften verschiedenen Kulturböden. *Z. PflBau.* 4, 177–181.

HOWELL S. F. and SUMNER J. B. (1934). The specific effects of buffers upon urease activity. *J. Biol. Chem.* 104, 619–626.

HUGHES R. B., KATZ S. A. and STUBBINS S. E. (1969). Inhibition of urease by metal ions. *Enzymologia* 36, 332–334.

HUTCHINSON G. L. and VIETS F. G. (1969). Nitrogen enrichment of surface water by absorption of ammonia volatilized from cattle feedlots. *Science* 166, 514–515.

HYSON A. M. (1963). Fertilizer composition comprising urea and dithiocarbamates, US Patent Office 3,073,694.

IRVING C. J. G. and COSGROVE D. J. (1976). The kinetics of soil acid phosphatase. *Soil Biol. Biochem.* 8, 335–340.

JONES H. W. (1932). Some transformations of urea and their resultant effects on the soil. *Soil Sci.* 34, 281–299.

KALINKEVICH A. F. (1961). Transformation of urea in the soil. *Soviet Soil Sci.* 4, 463–466.

KISS S. and BOARU M. (1965). Some methodological problems of soil enzymology. *In* "Symposium on Methods in Soil Biology", pp. 115–127. Rumanian National Society of Soil Science, Bucharest.

KISS S., DRĂGAN-BULARDA M. and RADULESCU D. (1975). Biological significance of enzymes accumulated in soil. *Adv. Agron.* 27, 25–87.

KISTIAKOWSKY G. B. and LUMRY R. (1949). Anomalous temperature effects in the hydrolysis of urea by urease. *J. Am. chem. Soc.* 71, 2006–2013.

KISTIAKOWSKY G. B. and ROSENBERG A. J. (1952). The kinetics of urea hydrolysis by urease. *J. Am. chem. Soc.* 74, 5020–5025.

KOEPF H. (1954). Experimental study of soil evaluation by biochemical reactions. I. Enzyme reactions and carbon dioxide evolution in different soils. *Z. Pflanzenernähr. Düng. Bodenkd.* 67, 262–270.

KOZLOVSKAYA N. A., NIKITINA G. D. and RUNKOV S. V. (1972). A rapid microdiffusion method for the determination of the urease activity of peat-bog soils. *Agrokhimiya* 5, 144–149.

KUPREVICH V. F. and SHCHERBAKOVA T. A. (1971a). "Soil Enzymes", Indian National Scientific Documentation Centre, New Delhi.

KUPREVICH V. F. and SHCHERBAKOVA T. A. (1971b). Comparative enzymatic activity in diverse types of soil. *In* "Soil Biochemistry", Vol. 2 (A. D. McLaren and J. J. Skujiņš, Eds) pp. 167–201. Marcel Dekker, New York.

LADD J. N. and BUTLER J. H. A. (1972). Short-term assays of soil proteolytic enzyme activities using proteins and dipeptide derivatives as substrates. *Soil Biol. Biochem.* **4**, 19–30.

LAIDLER K. J. and HOARE J. P. (1949). The molecular kinetics of the urea-urease system. I. The kinetic laws. *J. Am. Chem. Soc.* **71**, 2699–2702.

LAIDLER K. J. and HOARE J. P. (1950). The molecular kinetics of the urea-urease system. III. Heats and entropies of complex formation and reaction. *J. Am. Chem. Soc.* **72**, 2489–2494.

LARSON A. D. and KALLIO R. E. (1954). Purification and properties of bacterial urease. *J. Bacteriol.* **68**, 67–73.

LAUGESEN K. (1972). Urease activity in Danish soils. *Tidsskr. Pl.* **76**, 221–223.

LETHBRIDGE G. and BURNS R. G. (1976). Inhibition of soil urease by organo-phosphorus insecticides. *Soil Biol. Biochem.* **8**, 99–102.

LISTER A. J. (1956). The kinetics of urease activity in *Corynebacterium renale*. *J. Gen. Microbiol.* **14**, 478–484.

LLOYD A. B. (1975). Extraction of urease from soil. *Soil Biol. Biochem.* **7**, 357–358.

LLOYD A. B. and SHEAFFE M. J. (1973). Urease activity in soils. *Pl. Soil* **39**, 71–80.

LYNN K. R. (1967). Some properties and purifications of urease. *Biochim. Biophys. Acta* **146**, 205–218.

LYNN K. R. and YANKWICH P. E. (1962). ^{13}C kinetic isotope effects in the urease-catalyzed hydrolysis of urea. I. Temperature dependence. *Biochim. Biophys. Acta* **56**, 512–530.

MALHOTRA O. P. and RANI I. (1969). Purification and properties of urease of *Cajanus indicus*. *Indian J. Biochem.* **6**, 15–20.

MAY P. B. and DOUGLAS L. A. (1975). Germination of wheat and alfalfa seeds as affected by some soil urease inhibitors. *Agron. J.* **67**, 718–720.

MAY P. B. and DOUGLAS L. A. (1976). Assay for soil urease activity. *Pl. Soil* **45**, 301–305.

MCGARITY J. W. and MYERS M. G. (1967). A survey of urease activity in soils of northern New South Wales. *Pl. Soil* **27**, 217–238.

MCLAREN A. D. (1963). Enzyme activity in soils sterilized by ionizing radiation and some comments on micro-environments in nature. *In* "Recent Progress in Microbiology", Vol. 8 (N. E. Gibbons, Ed.) pp. 221–229. University of Toronto Press.

MCLAREN A. D. (1969). Radiation as a technique in soil biology and biochemistry. *Soil Biol. Biochem.* **1**, 63–73.

MCLAREN A. D. (1975). Soil as a system of humus and clay immobilized enzymes. *Chemica Scripta* **8**, 97–99.

MCLAREN A. D. and PACKER L. (1970). Some aspects of enzyme reactions in heterogeneous systems. *Adv. Enzymol.* **33**, 245–308.

MCLAREN A. D. and PUKITE A. (1972). Ubiquity of some soil enzymes and isolation of soil organic matter with urease activity. *In* "Humic Substances: Their Structure and Function in the Biosphere" (D. Povoledo and H. L. Golterman, Eds) pp. 187–193. Centre for Agricultural Publishing and Documentation, Wageningen.

MCLAREN A. D., PUKITE A. H. and BARSHAD I. (1975). Isolation of humus with enzymatic activity from soil. Soil Sci. 119, 178–180.

MCLAREN A. D., RESHETKO L. and HUBER W. (1957). Sterilization of soil by irradiation with an electron beam, and some observations on soil enzyme activity. Soil Sci. 83, 497–502.

MILLER R. J., PINKHAM C., OVERMAN A. R. and DUMFORD S. W. (1968). Temperature effects on the acid hydrolysis of glucose-1-phosphate and urease hydrolysis of urea. Biochim. Biophys. Acta 167, 607–609.

MOE P. G. (1967). Nitrogen losses from urea as affected by altering soil urease activity. Soil Sci. Soc. Am. Proc. 31, 380–382.

MULVANEY R. L. and BREMNER J. M. (1977). Evaluation of antimetabolites for retardation of urea hydrolysis in soils. Soil Sci. Soc. Am. J. 41, 1024–1027.

MUSA M. M. and MUKHTAR N. O. (1969). Enzymatic activity of a soil profile in the Sudan Gezira. Pl. Soil 30, 153–156.

MYERS M. G. and MCGARITY J. W. (1968). The urease activity in profiles of five great soil groups from northern New South Wales. Pl. Soil 28, 25–37.

NANNIPIERI P., CECCANTI B., CERVELLI S. and SEQUI P. (1974). Use of 0·1M pyrophosphate to extract urease from a podzol. Soil Biol. Biochem. 6, 359–362.

NANNIPIERI P., CERVELLI S. and PEDRAZZINI F. (1975). Concerning the extraction of enzymatically active organic matter from soil. Experientia 31, 513–515.

NIEKERK P. E. LE R. VAN. (1964). A modified method for the determination of urease activity in the soil. S. Afr. J. Agr. Sci. 7, 131–134.

NORSTADT F. A., FREY C. R. and SIGG H. (1973). Soil urease: Paucity in the presence of the fairy ring fungus Marasmius oreades (Bolt.) Fr. Soil Sci. Soc. Am. Proc. 37, 880–885.

OVERREIN L. N. (1963). The chemistry of urea nitrogen transformations in soil. Ph.D. Thesis, Purdue University, West Lafayette.

OVERREIN L. N. and MOE P. G. (1967). Factors affecting urea hydrolysis and ammonia volatilization in soil. Soil Sci. Soc. Am. Proc. 31, 57–61.

PANCHOLY S. K. and RICE E. L. (1972). Effect of storage conditions on activities of urease, invertase, amylase, and dehydrogenase in soil. Soil Sci. Soc. Am. Proc. 36, 536–537.

PANCHOLY S. K. and RICE E. L. (1973). Soil enzymes in relation to old field succession: amylase, cellulase, invertase, dehydrogenase, and urease. Soil Sci. Soc. Am. Proc. 37, 47–50.

PAULSON K. N. and KURTZ L. T. (1969a). Locus of urease activity in soil. Soil Sci. Soc. Am. Proc. 33, 897–901.

PAULSON K. N. and KURTZ L. T. (1969b). Evaluation of urea-hydrocarbon complexes as "slow release" nitrogen carriers. Soil Sci. Soc. Am. Proc. 33, 973.

PAULSON K. N. and KURTZ L. T. (1970). Michaelis constant of soil urease. Soil Sci. Soc. Am. Proc. 34, 70–72.

PEL'TSER A. S. (1972a). Kinetics of ^{14}C-urea decomposition in soil depending on soil moisture, addition of toxic chemicals, liming and soil pH. Soviet Soil Sci. 5, 571–576.

PEL'TSER A. S. (1972b). Kinetics of ^{14}C-urea decomposition in soil in relation to the amount of urea, microbiological activity, and soil group. Agrokhimiya 9, 32–37.

PETERSON A. F. and WALTER C. R. Jr. (1970). Regulation of urea hydrolysis in soil, US Patent Office 3,547,614.

PETERSON J., HARMON K. M. and NIEMANN C. (1948). The dependence of the specific activity of urease upon the apparent absolute enzyme concentration. *J. Biol. Chem.* **176**, 1–7.

PETTIT N. M., SMITH A. R. J., FREEDMAN R. B. and BURNS R. G. (1976). Soil urease: activity, stability and kinetic properties. *Soil Biol. Biochem.* **8**, 479–484.

PINCK L. A. and ALLISON F. E. (1961). Adsorption and release of urease by and from clay minerals. *Soil Sci.* **91**, 183–188.

PORTER L. K. (1965). Enzymes. In "Methods of Soil Analysis", Part 2 (C. A. Black, Ed.) pp. 1536–1549. American Society of Agronomy, Madison, Wisconsin.

PUGH K. B. and WAID J. S. (1969a). The influence of hydroxamates on ammonia loss from an acid loamy sand treated with urea. *Soil Biol. Biochem.* **1**, 195–206.

PUGH K. B. and WAID J. S. (1969b). The influence of hydroxamates on ammonia loss from various soils treated with urea. *Soil Biol. Biochem.* **1**, 207–217.

RABINOWITZ J. L., SALL T., BIERLY J. N. and OLEKSYSHYN O. (1956). Carbon isotope effects in enzyme systems. I. Biochemical studies with urease. *Arch. Biochem. Biophys.* **63**, 437–445.

RACHINSKIY V. V. and PEL'TSER A. S. (1965). The kinetics of urea decomposition in soil. *Dokl. TSKhA* **109**, 75–78.

RACHINSKIY V. V. and PEL'TSER A. S. (1967). Effect of temperature on rate of decomposition of urea in soil. *Agrokhimiya* **10**, 75–77.

RAMACHANDRAN K. B. and PERLMUTTER D. D. (1976). Effects of immobilization on the kinetics of enzyme-catalyzed reactions. II. Urease in a packed-column differential reactor system. *Biotechnol. Bioeng.* **18**, 685–699.

REITHEL F. J. (1971). Ureases. In "The Enzymes", Vol. 4 (P. D. Boyer, Ed.) pp. 1–21. Academic Press, New York.

REITHEL F. J., ROBBINS J. E. and GORIN G. (1964). A structural subunit molecular weight of urease. *Arch. Biochem. Biophys.* **108**, 409–413.

ROBERGE M. R. (1968). Effects of toluene on microflora and hydrolysis of urea in a black spruce humus. *Can. J. Microbiol.* **14**, 999–1003.

ROBERGE M. R. (1970). Behavior of urease added to unsterilized, steam-sterilized, and gamma radiation-sterilized black spruce humus. *Can. J. Microbiol.* **16**, 865–870.

ROBERGE M. R. and KNOWLES R. (1968a). Factors affecting urease activity in a black spruce humus sterilized by gamma radiation. *Can. J. Soil Sci.* **48**, 355–361.

ROBERGE M. R. and KNOWLES R. (1968b). Urease activity in a black spruce humus sterilized by gamma radiation. *Soil Sci. Soc. Am. Proc.* **32**, 518–521.

ROSS D. J. (1965). Effect of air-dry, refrigerated and frozen storage on activities of enzymes hydrolysing sucrose and starch in soils. *J. Soil Sci.* **16**, 86–94.

ROTINI O. T. (1935). La transformazione enzimatica dell'-urea nell terreno. *Ann. Labor. Ric. Ferm. Spallanzani* **3**, 143–154.

SANKHAYAN S. D. and SHUKLA U. C. (1976). Rates of urea hydrolysis in five soils of India. *Geoderma* **16**, 171–178.

SEARLE P. L. and SPEIR T. W. (1976). An automated colorimetric method for the determination of urease activity in soil and plant material. *Commun. Soil Sci. Pl. Anal.* **7**, 365–374.

194   J. M. BREMNER AND R. L. MULVANEY

SEQUI P. (1974). Enzymes in soil. *Ital. Agric.* 111, 91–109.

SHAW W. H. R. (1954). The inhibition of urease by various metal ions. *J. Am. chem. Soc.* 76, 2160–2163.

SHAW W. H. R. and RAVAL D. N. (1961). The inhibition of urease by metal ions at pH 8·9. *J. Am. chem. Soc.* 83, 3184–3187.

SILVA C. G. and PERERA A. M. A. (1971). A study of the urease activity in the rubber soils of Ceylon. *Quart. J. Rubber Res. Inst. Ceylon* 47, 30–36.

SIMPSON J. R. (1968). Losses of urea nitrogen from the surface of pasture soils. *Trans. 9th Int. Congr. Soil Sci., Adelaide* 2, 459–466.

SIMPSON D. M. H. and MELSTED S. W. (1963). Urea hydrolysis and transformation in some Illinois soils. *Soil Sci. Soc. Am. Proc.* 27, 48–50.

SIZER I. W. (1939). Temperature activation of the urease-urea system using crude and crystalline urease. *J. Gen. Physiol.* 22, 719–741.

SIZER I. W. (1940). The activation energy of urea hydrolysis catalyzed by soybean urease. *J. Biol. Chem.* 132, 209–218.

SIZER I. W. (1941). Temperature activation of the urease-urea system using urease of *Proteus vulgaris. J. Bacteriol.* 41, 511–527.

SKUJIŅŠ J. J. (1965). $^{14}CO_2$ detection chamber for studies in soil metabolism. *Biol. du Sol* 4, 15–17.

SKUJIŅŠ J. J. (1967). Enzymes in soil. *In* "Soil Biochemistry", Vol. 1 (A. D. McLaren and G. H. Peterson, Eds) pp. 371–414. Marcel Dekker, New York.

SKUJIŅŠ J. J. (1976). Extracellular enzymes in soils. *CRC Crit. Rev. Microbiol.* 4, 383–421.

SKUJIŅŠ J. J. and MCLAREN A. D. (1967). Enzyme reaction rates at limited water activities. *Science* 158, 1569–1570.

SKUJIŅŠ J. J. and MCLAREN A. D. (1968). Persistence of enzymatic activities in stored and geologically preserved soils. *Enzymologia* 34, 213–225.

SKUJIŅŠ J. J. and MCLAREN A. D. (1969). Assay of urease activity using ^{14}C-urea in stored, geologically preserved, and in irradiated soils. *Soil. Biol. Biochem.* 1, 89–99.

SOR K. M. (1968). Fertilizer composition consisting of urea, a urease inhibitor, and a hydrocarbon binder, US Patent Office 3,388,989.

SOR K. M. (1969). Inhibition of urea hydrolysis in fertilizers, British Patent No. 1,157,400.

SOR K. M., PELISSIER J. A. and LATHAM R. (1968). Urea fertilizer with slow ammonia release, French Patent No. 1,519,208 (*Chem. Abstr.* 71, 21311h).

SOR K. M., STANSBURY R. L. and DE MENT J. D. (1966). Agricultural nutrient containing urea, US Patent Office 3,232,740.

SOR K. M., PELISSIER J. A. and LATHAM R. (1971). Urease inhibited urea-containing compositions, US Patent Office 3,565,599.

SPEIR T. W. and ROSS D. J. (1975). Effects of storage on the activities of protease, urease, phosphatase, and sulphatase in three soils under pasture. *New Zealand J. Sci.* 18, 231–237.

STOJANOVIC B. J. (1959). Hydrolysis of urea in soil as affected by season and added urease. *Soil Sci.* 88, 251–255.

SUMNER J B. (1951). Urease. *In* "The Enzymes", Vol. 1 (J. B. Sumner, Ed.) pp. 873–892. Academic Press, New York.

SUMNER J. B. and SOMERS G. F. (1953). "Chemistry and Methods of Enzymes", 3rd ed. pp. 156–161. Academic Press, New York.

TABATABAI M. A. (1973). Michaelis constants of urease in soils and soil fractions. *Soil Sci. Soc. Am. Proc.* 37, 707–710.

TABATABAI M. A. (1977). Effects of trace elements on urease activity in soils. *Soil Biol. Biochem.* 9, 9–13.

TABATABAI M. A. and BREMNER J. M. (1970). Factors affecting soil arylsulfatase activity. *Soil Sci. Soc. Am. Proc.* 34, 427–429.

TABATABAI M. A. and BREMNER J. M. (1972). Assay of urease activity in soils. *Soil Biol. Biochem.* 4, 479–487.

TAGLIABUE L. (1958). Cryoenzymological research on urease in soils. *Chimica Milano* 34, 488–491 (*Chem. Abstr.* 53, 10618).

TALSKY G. and KLUNKER G. (1967). Über die Temperaturabhängigkeit enzymatischer Reaktionen. IV. Hydrolyse von Harnstoff durch Urease. *Z. Physiol. Chem.* 348, 1372–1376.

TAN K. H. and BREMNER J. M. (1977). Urease activity in Malaysian soils. Manuscript in preparation.

TANABE I. and ISHIZAWA S. (1969). Microbial activity of soil. *Bull. Natn. Inst. Agr. Sci., Tokyo, Sect.* B 21, 248–253.

THENTE B. (1970). Effects of toluene and high energy radiation on urease activity in soil. *Lantbr. Högsk. Annlr* 36, 401–418.

TOMLINSON T. E. (1967). Controlling urea hydrolysis in soils, British Patent No. 1,094,802.

TOMLINSON T. E. (1970). Urea—agronomic applications. *Proc. Fertil. Soc.* 113, 1–76.

TOREN E. C. and BURGER F. J. (1968). Trace determination of metal ion inhibitors of the urea-urease system by a pH-stat kinetic method. *Michrochim Acta* 5, 1049–1058.

VALMIKINATHAN K., VISVANATHA RAO V. N. and VERGHESE N. (1968). A study of urease from *Glyciridia maculata. Enzymologia* 34, 257–268.

VAN SLYKE D. D. and CULLEN G. E. (1914). The mode of action of urease and of enzymes in general. *J. Biol. Chem.* 19, 141–180.

VARNER J. E. (1960). Urease. *In* "The Enzymes", Vol. 4 (P. D. Boyer, H. Lardy and K. Myrbäck, Eds) pp. 247–256. Academic Press, New York.

VASILENKO Y. S. (1962). Urease activity in the soil. *Soviet Soil Sci.* 11, 1267–1272.

VELA G. R. and WYSS O. (1962). The effect of gamma radiation on nitrogen transformations in soil. *Bacteriol. Proc.* 62, 24.

VOETS J. P. and DEDEKEN M. (1966). Soil enzymes. *Meded. Rijksfac. Landbouwwet., Gent* 31, 177–190.

VOLK G. M. (1961). Gaseous loss of ammonia from surface applied nitrogenous fertilizers. *J. Agric. Fd Chem.* 9, 280–283.

WAID J. S. and PUGH K. B. (1967). Acetohydroxamate inhibition of the urease activity of an acid soil. *Chemy Ind.* 71–73.

WALL M. C. and LAIDLER K. J. (1953). The molecular kinetics of the urea-urease system. IV. The reaction in an inert buffer. *Arch. Biochem. Biophys.* 43, 299–306.

196　　　J. M. BREMNER AND R. L. MULVANEY

WANG C. H., TSENG Y. I. and PUH Y. S. (1966). A study on the behavior of urea in Taiwan soils. *Soils Fertil. Taiwan* **12**, 14–25.

ZANTUA M. I. and BREMNER J. M. (1975a). Comparison of methods of assaying urease activity in soils. *Soil Biol. Biochem.* **7**, 291–295.

ZANTUA M. I. and BREMNER J. M. (1975b). Preservation of soil samples for assay of urease activity. *Soil Biol. Biochem.* **7**, 297–299.

ZANTUA M. I. and BREMNER J. M. (1976). Production and persistence of urease activity in soils. *Soil Biol. Biochem.* **8**, 369–374.

ZANTUA M. I. and BREMNER J. M. (1977a). Stability of urease in soils. *Soil Biol. Biochem.* **9**, 135–140.

ZANTUA M. I. and BREMNER J. M. (1977b). Factors affecting urease activity in soils. Manuscript in preparation.

ZANTUA M. I., DUMENIL L. C. and BREMNER J. M. (1977). Relationships between soil urease activity and other soil properties. *Soil Sci. Soc. Am. J.* **41**, 350–352.

6

Soil Phosphatase and Sulphatase

T. W. SPEIR and D. J. ROSS

Soil Bureau, Department of Scientific and Industrial Research, Lower Hutt, New Zealand

I. Introduction

The elements phosphorus and sulphur are essential for plant growth and metabolism and they have, therefore, received a great deal of study by agronomists and soil scientists throughout the world. The transformations of these elements in the soil system are not well understood, but it is generally accepted that they are taken up by plant roots as inorganic phosphate and inorganic sulphate. Since a large proportion of the P and S in many soils is organically bound, the mineralization of these organic fractions is of major agricultural and economic importance. Soil phosphatase and sulphatase enzymes have been accorded a major role in this mineralization process; namely, catalysis of the hydrolytic cleavage of ester-phosphate and ester-sulphate bonds. It is the apparent similarity of their roles that has led to soil phosphatase and sulphatase being included together in this review, although very few attempts to compare, or contrast them will be made.

A considerable literature has accumulated on soil phosphatase activity and, consequently, this enzyme has been given a prominent place in a number of soil biochemistry and enzymology reviews (Cosgrove, 1967; Skujiņš, 1967; Ramírez-Martínez, 1968; Kuprevich and Shcherbakova, 1971; Halstead and McKercher, 1975; Kiss et al., 1975a). In contrast, soil sulphatase, because of its relatively recent discovery, has seldom been mentioned in review articles.

II. Soil Organic Phosphorus

The role of phosphorus in living systems and, in particular, in plant nutrition and physiology, has been outlined in a recent review (Halstead and McKercher, 1975). This element has been the subject of a large number of soil–plant studies in the past half-century, many of which have been concerned with the transfer of phosphorus from soil organic matter to the plant. A large proportion of total soil phosphorus is bound up in complex organic compounds, whose exact chemical nature has not been elucidated. Soil extraction techniques have yielded a number of relatively low molecular weight phosphate esters (Anderson, 1975a; Cheshire and Anderson, 1975; Halstead and McKercher, 1975). The principal esters identified in extracts are those of inositol, which may contain up to 60% of the organic phosphorus (Halstead and McKercher, 1975); nucleotides, which are thought to be derived from humic-bound nucleic acids and contain up to 2·4% of the organic phosphorus (Halstead and McKercher, 1975); and phospholipids, predominantly phosphatidyl choline (lecithin) and phosphatidyl ethanolamine, which are thought to be derived from glycerophosphatides (Anderson, 1975a), and contain up to 1% of the organic phosphorus (Halstead and McKercher, 1975). There is evidence that some of the remaining organic phosphorus is in the form of tightly bound phosphoprotein, and a small proportion may exist as glucose-1-phosphate (Anderson, 1975a). Traces of other sugar phosphates and of glycerophosphate have also been identified (Cheshire and Anderson, 1975). In the soil, small amounts of these compounds may exist in solution but in the main they occur as insoluble organic complexes, as relatively insoluble calcium and ferric salts, or as adsorbates with clay minerals (Halstead and McKercher, 1975).

III. Plant Root and Microbial Phosphatases

Plants grown in absence of soil have a recognized ability to obtain all of their phosphorus requirements from organic phosphorus compounds of the type found in soil extracts, when these compounds are provided in the nutrient solution (Pearson and Pierre, 1940; Hayashi and Takijima, 1951; Szember, 1960b). There is some evidence that organic phosphorus may be taken up directly by plant roots (Rogers et al., 1940; Totarskaya-Merenova, 1956; Thompson and Black, 1970b); however, recent work suggests that this mechanism is most unlikely and that the organic phosphorus compounds are "mineralized" in the nutrient medium, or on the root surface, to inorganic phosphorus which is subsequently absorbed (Bartlett and Lewis, 1973; Appiah and Thompson, 1974). This mineralization appears to be catalysed by root exocellular enzymes which may be either liberated into the nutrient solution (Pearson and Pierre, 1940; Rogers et al., 1942; Stefan, 1963; Chang and Bandurski, 1964; Dubovenko, 1966; Floyd and Ohlrogge, 1970; Guede Sueiro and De Felipe Anton, 1974), or bound to the root epidermal cells but outside their permeability barrier (Bieleski and Johnson, 1971; Hall and Davie, 1971; Ridge and Rovira, 1971; Bieleski, 1974; Shaykh and Roberts, 1974; Clark, 1975; Pammenter and Woolhouse, 1975). The enzymes have generally been given trivial names according to their substrate but are either phosphoric monoester hydrolases (EC 3.1.3) or phosphoric diester hydrolases (EC 3.1.4). To the first group belong such enzymes as phytase, which catalyses the hydrolytic removal of all six phosphate groups from inositol hexaphosphate, nucleotidases, sugar phosphatases and glycerophosphatase, whose catalytic functions are obvious. To the second group belong nucleases, which catalyse hydrolysis of ribo- and deoxyribonucleic acids to their individual nucleotides, and phospholipases, which catalyse hydrolysis of phospholipids.

Pure cultures of many microorganisms commonly found in soils and soil microbial isolates, have also been shown to produce exocellular phosphatases (Casida, 1959; Greaves et al., 1963; Chunderova, 1964; Mazilkin and Kuznetsova, 1964; Banerjee and Nandi, 1966; Shieh and Ware, 1968; Agre et al., 1969; Bezborodova and Il'ina, 1969; Cosgrove et al., 1970; Ko and Hora, 1970; Antheunisse, 1972; Jayaram and Prasad, 1972) and to be excellent mineralizers of organic phosphorus in culture media (Pacewic-zowa, 1960; Szember, 1960a; Kobus, 1961; Greaves and Webley, 1964). Mycorrhizal fungi, in pure culture, produce extracellular phytase which catalyses hydrolysis of calcium and sodium phytate but not ferric phytate (Theodorou, 1968). The root-mycorrhizal fungi system of beech has a high phosphatase activity (Bartlett and Lewis, 1973; Williamson and Alexander, 1975); this activity is primarily located in the fungal sheath and appears to be firmly bound.

The role of these plant and microbial enzymes and their ability to function in the soil environment, are a matter of speculation. Although the growth of plants does reduce the organic phosphorus content of soil (Thompson and Black, 1970a; Stone and Thorp, 1971; Appiah and Thompson, 1974), Appiah and Thompson (1974) found that addition of plant-root phosphatase extracts to soil had no effect on the organic phosphorus and Jackman and Black (1952b) showed that added phytase caused no detectable hydrolysis of soil phytate. On the contrary, Thompson and Black (1970b) showed that the addition of corn-root exocellular phosphatase preparations and commercial acid and alkaline phosphatases, actually caused an increase in soil organic phosphorus. Incubation studies in the absence of plants have shown that soil mineralizes native organic phosphorus (Thompson and Black, 1947; Kaila, 1949; Adebayo, 1973). Addition of an energy source containing readily assimilable carbon increases the rate of mineralization (Furukawa and Kawaguchi, 1969; Halm et al., 1971), indicating that microbial proliferation is at least partly responsible. Bacillus cereus and Bacillus megaterium have both been shown to improve crop yields when inoculated into soil. This effect was assumed to be due to exocellular enzymes decomposing soil organic compounds, thereby providing inorganic phosphorus for plant growth (Menkina, 1950; Mal'tseva, 1960; Kudzin and Yaroshevich, 1961; Mazilkin and Kuznetsova, 1964; Burangulova and Khaziev, 1965b). However, with Bacillus megaterium at least, improved plant yields are now believed to be due to production, by the bacteria, of plant growth inducers, such as auxin and gibberellin (Naumova et al., 1962; Samtsevich, 1962; Mishustin, 1963; Brown, 1974). There appears to be no clear evidence that plant and microbial exocellular phosphatase enzymes play a prominent role in the mineralization of soil organic phosphorus compounds. This is not wholly surprising when the complexity and heterogeneity of soil organic matter are taken into account. The fact that microorganisms require phosphorus for their own metabolism must not be overlooked. If the phosphorus concentration in the organic matter being assimilated by microorganisms is below 0·2–0·3%, no phosphorus will be mineralized; it will, rather, all be immobilized in microbial tissue (Kaila, 1949; Nilsson, 1957).

The mineralization of soil organic phosphorus is therefore intimately associated with the mineralization of organic matter as a whole. This subject involves all of the disciplines of soil science and, as such, is beyond the scope of this review.

IV. Soil Phosphatases

1. Terminology

This review is concerned with the phosphomonoesterases (phosphoric monoester hydrolases) and phosphodiesterases (phosphoric diester hydrolases) (EC 3.1.3 and 4) found in soil and covers the enzymes mentioned in the previous section plus the phosphatase activities measured by numerous workers using a variety of artificial substrates. Generally, the enzymes will be accorded the names used by the authors cited in the references; collectively, however, they will often be referred to as phosphatases.

The enzymes adenosine triphosphatase (Floyd and Ohlrogge, 1970; Galstyan and Abramyan, 1975) and pyrophosphatase (Khaziev, 1972a; Racz and Savant, 1972; Arutyunyan and Galstyan, 1974) have been detected in soils, but because they belong to class EC 3.6.1, namely hydrolases acting on acid anhydride bonds in phosphoryl-containing anhydrides, they have not been included here.

2. History

Parker (1927) suggested that enzymes may catalyse the decomposition of some of the organic phosphorus compounds in soil. Rogers (1942) demonstrated phosphatase activity in soil by separately incubating calcium glycerophosphate and nucleic acid solutions with soil and toluene; both substrates were extensively hydrolysed in 18 h under optimum pH and temperature conditions. The similarity of the nuclease pH and temperature optima to those of plant root nuclease prompted Rogers to propose, as had Parker, that the soil catalytic activity was of plant origin.

Throughout the 1940s, there were a number of studies on the mineralization of organic phosphorus compounds in soil (Chang, 1940; Scharrer and Keller, 1940; Dyer and Wrenshall, 1941; Pearson et al., 1941; Ghani and Aleem, 1942; Thompson and Black, 1947; Bower, 1949) but it was not until 1952 that specific reference was again made to a soil phosphatase enzyme. Jackman and Black (1952a, b) demonstrated phytase activity by incubating sodium phytate with soil and toluene in citrate buffer (pH 5) for 20 h at 45°C, and measuring the inorganic phosphate released; enzymic hydrolysis of soil phytate by added phytase was not detected. However, hydrolysis of soil organic phosphorus by native soil enzymes was demonstrated by Stefanic (1971), who incubated toluene-treated soil without added substrate for 24 h at 28°C and measured the increase in extractable inorganic phosphate.

The first study of soil phosphatase activity using an "artificial" substrate, i.e., a substrate neither identified as, nor even suspected of being, a compound of soil organic phosphorus, was that of Kroll et al. (1955), who used disodium phenylphosphate and measured both liberated phenol and inorganic phosphate. Since then, there has been an ever increasing number of investigations with so-called natural and artificial substrates, with most of the recent work being carried out in the Soviet Union. It is interesting to note that, although nucleic acids are initially hydrolysed by phosphodiesterase activity, no phosphodiester artificial substrate was used before 1974 when Ishii and Hayano used bis(p-nitrophenyl)phosphate. Phosphotriesterase activity was discovered by Kishk et al. (1976), when studying the degradation of methyl parathion, (O, O-dimethyl O-p-nitrophenyl) phosphorothionate, in soil.

3. Factors involved in the measurement of soil phosphatase activity

It is probable that many of the contradictions that have arisen from studies on soil enzymes are due to differences in the treatment of the soils before and during incubation and to the diversity of extraction and assay methods employed after incubation. Factors such as moisture status of the soil, type of buffer (if any), incubation temperature and time, have not received the amount of study they warrant. This is partly due to the fact that particular methods have been used by different workers for their particular soils, with or without a study of their suitability. Subsequently these methods may have been used by other workers for very different soils, for which they may be inadequate. The result is that no firm guidelines have been laid down to help the newcomer avoid the many pitfalls that exist in this field.

3.1. Incubation and assay methods

Rogers (1942), recognizing the slow rate of hydrolysis of nucleic acid in soil, incubated his soil with this substrate at the pre-determined temperature and pH optima (60°C, pH 7), for 18 h. Such conditions necessitated the use of toluene to prevent microbial proliferation during incubation. Other workers using natural substrates such as glycerophosphate, nucleic acid and phytate, when confronted with the same problem, have also usually employed higher temperatures and longer incubation times than would be used in enzyme studies on other tissue materials. For these reasons, they have been required to "sterilize" the soil by either chemical (toluene) or physical (irradiation) means (Jackman and Black, 1952a; Nilsson, 1957; Drobníková, 1961; McLaren et al., 1962; Burangulova and Khaziev, 1972).

Artificial substrates have superseded the natural ones in most of the soil phosphatase studies since the early 1960s, except in the Soviet Union where the use of natural substrates (primarily nucleic acids) has been continued (Kudzin et al., 1970; Lisoval, 1970; Kozlov et al., 1972; Blagoveschenskaya and Danchenko, 1974; Panikov and Aseeva, 1974), mainly by Khaziev and co-workers (Burangulova and Khaziev, 1965c; Khaziev and Burangulova, 1965; Khaziev, 1966, 1967, 1969, 1972b; Burangulova and Khaziev, 1972) and Galstyan and co-workers (Arutyunyan and Galstyan, 1974; Abramyan and Galstyan, 1975a, b; Arutyunyan and Abramyan, 1975). However, in many of these investigations, artificial, as well as the natural substrates have been used.

There are two major reasons for the change to artificial substrates. First, low molecular weight esters undergo relatively rapid hydrolysis in soil, compared with the natural substrates. Secondly, more importantly, the esters chosen have an organic moiety that is easily estimated; the assays do not then depend on the measurement of inorganic phosphate. With one exception (Kiss and Péterfi, 1961), all methods for determining soil phosphatase using natural substrates do require the measurement of the liberated inorganic phosphate. Because it was soon recognized that phosphate cannot be recovered from soil quantitatively, Kroll and colleagues (Kroll and Krámer, 1955; Kroll et al., 1955) were prompted to use phenylphosphate as substrate and measure the phenol produced, colorimetrically. This compound, mainly through the methods proposed by Hofmann and colleagues (Hofmann, 1963; Hofmann and Hoffmann, 1966; Hoffmann, 1967) Kramer and Erdei (1958) and Kramer and Yerdei (1959), has become the most widely used artificial substrate (e.g. Janossy, 1963; Halstead, 1964; Goian, 1968; Radulescu and Kiss, 1971; Stefanic et al., 1971; Kong and Dommergues, 1972; Gavrilova et al., 1973; Karki and Kaiser, 1974; Tyler, 1974; Verstraete et al., 1974; Voets et al., 1974). The other artificial substrates are phenolphthalein phosphate, first used by

Dubovenko (1964); *p*-nitrophenyl phosphate, first used by Bertrand and de Wolf (1968) and then independently by Tabatabai and Bremner (1969) and by Fauvel and Rouquerol (1970); bis(*p*-nitrophenyl)phosphate (Ishii and Hayano, 1974); α-naphthyl phosphate (Hochstein, 1962); and β-naphthyl phosphate (Ramírez-Martínez and McLaren, 1966a). In these methods, the products are assayed spectrophotometrically or fluorimetrically and incubation times are generally only 1–3 h. Because they contain the phosphate ion, these artificial substrates are adsorbed by clay minerals to some extent (Saalbach, 1956; Nannipieri *et al.*, 1972); however, under suitable extraction conditions, recovery of the organic moiety product is quantitative (Tabatabai and Bremner, 1969; Arutyunyan and Galstyan, 1975).

It was once believed that enzymes are most active at the natural pH of the soil (Kroll *et al.*, 1955). Although this may be true for phosphatase in some soils (Drobníková, 1961; Skujiņš *et al.*, 1962; Parks, 1974), it is more the exception than the rule (Drobníková, 1961; Halstead, 1964; Hoffmann and Elias-Azar, 1965; Khaziev and Burangulova, 1965; Khaziev, 1966; Ramírez-Martínez and McLaren, 1966b; Herlihy, 1972). Buffers are generally included in soil phosphatase incubations to enable activity to be determined at, or near, the pH optimum for the enzyme and to prevent pH changes during the course of the reaction. Care must be taken when choosing a buffer, Abramyan and Galstyan (1975a) having found that acetate and citrate–phosphate buffers inhibited soil phosphatase activity and that borate inhibited nuclease activity. Khaziev (1966) also found that phosphate inhibited nuclease activity.

3.2 Soil sterilization.

In soil enzyme studies requiring long incubation times, soil should ideally be pre-treated to meet the requirements laid down by McLaren (1963). However, no methods are yet available to achieve such requirements. The best that can be done is to "kill" all microbial activity, thereby preventing assimilation of substrates and products and the further synthesis of exoenzymes. The only techniques used in soil phosphatase investigations have been treatment with toluene or ionizing radiation; possible effects of these treatments on soil biochemical activities have been examined in some detail.

Rogers (1942) found no increase in carbon dioxide evolution when sucrose was incubated for 24 h with toluene-treated soil and thus believed that toluene effected soil sterilization during his 24 h phosphatase determination. For short incubation periods (1 h), phosphatase activity was identical in toluene-treated and untreated samples (cited in Kiss *et al.*, 1974); during longer incubations (3–24 h), activity was higher in untreated samples, presumably because of microbial proliferation. The use of toluene as a soil "sterilant" has, however, been the source of some controversy (Skujiņš, 1967) and it can have an effect on cell permeability (Jackson and DeMoss, 1965) and possibly soil structure (Černá, 1972). Thus, Jackson and DeMoss (1965) found that 5% toluene in the culture medium killed the cells of *Escherichia coli* but did not cause lysis; however, up to 25% of the intracellular protein was lost to the medium and the cell membranes became permeable to small molecules. Toluene treatment could therefore allow substrate molecules to come into contact with intracellular enzymes, whether they were liberated from, or still retained in, the cells. A possible effect of toluene on soil structure can be inferred from the results of Černá (1972), using soil aggregates of various sizes; he found that the addition of toluene had the same effect on phosphatase activity as would be expected from aggregate disruption. It is possible, then, that toluene may alter the surface structure of soil particles, thereby exposing more, or less, enzymes to substrates.

The use of ionizing radiation as a soil sterilant has recently been reviewed by Cawse (1975). McLaren *et al.* (1957) found that a dose of $2 \cdot 2 \times 10^6$ rep with an electron beam destroyed all microorganisms in a sample of air-dried soil but had no effect on urease activity. Such a dose caused a decrease in phosphatase activity in some air-dried soils (McLaren, 1963), although not in others (Voets *et al.*, 1965; Kiss *et al.*, 1974). Skujiņš *et al.* (1962), in a detailed study of the effect of ionizing radiation on soil β-glycerophosphatase, pre-irradiated their samples with 5×10^6 rep to sterilize them, and then further irradiated them at different doses before measuring enzyme activity. Activity was reduced to 37% of that present after the initial sterilizing dose by 17×10^6 rep of electrons, or a similar dose of γ-radiation. In the above studies, soil was irradiated in the air-dried state. Ramírez-Martínez and McLaren (1966b) studied the effect of 5×10^6 rep on soil samples at various stages of dryness and found that, as the moisture content decreased, the sensitivity of phosphatase to radiation also decreased, although the relationship was not a simple one. These measurements were made at pH 7, the optimum pH for phosphatase; at pH 5, in contrast, no significant difference between the phosphatase activity of irradiated and untreated soil was found. This pH effect could account for the conflicting results obtained by other workers concerning the influence of ionizing radiation on phosphatase activity (McLaren, 1963; Voets *et al.*, 1965; Kiss *et al.*, 1974).

According to Ramírez-Martínez and McLaren (1966b), the decrease of phosphatase activity caused by irradiation under some conditions may be due to inactivation of phosphatase in "killed" microbial cells, or, if there is a large pool of enzyme in soil from many generations of microorganisms, to the inactivation of free, unadsorbed enzyme.

Ionizing radiation at a minimum sterilizing dose appears to be a far superior soil sterilant to toluene and only about 10% of the phosphatase activity is thereby lost (Ramírez-Martínez and McLaren, 1966b). However, because the equipment required to produce such radiation is expensive and not readily accessible to many soil enzymologists, toluene continues to be used in most soil enzyme studies.

It would seem that, for short incubation periods, "sterilization" is not required at all (Kiss et al., 1974; Arutyunyan and Galstyan, 1975), especially if the temperature is not high. However, many workers are loath to change their techniques and soil "sterilization" will undoubtedly continue to be an integral part of phosphatase determinations for some time.

3.3. Air-drying and storage of soil

Because chemical measurements on soil were made with air-dry samples, early workers did not hesitate to use air-dried soil for enzyme studies (e.g. Rogers, 1942). Jackman and Black (1952a), however, showed that field-moist samples had greater phytase activity than did partially air-dried soils; the magnitude of the difference varied from soil to soil. They recommended that phytase activity be measured on field-moist samples immediately after collection. In spite of these results, most workers continued to use air-dry soils for phosphatase assays, and generally ignored any loss of activity due to drying (e.g. Halstead, 1964; Burangulova and Khaziev, 1965c; Tabatabai and Bremner, 1969; Khan, 1970; Voets and Vandamme, 1970; Stefanic, 1971; Cervelli et al., 1973; Arutyunyan and Galstyan, 1975; Karanth et al., 1975; Kelley and Rodriguez-Kabana, 1975); Hofmann and co-workers even recommended air-drying in their methods (Hofmann, 1963; Hofmann and Hoffmann, 1966; Hoffmann, 1967). Khaziev (1969), however, observed that air-drying reduced soil nuclease activity by 40–50%, and Ramírez-Martínez (1966) and Speir and Ross (1975) found that it reduced phosphatase activity by almost 30–40% and 50–55% respectively.

Because air-drying appears to be unsuitable and some form of storage is frequently required, refrigeration or freezing would seem to offer the best alternatives. Speir and Ross (1975), however, have shown variable effects with three soils stored at 4°C. One of them lost 36% of its activity in 21 days, but no more up to 216 days; the other two lost very little activity up to 21 days, and only 13% and 20% over 216 days. In samples kept at −20°C, phosphatase activity dropped by 20–30% over 24 days and by 50% or more over 200 days. Surprisingly, storage of the field-moist samples at room temperature (18–25°C) generally resulted in the smallest losses of activity, even though some of the changes were significant (Speir and Ross, 1975). Khaziev (1969) found that storage of field-moist soil in the cold resulted in a 20% loss of nuclease activity in 10 days but a more gradual loss thereafter; addition of toluene to the soil had no effect on this loss. The reduction of phosphatase activity during storage of field-moist soil may be due, in part at least, to proteolytic activity. In frozen soil, urease, phosphatase and sulphatase have been shown to be active, presumably due to unfrozen water around soil particles (Bremner and Zantua, 1975) and proteases could also well remain active under such conditions.

After the initial loss of about 50% of the phosphatase activity on air-drying of three soils, little change during subsequent storage at 4°C was found by Speir and Ross (1975). Other workers have recorded that soil stored air-dry for several years did not alter greatly in phosphatase activity (Halstead, 1964; Galstyan et al., 1967) and that some activity remained after even 60 years (Skujiņš and McLaren, 1967). However, Khaziev (1969) showed that, even in air-dry soil, nuclease activity declined by 10% in 60 days and by 40% in 13 months; this study was carried out at room temperature, and was confirmed by Speir and Ross (1975) who found a 19–41% loss of phosphatase activity over 208 days in soils stored under these conditions.

From the above results, it is obvious that the extent of change of phosphatase activity during storage of field-moist and air-dry soil depends to a large extent on the particular soil under consideration. Ideally, phosphatase activity should be assayed with field-moist samples as soon as possible after collection. If storage is necessary, a preliminary experiment to find the optimum conditions is very desirable. For long-term storage, air-drying, despite its initial drastic effect, is probably the best method for retaining activity.

4. Properties of soil phosphatases

4.1. Optimum pH

Soil phosphatase activity, as measured with artificial substrates, often appears to have two pH optima, one acid and the other alkaline (Hochstein, 1962; Hofmann and Kasseba, 1962, Hoffmann and Elias-Azar, 1965; Khaziev and Burangulova, 1965; Hofmann and Hoffmann, 1966). Acid and alkaline phosphatase are common in nature (Feder, 1973), and it is not surprising that both should occur in soil. In soils where they are both present (Halstead, 1964; Fauvel and Rouquerol, 1970; Arutyunyan and Galstyan, 1975), or where only one or the other is found (Skujiņš et al., 1962; Galstyan and Markosyan, 1965; Galstyan and Arutyunyan, 1966; Ramírez-Martínez, 1966; Chunderova, 1970; Herlihy, 1972; Parks, 1974), pH optima are generally within the ranges pH 4–6 and 8–10 respectively. Frequently, however, neither acid nor alkaline phosphatase is found, but rather a "neutral" phosphatase with an optimum around pH 7 (Halstead, 1964; Ramírez-Martínez, 1966; Ramírez-Martínez and McLaren, 1966a;

Thornton and McLaren, 1975; Kishk *et al.*, 1976). The peak may be broad, spanning several pH units, and represent a mixture of acid and alkaline phosphatases (McLaren, 1963).

Soil phosphatase activity, as measured with natural substrates, appears to have only one pH optimum, generally around pH 7 (Rogers, 1942; Drobníková, 1961; Khaziev and Burangulova, 1965; Khaziev, 1966), although an optimum at pH 3–5 has been reported for phytase (Khaziev and Burangulova, 1965). There are many factors that influence the optimum pH of soil phosphatases, and some of them will be discussed in subsequent sections.

4.2. Effect of temperature

As already mentioned, soil phosphatase is active at sub-zero temperatures, a property which reflects the nature of the soil rather than that of the enzyme. Activity was detectable at $-10°C$ and $-20°C$, but not at $-30°C$ and at best was less than 10% of that measured at $5°C$ (Bremner and Zantua, 1975).

Enzyme-catalysed reactions have somewhat lower temperature coefficient (Q_{10}) values, generally falling between 1 and 2, than do uncatalysed chemical reactions (Dixon and Webb, 1964). Coefficients for soil phosphatases have been determined in several chernozem soils (Galstyan, 1965; Khaziev, 1969, 1975); the values obtained depended upon the particular temperature interval and, to some extent, upon the type of chernozem. Typically, for a leached chernozem, Khaziev (1975) determined Q_{10} values of 1·38, 1·66, 1·43 and 1·16 for the temperature intervals 10–20°C, 20–30°C, 30–40°C and 40–50°C respectively. Apart from the 10–20°C interval, the Q_{10} decreased as temperature increased (Galstyan, 1965; Khaziev, 1969, 1975). At temperatures above 60°C, Q_{10} values were less than 1, indicating that temperature inactivation of the enzyme was occurring (Galstyan, 1965; Khaziev, 1969). Ramírez-Martínez and McLaren (1966b) determined a Q_{10} value of 2·5 for the interval 25–35°C in a clay loam soil. This is much higher than any value quoted by the Russian workers and it is possible that Q_{10} values may be more dependent upon soil type than is evident from the Russian results.

Khaziev (1975) also quotes activation energies and heats of activation for soil phosphatase and various other enzyme activities. The activation energies varied considerably and in some instances were very low. Khaziev suggested that at low temperatures, free and weakly bound soil enzymes are most active, but that with increasing temperature the immobilized enzymes become more active. He also proposed that the changes in thermodynamic characteristics indicate a heterogeneity in the composition and state of soil enzymes.

Optimum temperatures appear to be around 60–70°C for phosphatase (Galoppini *et al.*, 1962; Galstyan, 1965; Fauvel and Rouquerol, 1970) and nuclease (Rogers, 1942; Khaziev, 1969) activities in most soils studied, although an optimum of 42°C has been reported for nuclease (Drobníková, 1961). Drobníková (1961) found optima of 55°C for glycerophosphatase activity in three soils and 42°C in a fourth and Rogers (1942) found an optimum of 45°C for this enzyme. Phytase also appears to have an optimum temperature of 45°C (Jackman and Black, 1952a). The temperature optima for these enzymes may be dependent upon incubation time, since at high incubation temperatures a time-dependent inactivation could be taking place. This would explain the higher optima for "artificial-substrate" phosphatases (short incubation times) than those generally found for "natural-substrate" phosphatases (long incubation time). This proposition is supported by Parks (1974), who found that *Escherichia coli* phosphatase had an optimum temperature of 55°C for a 30 minute incubation, but only 50°C for a 60 minute incubation.

Temperature inactivation studies have shown that soil phosphatase, like other soil enzymes, is extraordinarily stable to heat treatment. Fauvel and Rouquerol (1970) measured soil alkaline phosphatase activity using incubation temperatures from 30–100°C. Activity was maximal at 60°C and rapidly decreased up to 100°C, but even at 100°C phosphatase was almost as active as at 30°C. Most studies have been with soil subjected to dry heat prior to incubation with substrate. Generally, activity was unaffected by heating to 60°C (Galstyan, 1965; Khaziev, 1969) but fell slightly at higher temperatures (Halstead, 1964; Galstyan, 1965; Ambroz, 1973b; Parks, 1974). Khaziev (1969) showed that 30–40% of the original nuclease activity was retained after heating soil at 100°C for 3 h. Kishk *et al.* (1976) pre-heated soil for 4 h at 160°C and found that 1–4% of the phosphotriesterase activity was retained. Wet heat appears to have a more drastic effect on soil enzymes. Parks (1974) found that soil phosphatase activity fell by 80–100% during autoclaving at 121°C for one hour, and other workers have reported complete inactivation of phosphatase during autoclaving (Geller and Dobrotvorskaya, 1961; Kishk *et al.*, 1976).

4.3. Stability during natural preservation

Phosphatase activity has been found in 9000-year-old peats that were subjected to permafrost conditions for most of this time (Skujiņš and McLaren, 1967; McLaren and Pukite, 1972). The levels of activity were, in fact, higher than in some of the fresh soils assayed. However, phosphatase activity was not detected in a 32 000-year-old buried silty soil, even though moderate levels of newly introduced microorganisms were present. Phosphatase activity has also been found in a Pliocene soil (Nannipieri *et al.*,

1973) and in lake sediments up to 13 000 years old (Radulescu and Kiss, 1971).

4.4. Inhibitors and activators

Soil phosphatase activity is strongly inhibited by inorganic phosphate. Kiss et al. (1974) describe a linear inverse relationship between phosphate concentration and phosphatase activity and Khaziev (1966) showed that soil nuclease was completely inhibited by 0·02 M KH_2PO_4. Abramyan and Galstyan (1975a) found that citrate–phosphate buffer inhibited phosphatase; this effect would probably have been due to phosphate, since citrate is a component of the modified universal buffer used by Tabatabai and Bremner (1969) and has no inhibitory action on phosphatase.

Fluoride appears to inhibit phosphatase activity in some soils (Halstead, 1964; Runkov and Kozlovskaya, 1974) but not in others (Halstead, 1964; Khaziev, 1966; Kiss et al., 1974) and azide has little effect (Kelley and Rodriguez-Kabana, 1975). Several cations inhibit soil phosphatase activity. Tyler (1974, 1976b) has shown that Cu and Zn ions, which were found together as pollutants in the organic mor of a podzolized spruce forest soil near a brass foundry, had a marked inhibitory effect on phosphatase activity. A linear inverse relationship ($P < 0·001$) was found between phosphatase activity and log (Cu+Zn) concentration (Tyler, 1976b). The Cu ion was more responsible for this inhibitory effect than was the Zn ion at equal concentration. Corroborative evidence has been provided by Andersson and Bengtsson (1975), who showed that the addition of Cu+Zn cations to soil caused an instantaneous inhibition of phosphatase, followed by a slow recovery of activity. Tyler (1976a) has also studied the effect on soil phosphatase of vanadium, a metal which can accumulate to high levels (50–100 mg kg^{-1}) in urban soils due to combustion of fuel oils. Concentrations of V, (as $NaVO_3$) of 30, 50, 100 and 1000 mg kg^{-1} added to spruce needle mor, inhibited phosphatase activity by 20, 40, 47 and 68% respectively. Other vanadium compounds inhibited phosphatase to the same extent. The mercuric ion also inhibits soil phosphatase activity but inhibition is only significant at relatively high concentrations (Kiss et al., 1974).

Miscellaneous compounds added to soil to test their effect on phosphatase activity include penicillin and streptomycin which inhibited it markedly (Ambrož, 1973b), formaldehyde and tannin which had some inhibitory action (Kiss et al., 1974) and benzene, xylene, chloroform, hydrogen peroxide and ether which had no effect at all (Ambrož, 1973b). Effluent from a starch-production plant decreased soil phosphatase activity (Ambrož, 1973a), whereas effluent from a PVC-plastic factory increased activity (Zayata and Kochkina, 1971).

4.5. Effect of fertilizers and trace elements

There is a wealth of contradictory evidence on the effect of fertilizers, generally concerning those containing inorganic phosphate, on soil phosphatases. Mineral fertilizers containing superphosphate or other phosphate compounds have been reported to increase soil phosphatase activity (Vlasyuk and Lisoval, 1964; Stefanic *et al.*, 1971; Kiss *et al.*, 1975b), to have no effect (Khaziev, 1967; Chunderova, 1970), or to decrease activity (Kotelev and Mekhtieva, 1961; Janossy, 1963; Burangulova and Khaziev, 1965b; Goian, 1968; Chunderova and Zubets, 1969; Muresanu and Goian, 1969; Mandrovskaya, 1971; Stebakova, 1972; Akimenko and Brid'ko, 1975; Stulin, 1975). Yaroshevich (1966) suggested that decreases in activity caused by superphosphate may be due to fluoride contamination. However, in a study of 544 derno-podzolic soils, Chunderova and Zubets (1969) found that soil phosphatase activity was increased by phosphorus fertilizers until the soluble-phosphorus content of the soil reached 20 mg 100 g^{-1}; it then declined and disappeared completely at a soluble-phosphorus content of 60–80 mg phosphorus 100 g^{-1} soil. Further evidence was obtained by Rankov and Dimitrov (1971), who showed that NP or NPK fertilizers, at 120 kg of each element ha^{-1}, increased soil phosphatase activity but that 360 kg phosphorus ha^{-1}, with various amounts of N and K, decreased activity. The stimulation of phosphatase activity by low levels of fertilizer phosphate would be due to increased microbial numbers in the soils and increased plant growth, which over a long period of time would cause a build-up of soil organic matter and enzyme levels. The effect of high levels of fertilizer phosphate is probably two-fold. First, the increased phosphate in soil solution would inhibit phosphatase activity and secondly, it would repress the synthesis of microbial phosphatase (Bezborodova and Il'ina, 1969; Shieh *et al.*, 1969; Adebayo, 1973; Feder, 1973); microbial numbers and soil organic matter content may therefore build up, but phosphatase activity would not.

Organic fertilizers (composts, manures) appear to stimulate soil phosphatase activity (Goian, 1968; Halstead and Sowden, 1968). They often enhanced the positive effect, or cancelled out the negative effect, of mineral fertilizers containing phosphate, when both were applied to soil (Vlasyuk *et al.*, 1957; Vlasyuk and Lisoval, 1964; Yaroshevich, 1966; Kiver, 1973; Blagoveshchenskaya and Danchenko, 1974; Lisoval *et al.*, 1974).

Application of fertilizers that do not contain phosphorous generally causes an increase in soil phosphatase activity. Urea, ammonium nitrate, with and without added potassium (Gerus, 1963; Khaziev, 1966; Goian, 1968; Kudzin *et al.*, 1968; Levchenko and Lisoval, 1974) and other N fertilizers (Burangulova and Khaziev, 1965b; Muresanu and Goian, 1969;

Stebakova, 1972), had stimulatory effects on soil phosphatase. However, Verstraete *et al.* (1964) found that slow-release N fertilizers had an immediate inhibitory effect on soil phosphatase and other enzyme activities; after twenty weeks, these activities recovered to be equal to control levels.

Fertilizers may affect different soil phosphatases in different ways. Thus, Kudzin *et al.*, (1970) found that the application of N and K fertilizers over a thirty-five-year period increased acid phosphatase, but did not alter neutral and alkaline phosphatases, glycerophosphatase, nuclease, or phytase activities.

Although liming is often included in fertilizer treatments, the only clear effect of liming on soil phosphatase comes from the study by Halstead (1964), who found that it reduced the phosphatase activity of an acid soil. This treatment was associated with an increase in the numbers of bacteria and actinomycetes and a decrease in fungi in the soil, as well as the expected rise in pH. A combination of the change in microbial populations and the rise in pH is the probable cause of the drop in phosphatase activity.

Khaziev (1966) studied the effect on soil nuclease of several ions commonly used in fertilizers. He included them in his incubations, along with toluene as a "sterilant"; they were not added to the soil prior to nuclease assay. Borate, Mg and Mn ions had little, if any, effect on nuclease activity, but Zn activated and molybdate strongly inhibited it. Khaziev warned against extrapolation of these results to pot or field situations, because the ions may affect microbial populations under such conditions. Indeed, Gorbanov (1975) found that molybdate, at $0 \cdot 1$–10 mg Mo kg soil^{-1}, raised phosphatase activity in all types of Bulgarian soils, presumably due to such an effect.

Generally, the changes of soil phosphatase activity caused by fertilizers, apart from the effect of phosphate, are probably due to increased microbial numbers and increased plant growth with a concomitant increase in soil organic matter and enzyme levels in general. The effect is thus indirect, and does not originate from activation of existing phosphatase, but rather from increased production of new enzyme.

4.6. Effect of other soil additives

Various pesticides and other agricultural chemicals have been evaluated to determine their effects on soil microorganisms and enzyme activities. Their effects on soil phosphatase vary considerably; some of the pesticides increased activity (Goian, 1969; Verstraete and Voets, 1974), some had no effect (Gruzdev *et al.*, 1973; Karki and Kaiser, 1974; Tyunyayeva *et al.*, 1974) but most reduced activity (Goian, 1969; Manorik and Malichenko, 1969; Voets and Vandamme, 1970; Zubets, 1973; Verstraete and Voets, 1974; Voets *et al.*, 1974; Karanth *et al.*, 1975; Abdel'Yussif *et al.*, 1976).

Voets *et al.* (1974) found the herbicide atrazine had significantly reduced phosphatase and other enzyme activities in soil samples eleven months after application but that renewed application caused no further loss. They suggested that the reduction of activity resulted, in part at least, from the elimination of vegetative cover and the associated reduction of soil organic matter. Zubets (1973) implicated a similar mechanism to account for the effect of simazine and atrazine. The reasons for any deleterious effects of other kinds of agricultural chemicals are less apparent. Fungicides may reduce soil phosphatase activity by eliminating one class of source organisms. Any effect of insecticides or nematocides may be due to direct inhibition of phosphatase activity, or to an indirect influence on microbial populations.

Voets *et al.* (1973) investigated the effect of the application of a bituminous emulsion soil conditioner on soil microorganisms and enzymes in a number of soils. The study contained too many variables to be interpreted easily, but it appears that phosphatase activity was reduced in several soils over a three-year period, whereas microbial numbers were unaffected or increased. The inhibition of phosphatase activity was attributed to complexing of the enzyme by hydrophobic bitumen polymers.

5. The effects of soil properties on phosphatases

5.1 Soil type.

Several reports have compared phosphatase and other enzyme activities in different soils (Hofmann and Kasseba, 1962; Keilling, 1964; Hoffmann and Elias-Azar, 1965; Burangulova and Mukatanov, 1973; Arutyunyan and Galstyan, 1974; Kiss *et al.*, 1974; Arutyunyan and Abramyan, 1975), mainly in the Soviet Union where chernozems generally appear to have high levels of phosphatase activity. Burangulova and Khaziev (1965c) compared different chernozems and found that nuclease activity increased in the following order: leached < typical thick < ordinary < calcareous. Galstyan and Tatevosyan (1964) determined various hydrolytic enzyme activities, including phosphatase, in a wide range of Armenian soils. They found that the saline and alkaline soils (solonchak+solonetz) had no hydrolytic activity, and that activity increased in the following sequence: semi-desert brown soils (low activity) < chestnut soils of the dry steppe < chernozems of the mountain steppe < prairie steppe chernozems < mountain prairie brown soils < mountain prairie peaty soils. The increasing activity followed the trend of increasing precipitation and altitude. Vukhrer and Shamshiyeva (1968) also observed increasing enzyme activities with increasing precipitation and altitude and related this to increasing organic matter content of the soil. Speir (1977b) found that phosphatase activity was generally not significantly different in a climosequence of New Zealand soils but that activity tended to increase with increasing organic matter content. Herlihy (1972) found that sphagnum peat had a very much higher phosphatase activity than did mineral soil. Other studies have also shown that peats, or soils tending towards peats, appear to have the highest levels of phosphatase activity compared with other soils (Galstyan and Tatevosyan, 1964; Vukhrer and Shamshiyeva, 1968; Chunderova and Zubets, 1969; Gavrilova et al., 1974).

5.2. Soil pH

Herlihy (1972), in a study of peat at various stages of cultivation, measured phosphatase activity at its optimum pH and at the natural pH of the peat. He discovered that the further the natural pH differed from the optimum pH, the smaller the proportion of optimum pH activity found. Chunderova (1970) found that phosphatase activity in sod-podzolic soils decreased as soil pH increased, with maximum activity occurring at a soil pH of 4·2–4·5. These results indicate that soil phosphatase may be acting at a level well below its potential if the soil pH is far removed from the optimum pH of the enzyme, as commonly occurs. There is also evidence that soil enzymes can be inactivated when there is a large difference between soil pH and the enzyme optimum pH (cited in Khaziev, 1966).

5.3. Soil moisture

Phosphatase activity often appears to be higher in saturated soils than in dry soils, or soils with normal amounts of moisture (Gavrilova and Shimko, 1969; Daraseliya et al., 1975). Goian (1968) found, with a leached chernozem, that the phosphatase activity of samples kept dry for eight months did not differ significantly from those intermittently wetted, whereas activity was 27% (significantly) greater in a sample kept wet at minimum water capacity (field capacity).

The distribution of acid and alkaline phosphatases may differ in soils with different moisture contents. Fauvel and Rouquerol (1970) found that phosphatase activity had two pH optima (pH 6 and 9–10) in a wet soil under rice, whereas it had only one pH optimum (pH 6) in an adjacent plot of dry, uncultivated soil. The differences noted in the distribution of acid and alkaline phosphatases in German, Mediterranean and Iranian (Persian) alkaline soils may be due, in part at least, to a difference in the sensitivity of these enzymes to aridity (Hofmann and Kasseba, 1962; Hoffmann and Elias-Azar, 1965). The wetter German soils, as a rule, contained more acid than alkaline phosphatase, whereas the ratio was reversed in the drier Spanish, Egyptian and Iranian soils. In contrast, Arutyunyan and Galstyan

(1975) found alkaline phosphatase was present in saturated soils and acid phosphatase in unsaturated soils.

5.4. Soil depth

Phosphatase activity, like most other enzyme activities, decreases with soil depth (Keilling *et al.*, 1960; Burangulova and Khaziev, 1965a, c; Hoffmann and Elias-Azar, 1965; Skujiņš and McLaren, 1967; Goian, 1968, 1970; Vukhrer and Shamshiyeva, 1968; Fauvel and Rouquerol, 1970; Roizin and Egarov, 1972; Stebakova, 1972; Samokhalov, 1973; Arutyunyan and Galstyan, 1974; Kiss *et al.*, 1974; Minenko, 1974; Pukhidskaya and Kovrigo,

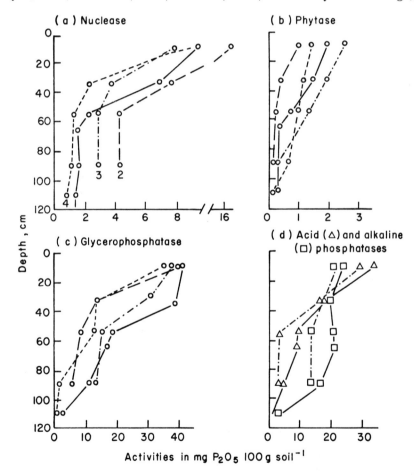

Fig. 1. Changes in the activity of phosphatase enzymes in the profile of different chernozems. (1) typical fertile chernozem; (2) carbonate chernozem; (3) leached chernozem; (4) ordinary chernozem. From Khaziev and Burangulova (1965).

1974a). Khaziev and Burangulova (1965) measured the activities of acid and alkaline phosphatase, nuclease, phytase and glycerophosphatase in four chernozem profiles. Fig. 1 shows the distribution of these activities in a typical fertile chernozem. The rate of decline of activity was much greater in the upper part of the profile than in the lower part. This is even more evident in soils with shallower A horizons than are found in chernozems (see Arutyunyan and Galstyan, 1974 and 1975 for comparisons of different soil types).

Khaziev and Burangulova (1965) found that the distribution of phosphatase activities corresponded with the distribution of microorganisms in the profiles. Definite positive relationships between phosphatase activity and organic matter content in soil profiles have also been found (Galstyan and Tatevosyan, 1964; Arutyunyan and Galstyan, 1975). The decrease in phosphatase activity with depth can be mainly attributed to the diminution of biological activity down the profile. Inactivation of enzymes by clay minerals in the deeper horizons may be partly responsible for the different distribution patterns of the different enzymes with depth (Khaziev and Burangulova, 1965).

5.5. Microbial numbers

Although many of the properties of phosphatases discussed in previous sections have been attributed, at least in part, to changes in microbial numbers and populations, attempts to relate soil phosphatase activity to microbial numbers have yielded contradictory results. Burangulova and Khaziev (1965c) and Khaziev and Burangulova (1965) proposed that the decline in phosphatase activity down a soil profile can, to some extent, be explained by a drop in microbial numbers. Pukhidskaya and Kovrigo (1974a) found a significant positive correlation between phosphatase activity and microbial numbers in soil profiles, whereas Roizin and Egarov (1972) did not. In comparative studies on different soils, Aliev and Gadzhiev (1973) and Panikov and Aseeva (1974) reported that phosphatase activity increased with increasing numbers of microorganisms but Herlihy (1972), Sanikidze et al. (1973) and Samokhalov (1973) found no such relationship. Hoffmann and Elias-Azar (1965) and Hofmann and Kasseba (1972) considered without evidence, that different microbial populations could explain, in part, the different distribution of acid and alkaline phosphatases in German, Mediterranean, and Iranian alkaline soils. Beck (1974) and Ramírez-Martínez and McLaren (1966b) found no relationship between microbial numbers and phosphatase activity throughout the plant growth season. Microbial numbers fluctuated strongly, whereas phosphatase activity, although it varied somewhat, was much more uniform (Fig. 2.).

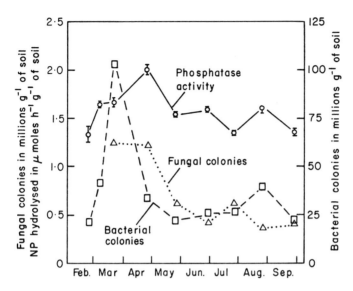

Fig. 2. Microbial plate counts and phosphatase activity of Oxford Tract soil as a function of sampling date. From Ramírez-Martínez and McLaren (1966b).

Kudzin *et al.* (1968) recorded that the application of ammonium nitrate and potassium chloride to soils for thirty-five years increased phosphatase activity and the numbers of microorganisms attacking organic phosphorus and Stefanic *et al.* (1971) found that the addition of lime and fertilizer also increased both phosphatase activity and microbial numbers. However, Verstraete *et al.* (1974) showed that certain slow release N-fertilizers reduced soil phosphatase activity, even though they stimulated microbial growth. Kramer and Yerdei (1959) found that, when soil was incubated with sucrose and ammonium nitrate, microbial numbers increased but phosphatase activity was unchanged, whereas Panikov and Aseeva (1974) found that the application of easily available compounds, such as glucose or peptone, to soils increased both microbial counts and nuclease activity.

5.6. Chemical properties

Since phosphate inhibits soil phosphatase activity and phosphate fertilizer can, as stated previously, cause a decrease in soil phosphatase activity, an inverse relationship between soil phosphate content and phosphatase activity might be expected. Correlation coefficients between phosphatase activity and total soil inorganic phosphate have not been measured, probably because of the difficulty of measuring soil phosphate. However, correlations between phosphatase activity and the so-called plant available fraction of inorganic phosphate, variously called available-P, mobile-P and soluble-P, have been measured, with contradictory results. Hofmann and Kasseba (1962) found a highly significant negative correlation between phosphatase activity and soluble-P in German soils but an equally significant positive correlation between these factors in Egyptian soils. Inverse relationships have been found by other workers (Kramer, 1957; Mamytov et al., 1974) and Kramer (1957) proposed using the inverse relationship between phosphatase activity and available-P to estimate biologically useful P in soil. However, other positive correlations between soil phosphatase activity and available- or mobile-P have also been found (Khaziev, 1972b; Ponomareva et al., 1972; Arutyunyan and Galstyan, 1974), while Burangulova and Mukatanov (1973) found no correlation at all.

Soil phosphatase activity is generally directly related to soil organic matter (organic carbon) content in soil profiles (Galstyan and Tatevosyan, 1964; Kanivets et al., 1973) and in different soils (Keilling et al., 1960; Galstyan and Tatevosyan, 1964; Gavrilova and Shimko, 1969; Khan, 1970; Vasyuk, 1971; Aliev and Gadzhiev, 1973; Laugesen and Mikkelsen, 1973; Kiss et al., 1974; Arutyunyan and Galstyan, 1975). Hoffmann and Elias-Azar (1965) found significant positive correlations ($P < 0.001$) between both alkaline and acid phosphatases and soil organic matter content in North Iranian soils, and Nannipieri et al. (1973) found a significant positive correlation ($P < 0.001$) between phosphatase activity and organic carbon content in thirty-two Italian soils. Speir (1977b) showed that phosphatase activity correlated highly significantly with organic carbon content, both within and between sites, in a climosequence of soils in New Zealand. There have been reports, however, of no significant correlation (Sanikidze et al., 1973), or even a negative relationship (Kozlov et al., 1973) between phosphatase activity and organic carbon content.

The relationship of phosphatase activity to total nitrogen in soil follows that for organic carbon content (Keilling et al., 1960; Hoffmann and Elias-Azar, 1965; Speir, 1977b), presumably because nitrogen is closely related to organic content rather than to the enzyme itself.

The relationship of phosphatase activity to soil organic phosphorus content has been the subject of several studies. Khaziev and Burangulova (1965) recorded inverse relationships between organic phosphorus content and soil nuclease, phytase, glycerophosphatase and phosphatase activities. Mamytov et al. (1974) found, in certain Russian cultivated soils, that alkaline phosphatase activity was inversely related and that acid and neutral phosphatase activities were directly related, to organic phosphorus content. Burangulova and Mukatanov (1973) found no correlation between phosphatase activity and organic phosphorus content of mountain chernozem soils of the Southern Urals and Burangulova and Khaziev (1965a) found that the nuclease activity of chernozems was not related to their organic phosphorus content. In most studies, however, significant positive correlations have been found between phosphatase activity and soil organic phosphorus content (Hoffmann and Elias-Azar, 1965; Gavrilova *et al.*, 1973; Arutyunyan and Galstyan, 1974, 1975; Gavrilova *et al.*, 1974).

Halm *et al.* (1971) found a seasonal relationship between phosphatase activity and the $NaHCO_3$-soluble organic phosphorus fraction. Increases of this fraction coincided with periods of maximum plant growth and highest microbial activity and the authors suggest that $NaHCO_3$-soluble organic phosphorus might represent a form of organic phosphorus which is available to plants. Khaziev (1967) demonstrated a definite positive relationship between nuclease activity and the loss of soil phosphorus to plants during a growing season and concluded that levels of nuclease activity may indicate the direction and rate of biochemical processes associated with mobilization of organic phosphorus in soil.

6 The effects of plants on soil phosphatase

There is no doubt that the presence of plants, and the type of plants, grown in a soil have a marked effect on its enzyme activities, including phosphatase (Voets and Dedeken, 1966; Gavrilova and Shimko, 1969; Khaziev, 1969; Vlasyuk and Lisoval, 1969; Khan, 1970; Soreanu, 1972; Gavrilova *et al.*, 1973; Neal, 1973; Beck, 1974; Dudchenko *et al.*, 1974; Kiss *et al.*, 1974). This effect is indirect and is caused by many different factors. Basically, however, changes in phosphatase activity are related to changes in soil organic matter content and microbial populations, brought about by the plants.

Rhizosphere phosphatase activities tend to be higher than are activities remote from the plant roots (Goian, 1969; Lisoval, 1970; Kiver, 1973; Kiss, *et al.*, 1975b), because of increased microbial numbers in the rhizosphere (Greaves and Webley 1964), and possibly because of higher phosphatase activities of the rhizosphere microorganisms (Petrenko, 1973; Teslinova and Pervushina-Grosheva, 1973) and the excretion of plant root enzymes (Neal, 1973). The stage of plant growth can affect soil phosphatase activity, with activity being high when growth is most intensive. However, this may be primarily a climatic effect on overall biological activity. Several workers have found positive correlations between crop yield and soil phosphatase activity (Khaziev, 1967; Stebakova, 1972; Samokhalov, 1973; Gavrilova *et al.*, 1975), but negative correlations have also been reported (Gubenko *et al.*, 1973; Kiss *et al.*, 1975 b).

7. Miscellaneous factors influencing soil phosphatase

7.1. Soil erosion

The work of Simonyan and colleagues has shown that erosion results in decreased soil phosphatase and other enzyme activities (Simonyan and Galstyan, 1974a, b; Arutyunyan and Simonyan, 1975), because of the loss of the enzyme in the soil surface layers. Arutyunyan and Simonyan (1975) have suggested that alkaline phosphatase activity is a good indicator of soil erosion in carbonate-containing chernozems.

7.2. Cultivation

Shcherbakova *et al.* (1974) found that phosphatase activity increased with roto-tilling and ploughing of peats, whereas Herlihy (1972) found that it decreased in cut-over sphagnum peats; if the cut-over peat was subsequently cultivated, phosphatase activity fell even more substantially. Herlihy also found that the optimum pH for phosphatase activity differed from the original in the cut-over peat.

7.3. Season

Ramírez-Martínez and McLaren (1966b) observed that phosphatase activity in Oxford Tract soil was highest in spring (April/May) (see Fig. 2). Other reports record that soil phosphatase activity can increase through spring to a maximum in the summer months and then fall in autumn (Aseeva and Vanyarkho, 1969; Ponomareva *et al.*, 1972; Samokhalov, 1973; Pukhidskaya and Kovrigo, 1974b; Kiss *et al.*, 1975). This seasonal pattern seems to depend upon many factors, such as aeration, soil moisture content, vegetation and microflora (Aseeva and Vanyarkho, 1969). A significant correlation between soil nuclease activity and seasonal temperature was found by Khaziev (1969).

V. Soil Organic Sulphur

Sulphur is no less important as a plant nutrient than is phosphorus, although it has received far less attention in soil–plant studies. Its role in plant metabolism has been reviewed recently by Anderson (1975c). A large proportion of soil sulphur may be organically bound (Anderson 1975b), especially in non-calcareous soils of the humid regions (Scott and Anderson, 1976). Virtually nothing is known about the structure of soil organic sulphur, and it is possible, unlike organic phosphorus, that it may exist predominantly as an integral part of the soil organic matter, and consequently be very resistant to extraction and degradation (Scott and Anderson, 1976). Although soil organic phosphorus is thought to be exclusively in the form of phosphate, not all organic sulphur is sulphate (Anderson, 1975b); under aerobic conditions, however, sulphate may well be the major form.

There are two "fractions" (not extractable as such from soil) that together account for most of the soil organic sulphur; the "HI-reducible" fraction (the reductant is actually a mixture of hydriodic, formic and hypophosphorous acids) and the "Raney-nickel reducible" fraction. The former is predominantly O-bonded sulphur (ester-sulphate), with possibly some N-bonded sulphur, and the latter is C-bonded sulphur (Anderson, 1975b). The "HI-reducible" fraction constitutes 50% (range 30–70%), 64% (range 44–88%) and 50% (range 31–63%) of the total organic sulphur in some soils from Australia (Freney, 1961; Williams, 1975), Scotland (Scott and Anderson, 1976) and U.S.A. (Tabatabai and Bremner, 1972a), respectively. The metabolism of sulphate esters has recently been reviewed (Fitzgerald, 1976). Compounds such as sulphated polysaccharides, choline sulphate and ethereal sulphates have been tentatively suggested as components of this fraction (Barrow, 1960; Freney, 1961; Anderson, 1975b).

VI. Plant Root and Microbial Sulphatases

It is generally accepted that plants take up their sulphur as inorganic sulphate and the availability of soil organic sulphur thus depends upon its mineralization (Williams, 1975). Freney et al. (1975) found that plants were able to obtain their sulphate from both ester-sulphate and C-bonded sulphur in soils. When the plants were utilizing indigenous organic sulphur, 60–63% was obtained from the C-bonded sulphur fraction. However, they showed, in an incubation study, that ^{35}S-labelled inorganic sulphate was largely incorporated into ester-sulphate and it was from this fraction that plants later derived all of the labelled-S taken up. Obviously, then, soil ester-sulphate is a major source of plant-available sulphur and a mechanism for its hydrolysis is therefore required.

Plant roots possess arylsulphatase activity (Poux, 1966; Hall and Davie, 1971) and there is evidence that some, at least, is located extracellularly, i.e., outside the permeability barrier of the root cells (Hall and Davie, 1971). Pure cultures of many microorganisms found in soil have also been shown to produce sulphatase (Whitehead et al., 1952; Harada and Spencer, 1962; Harada, 1964; Benkovic et al., 1971; Houghton and Rose, 1976).

There are many interacting factors involved in the transformations of sulphur, as of phosphorus in soil, and mineralization of organic sulphur to a form available for plant uptake involves far more than a straightforward hydrolysis to inorganic sulphate. The sulphatase-catalysed hydrolysis of ester-sulphate, like the hydrolysis of organic phosphorus, probably requires an initial breakdown of soil organic matter, necessitating an active microbial population producing other hydrolytic enzymes.

VII. Soil Sulphatases

1. Terminology

This part of the review will be concerned with sulphuric acid hydrolases (EC 3.1.6) found in soil. Most work reported has been on arylsulphatase (EC 3.1.6.1), since, apart from the study by Houghton and Rose (1976), the only substrate used has been p-nitrophenyl sulphate. Houghton and Rose used a number of substrates, and called the enzymes catalysing their hydrolysis, "sulphohydrolases". While there is no objection to this terminology, we prefer to call the enzymes "sulphatases".

2. History

Soil sulphatase activity was discovered by Tabatabai and Bremner (1970a), who were prompted to look for this enzyme when they realized that much of the mineralizable soil organic sulphur is in the ester-sulphate form. They described a simple incubation and assay procedure for measuring soil arylsulphatase activity, using p-nitrophenyl sulphate as substrate and presented such a comprehensive analysis of their method that, in almost all subsequent studies, it has been used with only minor, if any, modifications (e.g., Cooper, 1972; Galstyan and Bazoyan, 1974; Bremner and Zantua, 1975; Kowalenko and Lowe, 1975; Speir and Ross, 1975; Thornton and McLaren, 1975). Houghton and Rose (1976) studied soil sulphatase activities using a number of ^{35}S-labelled substrates and assaying activity either spectrophotometrically, or by scanning electrophoretograms for radioactivity.

3. Factors involved in the measurement of soil sulphatase activity

Tabatabai and Bremner (1970a) found that the pH optimum of soil sulphatase was pH 6·2 and they used an acetate buffer, pH 5·8 in their incubations. This pH was chosen because of the greater buffering capacity of acetate at pH 5·8; the amount of substrate hydrolysed at this pH was not significantly less than that hydrolysed at pH 6·2. They also showed that the $CaCl_2 + NaOH$ solution used in the assay allowed quantitative recovery of p-nitrophenol. However, Galstyan and Bazoyan (1974) do not appear to have accepted this result and they outlined a method in which the soil is removed by filtration before NaOH is added; they warned against over-titration with NaOH, recommending that the final pH be no higher than 8·5–8·6.

The incubation time recommended by Tabatabai and Bremner (1970a) was only one hour, so soil "sterilization" probably would have been un-necessary; however, toluene was included in the procedure and was found to increase sulphatase activity. All subsequent studies using this method have incorporated toluene, except that of Galstyan and Bazoyan (1974). Tabatabai and Bremner (1970a) found that 5 Mrad of γ-radiation reduced sulphatase activity in six air-dried soils by an average of 27%. Houghton and Rose (1976) did not use any form of soil "sterilization" in their study of soil sulphatase activities.

The effect of air-drying soil on its sulphatase activity has been studied by Tabatabai and Bremner (1970b) and Speir and Ross (1975). Tabatabai and Bremner found that air-drying increased sulphatase activity by an average of 43% in thirteen soils. Speir and Ross also found that activity increased, but only by 4–15%. This behaviour is in sharp contrast to that of soil phosphatase, and emphasizes that the effect of air-drying depends very much upon the particular enzyme. A possible explanation for the increase in sulphatase activity is that re-wetting of air-dry soil causes a breakdown of aggregates, thus increasing accessibility of enzyme to the substrate (Tabatabai and Bremner, 1970b).

Tabatabai and Bremner (1970b) studied the effects on sulphatase activity of storage of thirteen soils field-moist at room temperature, 5°C or −10°C, or air-dry at room temperature. Of these treatments, only storage at −10°C did not result in considerable changes of sulphatase activity over three months and this was recommended as the best means of soil storage prior to assay of this enzyme. Speir and Ross (1975), however, in a similar storage study, found that storage of three soils at −20°C resulted in sub-stantial and highly significant losses of sulphatase activity over 220 days. They included an extra treatment, storage air-dry at 4°C, and subsequently recommended this as the best means of retaining the sulphatase activity of these soils.

4. Properties of soil sulphatases

The pH optimum of soil sulphatase appears to be in the range pH 5·4–6·2 in a number of soils (Tabatabai and Bremner, 1970a; Galstyan and Bazoyan, 1974; Thornton and McLaren, 1975).

Soil sulphatase, like phosphatase and urease, is active at $-10°$ and $-20°C$ but not at $-30°C$ (Bremner and Zantua, 1975). Its temperature optimum was about $67°C$ in six soils which was similar to the temperature optima of several other soil enzymes.

Tabatabai and Bremner (1970a, b) found that sulphatase activity was destroyed in six soils by steam sterilization ($120°C$ for 30 min) or by boiling water ($100°C$ for 2 h). However, activity fell by an average of only 54% in thirteen soils that were subjected to dry heat ($105°C$) for 24 h.

Tabatabai and Bremner (1970a) tested the effects of various ions on soil sulphatase activity. The mercuric ion (4 mM) completely inhibited, and phosphate (5 mM) substantially inhibited, sulphatase activity. Sulphite and cyanide ions also inhibited to some extent but only at much higher concentrations, while sulphate and chloride had little, if any, effect. The action of these ions on soil sulphatase is similar to their action on arylsulphatases from other sources (Tabatabai and Bremner, 1970a). Tabatabai and Bremner (1970b) also tested several reagents used as extractants in determinations of the sulphur status of soils. Most did not deactivate the enzyme. The effects of the strongly alkaline reagents could not be assessed, however, because the dark colour they extracted from soil interfered with the colorimetric assay of p-nitrophenol.

Houghton and Rose (1976) added different types of sulphate esters to unsterilized soil and found that not all were immediately hydrolysed. Only dodecyl sulphate and the arylsulphate esters appeared to be hydrolysed by enzymes already present in soil. There was a delay of several hours before an appreciable rate of hydrolysis of the other substrates occurred, indicating a requirement for synthesis of enzymes by microorganisms and a limited substrate specificity of the native sulphatases present in soil.

5. The effects of soil properties on sulphatase

Galstyan and Bazoyan (1974) found that a solonetz-solonchak soil had no sulphatase activity, and that activity increased through a sequence of eight Armenian soils to be highest in a mountain–meadow prairie soil. Activity was strongly correlated with organic matter content ($r=0·93$, $P < 0·001$) in these soils. Speir (1977b) found significant differences in sulphatase activity between members of a climosequence of New Zealand soils (cf. phosphatase). The drier members of the sequence had low activities and the wetter podzolized members had high activities. The trend, however, was not so much one of increasing sulphatase activity with increasing precipitation but one of increasing activity with increasing soil moisture and organic matter contents.

Sulphatase activity is significantly correlated with organic matter content (organic carbon) in different soils (Tabatabai and Bremner, 1970b;

Cooper, 1972; Galstyan and Bazoyan, 1974; Speir, 1977b). Speir (1977b) also reported that sulphatase activity was correlated significantly with soil moisture content and total nitrogen, whereas Tabatabai and Bremner (1970b) found no significant relationship with % nitrogen, % clay or % sand.

Sulphatase activity decreases with depth in soil profiles (Tabatabai and Bremner 1970b; Galstyan and Bazoyan, 1974). In six profiles, activity was significantly correlated with organic carbon content ($r=0.78$; $P < 0.001$) (Tabatabai and Bremner, 1970b).

Correlations of sulphatase activity with various soil sulphur fractions have been determined. Activity does not appear to correlate significantly with total sulphur (Tabatabai and Bremner, 1970b; Galstyan and Bazoyan, 1974). It gave a barely significant negative correlation with inorganic sulphate (Galstyan and Bazoyan, 1974); this latter correlation, however, was strongly influenced by the extremely high sulphate content of an inactive solonetz soil. Galstyan and Bazoyan (1974) found a highly significant positive correlation between soil sulphatase activity and organic sulphur content. Cooper (1972) also found a significant correlation between sulphatase activity and this fraction ($r=0.56$, $P < 0.01$) in twenty soils, but there was a far more significant correlation between activity and the "HI-reducible" (ester-sulphate) fraction of organic S ($r=0.88$; $P < 0.001$). This could indicate a very important role for soil sulphatase in the mineralization of organic sulphur. However, incubation studies of soils in the absence of plants have shown no significant relationship between sulphatase activity and the amount of sulphur mineralized (Tabatabai and Bremner, 1972b; Kowalenko and Lowe, 1975). In contrast, Speir (1977b) found, in a pot trial using six soils with low levels of available sulphate, that soil sulphatase activity may be a useful indicator of the mineralized sulphur that has been taken up by plants.

Speir (1977b) also found that soil sulphatase activity correlated very significantly with phosphatase and urease activities, both between and within sites; correlations with oxygen uptake and dehydrogenase, invertase and amylase activities at the same tussock–grassland sites were also positive, and mainly significant (Ross et al., 1975a; Ross 1975). Ladd and Butler (1972) but not Ross and McNeilly (1975), found that sulphatase activity correlated significantly with protease activities.

6. Seasonal changes in soil sulphatase activity

Cooper (1972) studied the relationship between sulphatase activity and rainfall in three Nigerian soils, kept free from plants, from February to December. Periods of wetting and drying caused by widely dispersed rainstorms resulted in a decrease in sulphatase activity. When rainfall became more frequent and soils remained moist, sulphatase activity increased, reaching a maximum towards the end of the rainy season. Activity declined again during the dry season. Cooper suggested that seasonal variations in soil conditions may be more important than soil chemical properties in controlling the level of sulphatase activity in soils. However, these tropical soils were subject to severe seasonal variations of moisture content; the seasonal effect may be less important for soils with a more uniform climate.

VIII. The State of Phosphatase and Sulphatase in Soil

There have been several investigations in recent years that have led to a better understanding of the state of enzymes in soil (see reviews, Skujiņš, 1967; Ramírez–Martínez, 1968; Hayano, 1973; Sequi, 1974; Ladd and Butler, 1975), but only a very small proportion of this work has involved phosphatase or sulphatase activities. Evidence concerning the nature of soil enzymes can be gathered in a number of ways but the major techniques that have relevance to phosphatase and sulphatase are: enzyme extraction from soil, studies on adsorption of enzymes by mineral and organic soil components and studies of kinetic parameters of soil enzymes.

1. Extraction of phosphatase and sulphatase activities from soil

Although some other enzymes have been partially extracted from soil, there have been no reports of successful extraction of phosphatase or sulphatase activities. Fractionation studies have, however, yielded soil fractions enriched in these activities (Ladd and Paul, 1973; Aliev et al., 1975; Speir, 1977a).

In a fractionation study of a soil that had been incubated, and which consequently contained enhanced enzyme activities, Ladd and Paul (1973) found that most of the 43% of the phosphatase activity recovered was located in the two coarsest fractions ($>$ 2 μm and 0·2–2 μm diameter). These fractions were thought to contain intact microbial cells, cell debris and microbial metabolites adsorbed to larger soil particles. Because only a small proportion ($<$ 10%) of the recovered activity was associated with the soil colloids (smaller particle size fractions), Ladd and Paul suggested that newly synthesized phosphatase may be acting intracellularly. Results with other enzymes, where fractionation recoveries were greater, also suggest that the larger soil particle-size fractions may well be the major site of enzyme activity. McLaren and Skujiņš (1971) reported the glycerophosphatase activity of various particle size fractions of a Californian soil. After correction of results for adsorbed phosphate, all of the enzyme activity was recovered in these fractions. The silt fraction (2–20 μm) contained the largest amount of activity and organic matter; the clay ($< 2 \mu$m) and silt fractions were the most active per unit weight. These fractions correspond in size to the coarse fractions observed by Ladd and Paul (1973) to contain most of the soil phosphatase activity. In addition, McLaren and Skujiņš (1971) reported the presence of a mucilaginous fraction, containing bacteria, which constituted only 1% of the fractions by weight and was no more active per unit weight than was the whole soil.

Aliev et al. (1975) used ultrasonic dispersion to remove 75% of the microbial cells from a sod-podzolic soil. Only 51% of the phosphatase activity was recovered by this treatment, 37% in the "cell-free" soil and 14% in the cells. These recoveries were corrected for the 25% of cells remaining in the soil fraction. There was no supernatant (soluble) phosphatase activity, although recovery of some of the other enzymes was quite high in this fraction (e.g. urease, 17·5%). Aliev et al. concluded that the enzyme activity of soil is largely due to extra-cellular enzymes strongly retained by soil particles. However, it is impossible to state which soil fraction(s) would have contained the labile phosphatase lost by the ultrasonic treatment.

Speir (1977a) separated by flotation a "light fraction" from soil that had been ultrasonically dispersed in a heavy liquid. This fraction comprised a large proportion of the plant fragments and cellular debris of the soil. In ten soils, the mean recovery of phosphatase activity was 79%, and of sulphatase activity 62%, in the light fraction and mineral residue. The light fraction was considerably enriched in phosphatase activity but usually only slightly enriched in sulphatase activity, compared with the unfractioned soil. However, most of the recovered activity was retained in the mineral soil residue.

2. Adsorption of phosphatase activity

Adsorption of proteins by clay minerals and organic complexes is well known (Ladd and Butler, 1975). Adsorption of non-soil phosphatases by clay minerals appears to result in a loss of activity (Mortland and Gieseking, 1952; Ramírez–Martínez, 1966; Kiss et al., 1974; Panikov and Aseeva 1974). However, Kroll and Krámer (1955) and Kroll et al. (1955) found that the addition of clay minerals to soils did not affect soil phosphatase activity. This may indicate that the soil phosphatase is already largely in an adsorbed state. Scheffer et al. (1962) observed that extremely low concentrations of synthetic humic acids inhibited potato and wheat acid phosphatase but activated alkaline phosphatase. Saalbach (1956) found that humic acids stimulated root phosphatase activity.

Parks (1974) added various clay minerals to *Escherichia coli* alkaline phosphatase and found that 76–100% of the activity was adsorbed. The complexed enzymes had higher temperature optima but lower temperature coefficients, than did the free enzyme. The complexes also generally had higher values of the Michaelis constant (K_m) than did the free enzyme.

3. Kinetics of soil phosphatase and sulphatase

The specific activity of an enzyme is almost always decreased as a result of becoming attached to, or entrapped within, an insoluble matrix (Ladd and Butler, 1975). This reduction in specific activity is manifested by an increase in the K_m value of the enzyme.

Michaelis constants of soil phosphatase and sulphatase have been determined in a number of studies (Tabatabai and Bremner, 1971; Cervelli *et al.*, 1973; Brams and McLaren, 1974; Cervelli *et al.*, 1975; Thornton and McLaren, 1975; Kishk *et al.*, 1976). Tabatabai and Bremner (1971) found that their data fitted the linear transformation of the Michaelis–Menten equation and gave K_m values for phosphatase and sulphatase that were quite similar in nine soils. The uniformity of the K_m values suggested that the phosphatases and also the sulphatases in these soils, were similar in type and origin or that their K_m values were affected by, and made more uniform through, association of enzymes with soil constituents. The K_m values did not correlate significantly with other soil properties (pH, cation-exchange capacity, organic carbon, clay or sand contents).

Kishk *et al.* (1976), in a study of soil phosphotriesterase activity towards methyl parathion, found a discontinuity in their Michaelis–Menten plot and the two lines thus derived gave K_m values of 0·125 and 0·5 mM. They suggested, without evidence, that these may be attributed to free enzyme and adsorbed enzyme, respectively.

Nannipieri *et al.* (1972) suggested that, since the substrate p-nitrophenyl phosphate is adsorbed by soil, Michaelis–Menten kinetics cannot be applied without a correction factor. Cervelli *et al.* (1973) used adsorption parameters derived from the Freundlich isotherm to determine substrate concentrations and calculated corrected K_m values for soil phosphatase. These values (K_{m_c}) were somewhat lower than those found if substrate adsorption was ignored and were close to values reported for acid phosphatases from plants and microorganisms. Cervelli *et al.* (1975) presented an enlightened discussion on the relationship between K_{m_c} values and the locus of the phosphatase enzyme in soil. They measured the K_m and K_{m_c} values and also the specific surface of the soil and used these to "locate" phosphatase activity in different soils. In nine soils, K_{m_c} ranged from 0·35–5·4 mM. It was assumed that 0·35 mM is the true value for soil phosphatase and that the higher values are caused by diffusion phenomena, a factor known to increase K_m values. Thus, the higher the K_{m_c} value, the more the substrate has to diffuse to have contact with the enzyme. If a soil has a high K_{m_c} value and a high or normal specific surface, then its "active" organic material (i.e., that with biological, including phosphatase, activity) must be occluded by minerals or "inactive" organic components. In other words, diffusion is not limited by physical surface phenomena but by masking of active organic material by minerals or inactive organic material. If the specific surface is low but K_{m_c} is normal (approximately 0·35 mM), there must be an external distribution of active organic material, that is, on the outside of the particles; diffusion is then not required. Intermediate situations were found in some of the other soils examined.

Irving and Cosgrove (1976) examined the graphical techniques used by other workers and concluded that the supposition that soil phosphatase activity follows Michaelis–Menten kinetics is based on doubtful premises. They have found evidence for a model capable of accounting for the deviation of Michaelis–Menten kinetics that they observed in a kraznozem soil.

4. Conclusions

The results of Ladd and Paul (1973) and Aliev et al. (1975) suggest that a considerable proportion of soil phosphatase activity may be contained in, or attached to, microorganism cells and cellular debris. However, Ramírez-Martínez and McLaren (1966b) have shown that, in order to have a phosphatase activity comparable to that usually found in 1 g of soil, it would be necessary to have huge amounts of specialized microorganisms producing the enzyme. The calculated amounts were 10^{10} bacteria or 1 g of fungal mycelia! Therefore, they proposed an accumulation, or enrichment, of phosphatase activity in soil and postulated a stable extracellular component in soil humus. By analogy with other soil enzymes and from the small amount of evidence available concerning the state of soil phosphatase and sulphatase, the extracellular components of these enzymes appear to be intimately associated with the organo–mineral complex of soil. A diagrammatic representation of a possible form of this enzyme–organo–mineral complex is given by Sequi (1974). The results of Cervelli et al. (1975) suggest that the nature of this complex may vary from soil to soil.

IX. The Origin of Soil Phosphatase and Sulphatase

It is obvious that the extracellular component of soil phosphatase and sulphatase must have originated in living tissue. This assumes that the hydrolysis of substrates used for their assay is not due to catalysis by nonprotein soil components, an assumption that may not be completely justified. The ability of microorganisms and plant roots to release phosphatase and sulphatase activities into the soil has been covered in previous sections. Earthworms produce acid and alkaline phosphatases (De Jorge and Sawaya, 1967), and other soil animals would undoubtedly synthesize both phosphatase and sulphatase. Thus, there is no shortage of source organisms for these enzymes in soil but their relative importance has yet to be assessed. It seems logical to assume that microorganisms supply most of the soil enzyme activity since, with their large biomass, high metabolic activity and short lifetime under favourable conditions, they should produce and release far more enzymes than other organisms. Microorganisms also have the capacity to produce enzymes that enable them to assimilate compounds not initially degraded, as found by Houghton and Rose (1976) with various sulphate esters added to soil. Microorganisms also have the ability to increase their numbers dramatically in a very short time under favourable conditions. Ladd and Paul (1973) incubated soil at 22°C with glucose and sodium nitrate and found that bacterial numbers increased by almost two orders of magnitude in only 1·5 days. This increase was accompanied by a 3·2-fold increase in phosphatase activity. This newly synthesized phosphatase activity, however, had a short half-life and rapidly declined to a low level after twenty-one days. This suggests that a great deal of the increased activity during microbial proliferation is subsequently lost as the microorganisms die and lyse, presumably by inactivation and digestion by other microorganisms. Therefore, a great many flushes of microbial activity may be required to cause a significant permanent increase in the level of extracellular phosphatase activity.

Casida (1959) found that *Aspergillus niger* phytase and nuclease activities had pH optima of 2·5 and 4·0 respectively; these optima are considerably lower than those observed for soil phytase and nuclease. However, McLaren (1960) has shown that the optimum pH of an enzyme adsorbed on kaolinite is about two units higher than that of the free enzyme. This pH shift, together with the observation that bacterial acid phosphatases have higher pH optima than do the fungal enzymes, have resulted in proposals by Khaziev and Burangulova (1965) and Fauvel and Rouquerol (1970) that soils in which acid phosphatase predominates have high fungal populations and that soils in which neutral or alkaline phosphatases predominate have high bacterial populations. Adebayo (1973) also suggested that the acid phosphatase produced by soil bacteria could be responsible for the neutral phosphatase response shown by many agricultural soils.

Although microorganisms appear to be the major producers of soil phosphatase and sulphatase activities, the possibility of an important contribution from plants cannot be discounted. Indirectly, of course, plants influence microbial populations, with higher numbers and often higher enzyme activities being found in the rhizosphere soil than in soil remote from the roots. Plants also, through their photosynthetic systems, provide the soil with carbon, essential for microbial growth, and they also provide other nutrients. The possible role of plants as direct producers of phosphatase and sulphatase activities was examined by Speir (1976) and Speir and Ross (1976) (see Fig. 3). Their results indicated that the sulphatase activity of tussock green leaves was negligible and that the increasing activity found with increasing age of dead leaves was caused by microbial proliferation. In contrast, phosphatase activity in the tussock green leaves

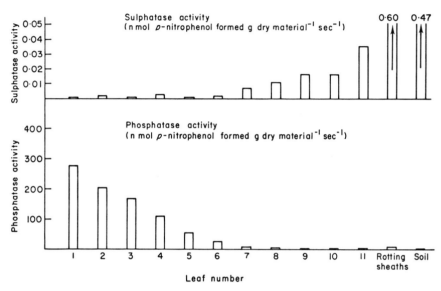

Fig. 3. Sulphatase and phosphatase activities of living and dead leaves of the snow-tussock *Chionochloa rigida*, and of soil from the same site. Leaves increased in age from numbers 1–11. Numbers 1–6 were living and numbers 7–11 were dead. "Rotting sheathes" comprised the wet, extensively decomposed material at the base of the tussock leaves.

was extremely high, and, on a dry weight basis, very much higher than in the soil. Phosphatase activity declined markedly with increasing leaf age. It was not possible to determine whether the phosphatase activity of the fallen litter was due to residual plant enzyme or to phosphatase activity of decomposer microorganisms. The roots of these plants were thought to be a far more likely source of soil phosphatase and sulphatase activities than were the leaves, because of their location within the soil at all times. Both phosphatase and sulphatase activities were quite high in the roots, phosphatase being higher than in the soil and sulphatase about the same or slightly lower on a dry weight basis. Again, there was no way of determining whether this activity was due to plant enzymes or to associated microbial enzymes. However, Speir (1976) concluded that the tussock root could contribute significantly to soil phosphatase, and possibly sulphatase, activities, after considering the results of Ridge and Rovira (1971), who found that microorganisms did not positively enhance root phosphatase activity.

X. The Role of Soil Phosphatase and Sulphatase

1. Function

Adebayo (1973) found that the amount of organic phosphorus mineralized in Nigerian soils which had been incubated at 35°C for four months, was not related to their phosphatase activity. Also, Tabatabai and Bremner (1972b) and Kowalenko and Lowe (1975) have observed no relationship between mineralized organic sulphur and sulphatase activity, in soil incubation studies. These results indicate that hydrolytic cleavage of the ester bond may not be the rate-limiting step in the mineralization process. Soeminto (1972) has suggested that the rate of mineralization of C-bonded sulphur might be the rate-limiting factor in the supply of sulphur to plants, from soils that have low levels of available sulphate.

Phosphatase preparations, added to soil, do not appear to release phosphate from soil organic phosphorus compounds (Jackman and Black, 1952b; Appiah and Thompson, 1974), although there is evidence that the native soil phosphatases do (Stefanic 1971). Hayashi and Takijima (1955) found that organic phosphorus extracted from soil was only partially hydrolysed by rice-bran phosphatase. Houghton and Rose (1976) showed that sulphatases from diverse species had no activity towards humic acid, although ester-sulphate could be demonstrated in this fraction.

After considering the complexity of soil organic matter, Appiah and Thompson (1974) proposed that the mineralization of organic phosphorus is principally a microbial phenomenon, rather than a purely chemical enzymic one. They suggested that phosphatase activity becomes important only after initial breakdown of soil organic matter, presumably catalysed by a host of microbial enzymes. It is probably this initial breakdown, or some part of it, that is the rate-limiting step of organic phosphorus mineralization, and probably also of organic sulphur mineralization. Halm et al. (1971) found that a large proportion of organic phosphorus in a calcareous soil could be mobilized if a suitable energy source, such as fresh plant litter, was available, and Thompson et al. (1954) showed that the organic phosphorus mineralized in a soil, correlated very significantly with the nitrogen and carbon mineralized concurrently. These results tend to support the hypothesis of Appiah and Thompson (1974).

The hydrolysis of ester-phosphate and sulphate is an essential part of the overall mineralization process, so, in spite of the above results, phosphatase and sulphatase activities have a vital, if only secondary, role to play. The relative importance of microbial, plant or cell-free soil enzymes in this secondary process cannot yet be assessed. Ridge and Rovira (1971) indicated that root surface enzymes, by acting in the immediate root environment, may be more important in organic phosphorus mineralization than the larger amounts of enzymes distant from the roots and bound to clays and organic matter. Other results indicate that microbial phosphatase and sulphatase, under conditions for microbial growth, may also be more important than the cell-free soil enzymes. However, under conditions that normally exist in soil, cell-free phosphatase and sulphatase appear to constitute the major proportion of these enzymes. They should, therefore, play a considerable role in the constant mineralization of organic matter which occurs as part of the nutrient cycle in the soil–plant system.

2. Diagnostic value

Evidence presented in the previous section would tend to negate any possibility of using soil phosphatase or sulphatase activities as simple indices of organic phosphorus and sulphur mineralization. However, Khaziev (1967) found a direct correlation between soil nuclease activity and the mineralization rate of the "ammonium-soluble" fraction of organic phosphorus, suggesting that some of the soil organic phosphorus may be susceptible to direct dephosphorylation. Khaziev (1972b) also showed that nuclease activity correlated positively with phosphorus uptake from soil by certain crops. Attempts to relate soil phosphatase activity to crop yields have given contradictory results. Significant positive correlations (Khaziev, 1967; Gavrilova et al., 1973, 1975), a significant negative correlation (Kiss et al. 1975b) and no significant correlation (Gubenko et al., 1973) have all been reported. However, soil phosphatase activity, in situations where it does correlate significantly with a fertility or productivity measurement, may be a useful predictive tool.

Kramer (1957) proposed that soil phosphatase activity may be useful as a negative indicator of plant-available inorganic phosphate. However, Adebayo (1973) felt that such a role has questionable validity, because the critical concentration of phosphate for derepression of microbial alkaline phosphatase biosynthesis was found to be appreciably lower than the levels likely to be P-limiting for plants.

Speir (1977b) found that soil sulphatase activity was not related to ryegrass yield, in a pot trial, but appeared to be very strongly related to the organic-matter derived sulphur that was taken up by these plants. Further work is required to see if this result is valid for a wider range of soils, and if it can be extrapolated to the field situation.

Khaziev (1972b) proposed that soil enzyme activities could be used for evaluating soil fertility and suggested the measurement of four activities, namely phosphatase, invertase, catalase and urease.

The measurement of a number of soil enzyme activities, including phosphatase, has also been used as a means of characterizing soil types (Galstyan and Tatevosyan, 1964; Kozlov, 1966). Ross *et al.*, (1975b) measured a number of biochemical properties, including phosphatase and sulphatase activities, of a climosequence of New Zealand soils in tussock grasslands and found that the soils could be grouped biochemically into arrangements consistent with either the pedological classification or the influence of a single dominant environmental factor.

Thornton and McLaren (1975) investigated the possibility of using enzyme activities to differentiate similar soils for forensic purposes. They found that the levels of phosphatase, sulphatase, urease, invertase and protease activities, together served to characterize a soil as having originated from a given location. Soils collected from close proximity, which were indistinguishable on pedological grounds, could be differentiated enzymically.

XI. Conclusions

The complexity, heterogeneity and sorptive properties of the soil organo-mineral complex, coupled with the diversity of populations and numbers of organisms in the soil, make soil enzyme research a more difficult problem than the study of enzymes from other sources. It is, therefore, not surprising, that, since the review of Ramírez-Martínez (1968), little progress has been made in understanding the nature, origin, and role of soil phosphatase. By extrapolation of the data available on soil urease and from the kinetic studies of Nannipieri and colleagues, it may be possible to propose a tentative location of the cell-free component of soil phosphatase. However, very little is known of the proportion of total activity that is cell-free and this situation may remain unless an efficient phosphatase extraction technique becomes available. Because of its recent discovery, soil sulphatase activity has been much less studied; however, there is no reason to suppose that the above comments do not equally apply to this enzyme.

Techniques for soil pre-treatment, and incubation and assay procedures for the study of soil phosphatase and sulphatase need to be standardized. Some of the apparent contradictions that are evident in the studies on these enzymes may then be eliminated.

A critical evaluation of the labile, low molecular weight esters generally used as substrates for soil phosphatase and sulphatase determinations is required. Are there a multitude of enzymes with a high degree of substrate specificity, or just one or two relatively nonspecific enzymes? Are the hydrolyses of the so-called "natural" substrates (phytate, glycerophosphate, nucleic acids, etc.) catalysed by the same enzymes as these artificial substrates? This is especially important for nucleic acid, which, because of its high molecular weight, may be unable to reach adsorbed or partially occluded phosphatases. Finally, what relevance do any of these substrates have to the true soil substrates for phosphatase and sulphatase?

There may be little profit in continuing to catalogue soil phosphatase and sulphatase activities in different soil types and to relate them to diverse soil properties and plant productivity measurements. Although they can lead to a greater understanding of these enzymes, such studies make little attempt to understand the processes in which they are involved. A more fruitful, although more complex approach, would be a multidisciplinary study involving agronomy, plant physiology, soil chemistry and microbiology, as well as soil biochemistry. From such an approach, a more comprehensive understanding should be gained of the relationships between phosphatase and sulphatase activities in soil, the processes involved in the mineralization of soil organic matter, and the supply of phosphorus and sulphur to plants. In this way, possible applications of these enzyme activities as simple indices of availability of some fraction(s) of soil phosphorus and sulphur, or of plant productivity, may become evident.

Acknowledgements

We would like to thank Mrs Jewel Davin for her help in obtaining reference material; and Prof. R. T. Truscoe and Mrs Anne Lee for translating Russian and German papers and abstracts.

References

ABDEL'YUSSIF R. M., ZINCHENKO V. A. and GRUZDEV G. S. (1976). The effect of nematocides on the biological activity of soil. *Izv Timiryazev. sel'skohoz. Akad.* (1), 206–214. (*Soils Fertil.* **39**, 3982).

ABRAMYAN S. A. and GALSTYAN A. SH. (1975a). Effect of the nature of buffer solutions on the activity of soil enzymes. *Biol. Zh. Arm.* (2), 25–31.

ABRAMYAN S. A. and GALSTYAN A. SH. (1975b). On the use of masking substances in studying soil enzymes. *Biol. Zh. Arm.* (10), 32–37 (*Chem. Abstr.* **84**, 104308g).

ADEBAYO A. A. (1973). Mineralization of organic phosphorus in Nigerian soils, Ph.D. Thesis, University of Wisconsin (*Diss. Abstr. Int. B.* **34**, 1822–1823).

AGRE N. S., TATARSKAYA R. I., DEMBO G. V. and PITRYUK A. P. (1969). Nuclease activity of thermophylic soil actinomycetes. *Microbiology* **38**, 932–936.

AKIMENO V. A. and BRID'KO YU. I. (1975). Effect of mineral fertilizers on the enzymic activity of peaty-gley soils under rice. *Agrokhimiya* (11), 81–84 (*Chem. Abstr.* **84**, 57799f).

236   T. W. SPEIR AND D. J. ROSS

ALIEV R. A., ZVYAGINTSEV D. G. and KOZHEVIN P. A. (1975). Determination of extracellular and intracellular enzymes in soil by treatment with ultrasound and subsequent differential centrifugation. *Moscow Univ. Soil Sci. Bull.* **30**, 1–3.

ALIEV S. A. and GADZHIEV D. A. (1973). Correlated changes of enzyme activity in soils of vertical zones. *Biol. Nauki* (5), 121–126 (*Chem. Abstr.* **79**, 52298d).

AMBROŽ Z. (1973a). Biological processes in soils irrigated with starch waste waters. *Acta Univ. Agric., Brno, Fac. Agron.* **20**, 575–580 (*Chem. Abstr.* **80**, 2614d).

AMBROŽ Z. (1973b). Effects of soil drying and wetting and of some antiseptics and antibiotics on soil enzyme activity. *Acta Univ. Agric., Brno, Fac. Agron.* **21**, 9–14 (*Chem. Abstr.* **80**, 132130d).

ANDERSON G. (1975a). Other organic phosphorus compounds. *In* "Soil Components" (J. E. Gieseking, Ed.) Vol. 1, pp. 305–331. Springer-Verlag, New York.

ANDERSON G. (1975b). Sulphur in soil organic substances. *In* "Soil Components" (J. E. Gieseking, Ed.) Vol. 1, pp. 333–341. Springer-Verlag, New York.

ANDERSON J. W. (1975c). The function of sulphur in plant growth and metabolism. *In* "Sulphur in Australasian Agriculture" (K. D. McLachlan, Ed.) pp. 87–97. Sydney University Press, Sydney.

ANDERSSON P. and BENGTSSON M. (1975). Effects of copper and zinc on urease and phosphatase activity of spruce needle mor. *Medd. Avd. Ekol. Bot. Lunds Univ.* **3**, No. 6.

ANTHEUNISSE, J. (1972). Decomposition of nucleic acids and some of their degradation products by microorganisms. *Antonie van Leeuwenhoek* **38**, 311–327.

APPIAH M. R. and THOMPSON E. J. (1974). The effect of successive croppings on soil organic phosphorus. *Ghana Jl. Agric. Sci.* **7**, 25–30.

ARUTYUNYAN E. A. and ABRAMYAN S. A. (1975). Activity of soil nucleases. *Izv. Sel'skokhoz. Nauk* (1), 49–53 (*Chem. Abstr.* **83**, 57301s).

ARUTYUNYAN E. A. and GALSTYAN A. SH. (1974). Forms of soil phosphorus and characteristics of soil phosphatase action. *Biol. Zh. Arm.* (8), 41–46. (*Chem. Abstr.* **82**, 30114a).

ARUTYUNYAN E. A. and GALSTYAN A. SH. (1975). Determination of the activity of alkaline and acid phosphatases in soils. *Agrokhimiya* (5), 128–133.

ARUTYUNYAN E. A. and SIMONYAN B. N. (1975). Forms of phosphorus and phosphatase activity in eroded chernozems. *Izv. Sel'skokhoz. Nauk* (2), 49–53 (*Chem. Abstr.* **83**, 130488q).

ASEEVA I. V. and VANYARKHO V. A. (1969). Enzymic activity of sod-podzols. *Biol. Nauk* (11), 128–131 (*Chem. Abstr.* **72**, 7799lt).

BANERJEE A. K. and NANDI P. (1966). Studies on extracellular phosphatases in actinomycetes. *Indian J. Exp. Biol.* **4**, 188–189.

BARROW N. J. (1960). The effects of varying the nitrogen, sulphur and phosphorus content of organic matter on its decomposition. *Aust. J. Agric. Res.* **11**, 317–330.

BARTLETT E. M. and LEWIS D. H. (1973). Surface phosphatase activity of mycorrhizal roots of beech. *Soil Biol. Biochem.* **5**, 249–257.

BECK TH. (1974). Effects of extended monoculture and crop rotation systems on microbiological activities in soil. *Landw. Forsch.* **31**, 268–276.

BENKOVIC S. J., VERGARA E. V. and HEVEY R. C. (1971). Purification and properties of an arylsulphatase from *Aspergillus oryzae. J. Biol. Chem.* **246**, 4926–4933.

BERTRAND D. and DE WOLF A. (1968). Effect of microelements applied as complementary fertilizers on the soil microflora. *C.R. Acad. Agric. Fr.* **54**, 1130–1133.

BEZBORODOVA S. I. and IL'INA T. V. (1969). Extracellular phosphomonoesterases of fungi belonging to the genus *Fusarium*. *Microbiology* **39**, 643–648.

BIELESKI R. L. (1974). Development of an externally-located alkaline phosphatase as a response to phosphorus deficiency. *In* "Mechanisms of Regulation of Plant Growth" (R. L. Bieleski, A. R. Ferguson and M. M. Cresswell, Eds) Bulletin 12, pp. 165–170. The Royal Society of New Zealand, Wellington.

BIELESKI R. L. and JOHNSON P. N. (1971). The external location of phosphatase activity in the phosphorus-deficient *Spirodela oligorrhiza*. *Aust. J. Biol. Sci.* **25**, 707–720.

BLAGOVESHCHENSKAYA Z. K. and DAUCHENKO N. A. (1974). Activity of soil enzymes after prolonged application of fertilizers to a corn monoculture and crops in rotation. *Soviet Soil Sci.* (5), 569–575.

BOWER C. A. (1949). Studies on the forms and availability of soil organic phosphorus. Research Bull. No. 362. Iowa State College of Agriculture, Ames.

BRAMS W. H. and MCLAREN A. D. (1974). Phosphatase reactions in a column of soil. *Soil Biol. Biochem.* **6**, 183–189.

BREMNER J. M. and ZANTUA M. I. (1975). Enzyme activity in soil at subzero temperatures. *Soil Biol. Biochem.* **7**, 383–387.

BROWN M. E. (1974). Seed and root bacterization. *A. Rev. Phytopath.* **12**, 181–197.

BURANGULOVA M. N. and KHAZIEV F. KH. (1965a). The nuclease activity of soils. (Preliminary communication). *Úst. Věd. Inf. MZLVH Rostl. Výroba* **37**, 579–584 (*Soils Fertil.* **29**, 349).

BURANGULOVA M. N. and KHAZIEV F. KH. (1965b). The phosphatase activity of the soil as affected by chemical fertilizers. *Agrokém. Talajt.* **14**, 101–110.

BURANGULOVA M. N. and KHAZIEV F. KH. (1965c). Nuclease activity of the soil *Soviet. Soil Sci.* (13), 1523–1526.

BURANGULOVA M. N. and KHAZIEV F. KH. (1972). The role of soil microorganisms in the transformation of organic phosphorus in soil. *Symp. Biol. Hung.* **11**, 271–275.

BURANGULOVA M. N. and MUKATANOV A. KH. (1973). Enzyme activity in mountain chernozems of the Southern Urals. *Biol. Nauki* (6), 121–124.

CASIDA L. E. Jr. (1959). Phosphatase activity of some common soil fungi. *Soil Sci.* **87**, 305–309.

CAWSE P. A. (1975). Microbiology and biochemistry of irradiated soils. *In* "Soil Biochemistry" (E. A. Paul and A. D. McLaren, Eds) Vol. 3, pp. 213–267. Marcel Dekker, New York.

ČERNÁ S. (1972). Phosphatase activity in structural soil. *Acta Univ. Carol.-Biologica* **1970**, 455–459.

CERVELLI S., NANNIPIERI P., CECCANTI B. and SEQUI P. (1973). Michaelis constant of soil acid phosphatase. *Soil Biol. Biochem.* **5**, 841–845.

CERVELLI S., NANNIPIERI P., GIOVANNINI G. and PERNA A. (1975). Concerning the distribution of enzymes in soil organic matter. *Studies about Humus. Trans. Int. Symp., Humus et Planta VI*, 291–296.

CHANG C. W. and BANDURSKI R. S. (1964). Exocellular enzymes of corn roots. *Pl. Physiol.* **39**, 60–64.

CHANG S. C. (1940). Assimilation of phosphorus by a mixed soil population and by pure cultures of soil fungi. *Soil Sci.* **49**, 197–210.

CHESHIRE M. V. and ANDERSON G. (1975). Soil polysaccharides and carbohydrate phosphates. *Soil Sci.* **119**, 356–362.

CHUNDEROVA A. I. (1964). Phosphatase activity of corn rhizosphere microflora. *Microbiology* **33**, 89–92.

CHUNDEROVA A. I. and ZUBETS T. P. (1969). Phosphatase activity in dernopodzolic soils. *Pochvovedenie* (11), 47–53 (*Soils Fertil.* **33**, 2092).

CHUNDEROVA A. N. (1970). Enzyme activity and pH of soil. *Soviet Soil Sci.* (3), 308–314.

CLARK R. B. (1975). Characterization of phosphatase of intact maize roots. *J. Agric. Fd. Chem.* **23**, 458–460.

COOPER P. J. M. (1972). Aryl sulphatase activity in Northern Nigerian soils. *Soil Biol. Biochem.* **4**, 333–337.

COSGROVE D. J. (1967). Metabolism of organic phosphates in soil. *In* "Soil Biochemistry" (A. D. McLaren and G. H. Peterson, Eds) pp. 216–228. Marcel Dekker Inc., New York.

COSGROVE D. J., IRVING G. C. J. and BROMFIELD S. M. (1970). Inositol phosphate phosphatases of microbiological origin. The isolation of soil bacteria having inositol phosphate phosphatase activity. *Aust. J. Biol. Sci.* **23**, 339–343.

DARASELIYA N. A., KALATOZOVA G. B. and DZADZAMIYA T. D. (1975). Microbiological and enzymic activity of podzolic-gley and kraznozem soils with various moisture contents. *Subtrop. Kul't.* (1), 103–105 (*Chem. Abstr.* **83**, 191765z).

DE JORGE F. B. and SAWAYA M. C. (1967). Comparative biochemical studies on the oligochaetes *Pheretima hawayana, Glossoscolex grandis,* and *Rhinodrilus sp. Comp. Biochem. Physiol.* **22**, 359–369.

DIXON M. and WEBB E. C. (1964). "Enzymes", Longmans, London.

DROBNÍKOVÁ V. (1961). Factors influencing the determination of phosphatases in soils. *Folia microbiol., Praha* **6**, 260–267.

DUBOVENKO E. K. (1964). Phosphatase activity of different soils. *Zhivlennya Udobr. Sil.-gospod. Kul't., Ukr. Nauk Dosl. Inst. Zemlerobstva,* pp. 29–32 (*Chem. Abstr.* **62**, 13794f).

DUBOVENKO E. K. (1966). Phosphatase activity of the roots of some plants and the effect of microorganisms upon their activity. *Soviet Pl. Physiol.* **13**, 469–471.

DUDCHENKO V. G., BESKROVNYI A. K., ULYASHOVA R. M. and IVANKEVICH N. P. (1974). Effect of crop rotation on the biological activity of peat-bog soils. *Trans. 10th Int. Congr. Soil Sci.* **10**, 332–338. (*Chem. Abstr.* **83**, 162867d).

DYER W. J. and WRENSHALL C. L. (1941). Organic phosphorus in soils. III. The decomposition of some organic phosphorus compounds in soil cultures. *Soil Sci.* **51**, 323–329.

FAUVEL B. and ROUQUEROL T. (1970). The phosphatase test considered as an index of soil activity and evolution. *Revue Écol. Biol. Sol* **7**, 393–406.

FEDER J. (1973). The phosphatases. *In* "Environmental Phosphorus Handbook" (E. J. Griffith, A. Beeton, J. M. Spencer and D. T. Mitchell, Eds) pp. 475–508. John Wiley and Sons, New York.

FITZGERALD J. W. (1976). Sulfate ester formation and hydrolysis: a potentially important yet often ignored aspect of the sulfur cycle of aerobic soils. *Bacteriol. Rev.* **40**, 698–721.

FLOYD R. A. and OHLROGGE A. J. (1970). Gel formation on nodal root surfaces of *Zea mays*. I. Investigation of the gel's composition. *Pl. Soil* **33**, 331–343.

FRENEY J. R. (1961). The nature of organic sulphur compounds in soil. *Aust. J. Agric. Res.* **12**, 424–432.

FRENEY J. R., MELVILLE G. E. and WILLIAMS C. H. (1975). Soil organic matter fractions as sources of plant-available sulphur. *Soil Biol. Biochem.* **7**, 217–221.

FURUKAWA H. and KAWAGUCHI K. (1969). Contribution of organic phosphorus to the increase of easily soluble phosphorus in waterlogged soils, especially related to phytic phosphorus (inositol hexaphosphate). *Nippon Dojo-Hiryogaku Zasshi* **40**, 141–148 (*Chem. Abstr.* **71**, 100848s).

GALOPPINI C., LOTTI G. and TOGNONI F. (1962). Investigations on phosphatase in the agricultural soil. *Chimica Ind., Milano* **44**, 255 (*Chem. Abstr.* **57**, 10247i).

GALSTYAN A. SH. (1965). Effect of temperature on the activity of soil enzymes. *Dokl. Akad. Nauk Arm. SSR* **40**, 177–181.

GALSTYAN A. SH. and ABRAMYAN S. A. (1975). On the adenosinetriphosphatase activity of soils. *Dokl. Akad. Nauk Arm. SSR* **61**, 298–300 (*Chem. Abstr.* **84**, 163395f).

GALSTYAN A. SH. and ARUTYUNYAN E. A. (1966). Determination of soil phosphatase activity. *Biol. Zh. Armenii* (3), 25–29 (*Chem. Abstr.* **65**, 7944f).

GALSTYAN A. SH. and BAZOYAN G. V. (1974). Activity of soil arylsulphatase. *Dokl. Akad. Nauk Arm. SSR* **59**, 184–187.

GALSTYAN A. SH. and MARKOSYAN L. V. (1965). The pH optima of enzymes of the soil. *Izv. Akad. Nauk Arm. SSR. Biol. Nauki* (7), 21–27 (*Soils Fertil.* **29**, 1127).

GALSTYAN A. SH. and TATEVOSYAN G. S. (1964). Enzyme activity as a characteristic of soil types. *Trans. 8th Int. Congr. Soil Sci.* **3**, 711–718.

GALSTYAN A. SH., MARKOSYAN L. V. and ARUTYUNYAN E. A. (1967). Principles for the study of soil enzymes. *Trudy Nauchno-issled. Inst. Pochv. Agrokhim., Yerevan* (3), 335–345 (*Chem. Abstr.* **70**, 114190n).

GAVRILOVA A. N. and SHIMKO N. A. (1969). Organophosphate level and phosphatase activity in some soils of the Belorussian SSR. *Vestsi Akad. Navuk BSSR, Ser. biyal. Navuk* (6), 35–41 (*Chem. Abstr.* **72**, 99556g).

GAVRILOVA A. N., SHIMKO N. A. and SAVCHENKO V. F. (1973). Dynamics of organic phosphorus compounds and phosphatase activity in pale yellow sod-podzolic soil. *Soviet Soil Sci.* (3), 320–328.

GAVRILOVA A. N., SAVCHENKO N. I. and SHIMKO N. A. (1974). Forms of phosphorus and phosphatase activity of the chief soil types of the Belorussian SSR. *Trans. 10th Int. Congr. Soil Sci.* **4**, 281–288.

GAVRILOVA A. N., SAVCHENKO N. I. and SHIMKO N. A. (1975). Content of organo-phosphates and the activity of phosphatase in soddy-pale-yellow-podzolic soils cultivated to different degrees. *Pochvovedenie* (1), 81–85 (*Chem. Abstr.* **82**, 123978a).

GELLER I. A. and DOBROTVORSKAYA K. N. (1961). Phosphatase activity of soils in beet root-seedling areas. *Trudy Inst. Mikrobiol., Mosk.* (11), 215–221 (*Chem. Abstr.* **56**, 13278d).

GERUS O. I. (1963). Effect of fertilizers on biological activity of soil. *Sb. Nauch. Rab. Sib. Nauchno-issled. Inst. Sel'skokhoz.* (8), 155–159 (*Soils Fertil.* **27**, 3545).

GHANI M. O. and ALEEM S. A. (1942). Effect of liming on the transformation of phosphorus in acid soils. *Indian J. Agric. Sci.* **12,** 873–882 (*Chem. Abstr.* **42,** 2703g).

GOIAN M. (1968). Studies on the phosphatase activity of soils. *Lucr. ştiinţ. Inst. agron. Timisoara, Ser. Agron.* **11,** 139–152.

GOIAN M. (1969). Studies on the phosphatase activity of soils. II. *Lucr. ştinţ. Inst. Agron. Timisoara, Ser. Agron.* **12,** 489–496.

GOIAN M. (1970). Phosphatase activity in saline soils. *Ştinţa Sol.* **8,** 13–17.

GORBANOV S. P. (1975). The effect of molybdenum on soil phosphatase activity. *Pochv. Agrokhim.* (3), 97–102.

GREAVES M. P. and WEBLEY D. M. (1964). A study of the breakdown of organic phosphates by microorganisms from the root region of certain pasture grasses. *J. Appl. Bact.* **28,** 454–465.

GREAVES M. P., ANDERSON G. and WEBLEY D. M. (1963). A rapid method for determining phytase activity of soil microorganisms. *Nature,* **200,** 1231–1232.

GRUZDEV G. S., MOZGOVOI A. F. and KUZ'MINA I. V. (1973). Effect of treflan on biological activity of soil under sunflower. *Izv. Timiryazev. Sel'skokhoz. Akad.* (4), 136–141 (*Soils Fertil.* **37,** 250).

GUBENKO V. A., SOKRUTA I. F. and STULIN A. F. (1973). Dynamics of nitrogen and the biological activity of the Azov region chernozem under corn following various crops. *Agrokhim. Gruntozn* **23,** 44–49 (*Chem. Abstr.* **82,** 15646g).

GUEDE SUEIRO S. and DE FELIPE ANTON M. R. (1974). Histochemical localization of acid phosphatase in the mucilage of young roots of *Zea mays* plants grown in nutrient solutions. *An. Edafol. Agrobiol.* **33,** 199–214 (*Chem. Abstr.* **82,** 135660d).

HALL J. L. and DAVIE C. A. M. (1971). Localization of acid hydrolase activity in *Zea mays* L. root tips. *Ann. Bot.* **35,** 849–855.

HALM B. J., STEWART J. W. B. and HALSTEAD R. L. (1971). The phosphorus cycle in a native grassland ecosystem. *In* "Isotopes and Radiation in Soil-Plant Relationships including Forestry. Proc. Symp.", pp. 571–586 I.A.E.A., Vienna.

HALSTEAD R. L. (1964). Phosphatase activity of soils as influenced by lime and other treatments. *Can. J. Soil Sci.* **44,** 137–144.

HALSTEAD R. L. and MCKERCHER R. B. (1975). Biochemistry and cycling of phosphorus. *In* "Soil Biochemistry" (E. A. Paul and A. D. McLaren, Eds) Vol. 4, pp. 31–63. Marcel Dekker Inc., New York.

HALSTEAD R. L. and SOWDEN F. J. (1968). Effect of long-term additions of organic matter on crop yields and soil properties. *Can. J. Soil Sci.* **48,** 341–348.

HARADA T. (1964). The formation of sulphatases by *Pseudomonas aeruginosa. Biochim. Biophys. Acta* **81,** 193–196.

HARADA T. and SPENCER B. (1962). The effect of sulphate assimilation on the induction of arylsulphatase synthesis in fungi. *Biochem. J.* **82,** 148–156.

HAYANO T. (1973). On soil enzymes. *Nippon Dojo-Hiryogaku Zasshi* **44,** 395–402.

HAYASHI T. and TAKIJIMA Y. (1951). Metabolism in the roots of crop plants. I. Dephosphorization of organic phosphorus by roots. *J. Sci. Soil Manure, Tokyo* **21,** 185–189 (*Chem. Abstr.* **45,** 10321a).

HAYASHI T. and TAKIJIMA Y. (1955). Utilization of soil organic phosphorus by crop plants. II. Mineralization of soil organic phosphorus. *J. Sci. Soil Manure, Tokyo* **26,** 135–138 (*Chem. Abstr.* **49,** 15141f).

HERLIHY M. (1972). Microbial and enzyme activity in peats. *Tech. Commun. Int. Soc. Hort. Sci., Acta Horticulturae* (26), 45–50.

HOCHSTEIN L. (1962). The fluorometric assay of soil enzymes. *NASA Rep. No. NASA CR-50919* (*Chem. Abstr.* **63**, 15164g).

HOFFMANN G. (1967). A photometric method for the determination of the phosphatase activity in soils. *Z. Pflanzenernähr Düng. Bodenkd.* **118**, 161–172. (*Chem. Abstr.* **69**, 18279g).

HOFFMANN G. and ELIAS-AZAR K. (1965). Various soil fertility factors of North Persian soils and their connections with the activity of hydrolytic enzymes. *Z. Pflanzenernähr. Düng. Bodenkd.* **108**, 199–217.

HOFMANN E. (1963). The analysis of enzymes in soils. *In* "Moderne Methoden der Pflanzenanalyse" (H. F. Linskens and M. V. Tracey, Eds) Vol. VI, 416–423. Springer-Verlag, Berlin.

HOFMANN E. and HOFFMANN G. (1966). Enzymic methods for estimation of the biological activity of soils. *Adv. Enzymol.* **28**, 365–390.

HOFMANN E. and KASSEBA A. (1962). Enzymes in Egyptian soils. *Z. Pflanzenernähr. Düng. Bodenkd.* **99**, 9–20 (*Chem. Abstr.* **58**, 4993d).

HOUGHTON C. and ROSE F. A. (1976). Liberation of sulphate from sulphate esters by soils. *Appl. Envir. Microbiol.* **31**, 969–976.

IRVING G. C. J. and COSGROVE D. J. (1976). The kinetics of soil acid phosphatase. *Soil Biol. Biochem.* **8**, 335–340.

ISHII T. and HAYANO K. (1974). Method for the estimation of phosphodiesterase activity in soil. *Nippon Dojo-Hiryogaku Zasshi* **45**, 505–508 (*Chem. Abstr.* **83**, 8023u).

JACKMAN R. H. and BLACK C. A. (1952a). Phytase activity in soils. *Soil Sci.* **73**, 117–125.

JACKMAN R. H. and BLACK C. A. (1952b). Hydrolysis of phytate phosphorus in soils. *Soil Sci.* **73**, 167–171.

JACKSON R. W. and DE MOSS J. A. (1965). Effects of toluene on *Escherichia coli. J. Bact.* **90**, 1420–1425.

JANOSSY G. (1963). Changes in phosphatase activity of soil microorganisms depending on status of phosphorus nutrition. *Agrokém. Talajt.* **12**, 285–292.

JAYAYAMAN K. N. and PRASAD N. N. (1972). Production of phosphatase by soil Aspergilli. *Madras agric. J.* **59**, 640–641.

KAILA A. (1949). Biological absorption of phosphorus. *Soil Sci.* **68**, 279–289.

KANIVETS V. I., MIKHNOVSKAYA A. D. and MAMCHENKO O. A. (1973). Biochemical properties of the main types of soils of the Ukraine. *In* "Agrokhim. Kharacter. Pochv. SSSR, Ukr. SSR (A. N. Sokolov, Ed.) pp. 92–106. Nauka, Moscow (*Chem. Abstr.* **80**, 132028b).

KARANTH N. G. K., CHITRA C. and VASANTHARAJAN V. N. (1975). Effect of fungicide, Dexon, on the activities of some soil enzymes. *Indian J. Exp. Biol.* **13**, 52–54.

KARKI A. B. and KAISER P. (1974). Effects of sodium chlorate on soil microorganisms, their respiration and enzyme activity. II. Laboratory incubation study. *Revue Écol. Biol. Sol.* **11**, 477–498.

KEILLING J. (1964). On the biology of soils. *C.R. Acad. Agric. Fr.* **50**, 1131–1138.

KEILLING J., CAMUS A., SAVIGNAC G., DAUCHEZ P. and BOITEL M. (1960). Contribution to the study of soil biology. *C.R. Acad. Agric. Fr.* **46**, 647–652.

KELLEY W. D. and RODRIGUEZ-KABANA R. (1975). Effects of potassium azide on soil microbial populations and soil enzymatic activities. *Can. J. Microbiol.* **21**, 565–570.

KHAN S. U. (1970). Enzymatic activity in a grey wooded soil as influenced by cropping systems and fertilizers. *Soil Biol. Biochem.* **2**, 137–139.

KHAZIEV F. KH. (1966). Dependence of the nuclease activity of soil on the pH and influence of various substances on it. *Soviet Soil Sci.* (13), 1547–1550.

KHAZIEV F. KH. (1967). Relationship between nuclease activity of soil and the biodynamics of organic phosphates. *Soviet Soil Sci.* (13), 1822–1826.

KHAZIEV F. KH. (1969). Relation of soil nuclease activity to temperature and soil conditions. *Biol. Nauki* (1), 109–113.

KHAZIEV F. KH. (1972a). Determination of pyrophosphatase activity of soil. *Pochvovedenie* (2), 67–73 (*Soils Fertil.* **35**, 2981).

KHAZIEV F. KH. (1972b). Soil enzymes and their role in soil fertility. *Biol. Nauki* (2), 114–119 (*Chem. Abstr.* **77**, 33479f).

KHAZIEV F. KH. (1975). Thermodynamic characteristics of enzymic reactions in soil. *Biol. Nauki* (10), 121–127 (*Chem. Abstr.* **84**, 42394w).

KHAZIEV F. KH. and BURANGULOVA M. N. (1965). Activity of enzymes which dephosphorylate organic phosphorus compounds of soil. *Prikl. Biokhim Mikrobiol.* **1**, 373–379.

KISHK F. M., EL-ESSAWAI T., ABDEL-GHAFAR S. and ABOU-DONIA M. B. (1976). Hydrolysis of methylparathion in soils. *J. Agric. Fd. Chem.* **24**, 305–307.

KISS S. and PETERFI S. (1961). Paper-chromatographic method for identifying the phosphomonoesterases of the soil. *Stud. Univ. Babeş-Bolyai, Ser. Chem.* **8**, 369–373 (*Chem. Abstr.* **61**, 11026a).

KISS S., ŞTEFANIC G. and DRĂGAN-BULARDA M. (1974). Soil enzymology in Romania (Part I). *Contrib. Bot. Cluj,* 207–219.

KISS S., DRĂGAN-BULARDA M. and RĂDULESCU D. (1975a). Biological significance of enzymes accumulated in soil. *Adv. Agron.* **27**, 25–87.

KISS S., ŞTEFANIC G. and DRĂGAN-BULARDA M. (1975b). Soil enzymology in Romania (Part II). *Contrib. Bot., Cluj,* 197–207.

KIVER K. F. (1973). Microbiological and enzymic activity of soil during irrigation and fertilization. *Mỹkrobiol. Zh.* **35**, 149–153 (*Chem. Abstr.* **79**, 41348u).

KO W.-H. and HORA F. K. (1970). Production of phospholipases by soil microorganisms. *Soil Sci.* **110**, 355–358.

KOBUS J. (1961). Role of microorganisms in transformations of phosphorus compounds in the soil. *Roczn. Naukro ln. Ser. D.* **91**, 5–102 (*Chem. Abstr.* **57**, 6417a).

KONG K. T. and DOMMERGUES Y. (1972). Limitation of cellulolysis in organic soils. II. —Study of soil enzymes. *Revue Écol. Biol. Sol.* **9**, 629–640.

KOTELEV V. V. and MEKHTIEVA E. A. (1961). Interdependence between the phosphatase activity of the microflora and the content of mobile P in soil. *Izv. Moldav. Fil. Akad. Nauk. SSSR* (7), 41–47 (*Chem. Abstr.* **62**, 8343d).

KOWALENKO C. G. and LOWE L. E. (1975). Mineralization of sulphur from four soils and its relationship to soil carbon, nitrogen and phosphorus. *Can. J. Soil Sci.* **55**, 9–14.

KOZLOV K. A. (1966). Biological activity of soils. *Izv. Akad. Nauk SSSR, Ser. Biol.* (5), 719–733.

KOZLOV K. A., KISLITSINA V. P., ZHDANOVA E. M., MARKOVA YU. and MIKHAILOVA E. N. (1972). On the problem of modelling in soil enzymology. *Symp. Biol. Hung.* **11**, 277–281.

KOZLOV K. A., KISLITSINA V. and MARKOVA YU. (1973). Enzymic activity and humus content in Eastern Siberian soils. *Zentralbl. Bakteriol. Parasitenkd. Abt. II* **128**, 144–148.

KRÁMER M. (1957). Phosphatase activity as an indicator of biologically useful phosphorus in soil. *Naturwissenschaften* **44**, 13.

KRÁMER M. and ERDEI G. (1958). Investigation of the phosphatase activity of soils with the use of disodium phenylphosphate. *Agrokém. Talajt.* **7**, 361–366.

KRÁMER M. and YERDEI G. (1959). Application of the method of phosphatase activity determination in agricultural chemistry. *Soviet Soil Sci.* (9), 1100–1103.

KROLL L. and KRÁMER M. (1955). Effect of clay minerals on the enzyme activity of soil phosphatase. *Naturwissenschaften* **42**, 157–158.

KROLL L., KRÁMER M. and LÖRINCZ E. (1955). The application of enzyme analysis with phenylphosphate to soils and fertilizers. *Agrokém. Talajt.* **4**, 173–182.

KUDZIN YU. K. and YAROSHEVICH I. V. (1961). The mobilization of organic phosphates in black soils and the P nutrition of plants. *Trudy Inst. Mikrobiol., Mosk.* (11), 252–259 (*Chem. Abstr.* **56**, 10609e).

KUDZIN YU. K., YAROSHEVICH I. V. and GUBENKO V. A. (1968). Mobilization of phosphorus reserves in chernozems during prolonged application of nitrogen and potassium fertilizers. *Dokl. Vses. Akad. Sel'skokhoz. Nauk* (12), 6–8 (*Soils Fertil.* **32**, 1703).

KUDZIN YU. K., GUBENKO V. A., PASHOVA V. T. and YAROSHEVICH I. V. (1970). Mobilization of phosphorus reserves in chernozem and "phosphatization" after prolonged fertilization. *Soviet Soil Sci.* (4), 409–416.

KUPREVICH V. F. and SHCHERBAKOVA T. A. (1971). "Soil Enzymes", INSDOC, New Delhi. Translated from *Soil Enzymes*, Nauka i Tekhnika, Minsk, 1966.

LADD J. N. and BUTLER J. H. A. (1972). Short-term assays of soil proteolytic enzyme activities using proteins and dipeptide derivatives as substrates. *Soil Biol. Biochem.* **4**, 19–30.

LADD J. N. and BUTLER J. H. A. (1975). Humus–enzyme systems and synthetic, organic polymer–enzyme analogs. *In* "Soil Biochemistry" (E. A. Paul and A. D. McLaren, Eds) Vol. 4, pp. 143–194. Marcel Dekker Inc., New York.

LADD J. N. and PAUL E. A. (1973). Changes in enzymic activity and distribution of acid-soluble amino acid-nitrogen in soil during nitrogen immobilization and mineralization. *Soil Biol. Biochem.* **5**, 825–840.

LAUGESEN K. and MIKKELSEN J. P. (1973). Phosphatase activity in Danish soils. *Tidsskr. Pl.* **77**, 352–356.

LEVCHENKO L. A. and LISOVAL A. P. (1974). Effect of nitrogen fertilizers on the behaviour of phosphorus in the roots of winter wheat. *Vest. Sel'skokhoz. Nauki, Mosk.* (6), 23–29 (*Chem. Abstr.* **81**, 76912s).

LISOVAL A. P. (1970). Effect of plants on soil ribonuclease activity. *Nauk. Pr., Ukr. Sil'.-gospod. Akad.* (24), 105–112 (*Chem. Abstr.* **75**, 97731y).

LISOVAL A. P., ULASEVICH E. YU. and LEVCHENKO L. A. (1974). Change in the activity of phosphate-decomposing bacteria and phosphatases of soil in crop rotation under the influence of fertilizers. *Din. Mikrobiol. Protsessov. Poche Obuslovlivayushchie Ee Faktory, Mater. Simp.* **2**, 34–38 (*Chem. Abstr.* **83**, 162724e).

MAL'TSEVA N. N. (1960). Phosphatase activity of *Bacillus megaterium*. *Mikrobiol. Zh., Akad. Nauk Ukr. RSR, Inst. Microbiol. im D. K. Zabolotnogo* (5) 25–30 (*Chem. Abstr.* **55**, 15608h).

MAMYTOV A. M., VORONOV S. I. and TYNAEV ZH. M. (1974). Composition of phosphorus compounds and phosphatase activity of soils of the territory brought under cultivation in the Issyk-Kul Basin of Kirgizia. *Dokl Vses. Akad. Sel'skokhoz. Nauk* (3), 5–7 (*Chem. Abstr.* **80,** 144651n).

MANDROVSKAYA N. M. (1971). Enzymic activity of peat soil during the first years of cultivation. *Fiziol. Biokhim. Kul't. Rast.* **3,** 176–179 (*Chem. Abstr.* **75,** 117566d).

MANORIK A. V. and MALICHENKO S. M. (1969). Effect of symmetric triazines on phosphatase and urease activities in soil. *Fiziol. Biokhim. Kul't Rast.* **1,** 173–178 (*Chem. Abstr.* **73,** 2895b).

MAZILKIN I. A. and KUZNETSOVA M. G. (1964). Nuclease and phosphatase activity of soil bacteria. *Izv. Akad. Nauk SSSR, Ser. Biol.* (4), 587–594 (*Chem. Abstr.* **61,** 16446a).

MCLAREN A. D. (1960). Enzyme action in structurally restricted systems. *Enzymologia* **21,** 356–364.

MCLAREN A. D. (1963). Enzyme activity in soils sterilized by ionizing radiation and some comments on micro-environments in nature. *"Recent Progress in Microbiology". Symp. 8th Int. Congr. Microbiol.,* Montreal, **1962,** 221–229.

MCLAREN A. D. and PUKITE A. (1972). Ubiquity of some soil enzymes and isolation of soil organic matter with urease activity. *Proc. Int. Meet. Humic Substances, Pudoc; Wageningen,* pp. 187–193.

MCLAREN A. D. and SKUJIŅŠ J. J. (1971). Trends in the biochemistry of terrestrial soils. *In* "Soil Biochemistry" (A. D. McLaren and J. J. Skujiņš, Eds) Vol. 2, pp. 1–15. Marcel Dekker Inc., New York.

MCLAREN A. D., RESHETKO L. and HUBER W. (1957). Sterilization of soil by irradiation with an electron beam, and some observations on soil enzyme activity. *Soil Sci.* **83,** 497–502.

MCLAREN A. D., LUSE R. A. and SKUJIŅŠ J. J. (1962). Sterilization of soil by irradiation and some further observations on soil enzyme activity. *Proc. Soil Sci. Soc. Am.* **26,** 371–377.

MENKINA R. A. (1950). Bacteria which mineralize organic phosphorus compounds. *Mikrobiologiya* **19,** 308–316 (*Soils Fertil.* **14,** 627).

MINENKO A. K. (1974). Phosphatase activity and number of phosphorus-mineralizing microorganisms in soils of the nonchernozem zone. *Din. Mikrobiol. Protsessov. Poche Obuslovlivayushchie Ee Faktory, Mater. Simp.* **1,** 179–181 (*Chem. Abstr.* **83,** 146327x).

MISHUSTIN E. N. (1963). Bacterial fertilizers and their effectiveness. *Microbiology* **32,** 774–778.

MORTLAND M. M. and GIESEKING J. E. (1952). The influence of clay minerals on the enzymatic hydrolysis of organic phosphorus compounds. *Proc. Soil Sci. Soc. Am.* **16,** 10–13.

MUREŞANU P. L. and GOIAN M. (1969). Investigation on phosphomonoesterase activity in soils treated with fertilizers and lime. *Agrokém. Talajt.* **18,** 102–106.

NANNIPIERI P., CECCANTI B. and CERVELLI S. (1972). Substrate absorption in the soil phosphatase activity determination using *p*-nitrophenylphosphate. *Agric. Ital., Pisa* **72,** 373–380.

NANNIPIERI P., CERVELLI S. and PERNA A. (1973). Enzyme activities in some Italian soils. *Agric. Ital., Pisa* **73,** 367–376.

NAUMOVA A. N., MISHUSTIN E. N. and MAR'ENKO V. M. (1962). Nature of the action of bacterial fertilizers (azotobacterin, phosphobacterin) on agricultural plants. *Izv. Akad. Nauk SSSR, Ser. Biol.* (5), 709–717 (*Soils Fertil.* **26**, 244).

NEAL J. R. Jr. (1973). Influence of selected grasses and herbs on soil phosphatase activity. *Can. J. Soil Sci.* **53**, 119–121.

NILSSON P. E. (1957). Influence of crop on biological activity of soil. *K. Lantbr Högsk. Annlr* **23**, 175–218.

PACEWICZOWA T. H. (1960). Influence of pH upon the decomposition of organic P compounds by soil microorganisms. *Roczn. Naukro ln., Ser. A.* **82**, 211–218 (*Chem. Abstr.* **55**, 20280g).

PAMMENTER N. W. and WOOLHOUSE H. W. (1975). The utilization of P-N compounds by plants. II. The role of extracellular root phosphatases. *Ann. Bot.* **39**, 347–361.

PANIKOV N. S. and ASEEVA I. V. (1974). Relation of the nuclease activity of soils to the action of physicochemical factors and the development of soil microorganisms. *Din. Mikrobiol. Protsessov. Poche Obuslovlivayushchie Ee Faktory, Mater. Simp.* **2**, 122–124 (*Chem. Abstr.* **83**, 162671k).

PARKER F. W. (1927). Soil phosphorus studies: III. Plant Growth and the absorption of phosphorus from culture solutions at different phosphate concentrations. *Soil Sci.* **24**, 129–146.

PARKS F. P. (1974). The sorption of *Escherichia coli* alkaline phosphatase by selected clay minerals, Ph. D. Thesis, University of Idaho.

PEARSON R. W. and PIERRE W. H. (1940). Assimilation of organic phosphorus by corn and tomato plants. *Rep. Iowa Agric. Exp. Stn*, pp. 95–96.

PEARSON R. W., NORMAN A. G. and CHUNG HO (1941). The mineralization of the organic phosphorus of various compounds in soil. *Soil Sci. Soc. Am. Proc.* **6**, 168–175.

PETRENKO M. B. (1973). Change of biochemical characteristics of bacteria as a result of adaptation to the plant. *Biol. Nauki* (6), 101–106 (*Chem. Abstr.* **79**, 134166y).

PINCK L. A., SHERMAN M. S. and ALLISON F. E. (1941). The behaviour of soluble organic phosphates added to soil. *Soil Sci.* **51**, 351–365.

PONOMAREVA N. S., PIROGOVA T. I., NIKULINA V. D. and ANTONOVA B. K. (1972). Phosphatase activity of high solonetzs of the wooded steppe of the Omsk region. *Agrokhimiya* (6), 102–108 (*Soils Fertil.* **35**, 5038).

POUX N. (1966). Ultrastructural localization of aryl sulphatase activity in plant meristematic cells. *J. Histochem. Cytochem.* **14**, 932–936.

PUKHIDSKAYA N. S. and KOVRIGO V. P. (1974a). Enzyme activity of basic types of soils of the Udmart Autonomous SSR. *Trudy Izhevsk. Sel'skokhoz. Inst.* **23**, 110–119 (*Chem. Abstr.* **84**, 16068y).

PUKHIDSKAYA N. S. and KOVRIGO V. P. (1974b). Enzyme activity of soils of the Udmart Autonomous SSR. *Trudy Izhevsk. Sel'skokhoz. Inst.* **23**, 119–124 (*Chem. Abstr.* **84**, 16069z).

RACZ G. J. and SAVANT N. K. (1972). Pyrophosphate hydrolysis in soil as influenced by flooding and fixation. *Proc. Soil Sci. Soc. Am.* **36**, 678–682.

RADULESCU D. and KISS S. (1971). Enzyme activities in the sediments of the Zanoaga and Zanoguta lakes (Retezat mountain mass). *Progrese în Palinologia Românească. Edit. Acad. RSR Bucuresti*, pp. 243–248.

RAMÍREZ-MARTÍNEZ J. R. (1966). Characterization and localization of phosphatase activity in soils, Ph.D. Thesis, University of California, Berkeley (*Diss. Abstr. Int. B* **27**, 3754).

RAMÍREZ-MARTÍNEZ J. R. (1968). Organic phosphorus mineralization and phosphatase activity in soils. *Folia Microbiol., Praha* **13**, 161–174.

RAMÍREZ-MARTÍNEZ J. R. and MCLAREN A. D. (1966a). Determination of soil phosphatase activity by a fluorimetric technique. *Enzymologia* **30**, 243–253.

RAMÍREZ-MARTÍNEZ J. R. and MCLAREN A. D. (1966b). Some factors influencing the determination of phosphatase activity in native soils and in soils sterilized by irradiation. *Enzymologia* **31**, 23–38.

RANKOV V. and DIMITROV G. (1971). Soil phosphatase activity resulting from the manuring of tomato and cabbage. *Pochv. Agrokhim.* (5), 93–98.

RIDGE E. H. and ROVIRA A. D. (1971). Phosphatase activity of intact young wheat roots under sterile and non-sterile conditions. *New Phytol.* **70**, 1017–1026.

ROGERS H. T. (1942). Dephosphorylation of organic phosphorus compounds by soil catalysts. *Soil Sci.* **54**, 439–446.

ROGERS H. T., PEARSON R. W. and PIERRE W. H. (1940). Absorption of organic phosphorus by corn and tomato plants and the mineralizing action of exoenzyme systems of growing roots. *Proc. Soil Sci. Soc. Am.* **5**, 285–291.

ROGERS H. T., PEARSON R. W. and PIERRE W. H. (1942). The source and phosphatase activity of exoenzyme systems of corn and tomato roots. *Soil Sci.* **54**, 353–366.

ROIZIN M. B. and EGOROV V. I. (1972). Biological activity of podzolic soils of the Kola peninsula. *Pochvovedenie* (3), 106–114 (*Soils Fertil.* **35**, 3087).

ROSS D. J. (1975). Studies on a climosequence of soils in tussock grasslands 5. Invertase and amylase activities of topsoils and their relationships with other properties. *N. Z. Jl Sci.* **18**, 511–518.

ROSS D. J. and MCNEILLY B. A. (1975). Studies of a climosequence of soils in tussock grasslands 3. Nitrogen mineralization and protease activity. *N.Z. Jl Sci.* **18**, 361–375.

ROSS D. J., MCNEILLY B. A. and MOLLOY L. F. (1975a). Studies on a climosequence of soils in tussock grasslands 4. Respiratory activities and their relationships with temperature, moisture and soil properties. *N. Z. Jl Sci.* **18**, 377–389.

ROSS D. J., SPEIR T. W., GILTRAP D. J., MCNEILLY B. A. and MOLLOY L. F. (1975b). A principal components analysis of some biochemical activities in a climosequence of soils. *Soil Biol. Biochem.* **7**, 349–355.

RUNKOV S. V. and KOZLOVSKAYA N. A. (1974). On a method for determining phosphatase activity in peat-bog soils. *Agrokhimiya* (4), 142–148 (*Chem. Abstr.* **81**, 12275d).

SAALBACH E. (1956). The influence of humic substances on the metabolism of plants. *Trans. 6th int. Congr. Soil Sci.* **D**, 107–111.

SAMOKHALOV A. N. (1973). Enzyme activity of sod-podzolic soil under sunflowers. *Vest. mosk Univ. Ser. Biol. Pochv. Geol. Geogr.* (3), 109–114 (*Chem. Abstr.* **79**, 114517k).

SAMTSEVICH S. A. (1962). Preparation, use and effectiveness of bacterial fertilizers in the Ukrainian SSR. *Microbiology* **31**, 747–755.

SANIKIDZE G. S., GOGORIKIDZE N. I. and SHONIYA N. K. (1973). Enzymic activity of basic soil types of the subtropical zone of the Western Georgian SSR. *Subtrop. Kul't.* (3), 166–170 (*Chem. Abstr.* **79**, 114397w).

SCHARRER K. and KELLER B. (1940). The distribution, mineralization and absorption of organic phosphoric acid compounds in the soil. *Bodenkd. Pflanzenernähr.* **19**, 109–124 (*Chem. Abstr.* **35**, 3019⁹).

SCHEFFER F., ZEICHMANN W. and ROCHUS W. (1962). The effect of synthetic humic acids on phosphatases. *Naturwissenschaften* **49**, 131–132 (*Chem. Abstr.* **59**, 5017i).

SCOTT N. M. and ANDERSON G. (1976). Organic sulphur fractions in Scottish soils. *J. Sci. Fd Agric.* **27**, 358–366.

SEQUI P. (1974). Soil enzymes. *Ital. Agric.* (10), 92–109.

SHAYKH M. M. and ROBERTS L. W. (1974). A histochemical study of phosphatases in root apical meristems. *Ann. Bot.* **38**, 165–174.

SHCHERBAKOVA T. A., BOROD'KO S. N. and SHIMKO N. A. (1974). Change in the initial level of the enzymic activity of shallow peat soils during cultivation. *Din. Mikrobiol. Protsessov. Poche Obuslovlivayushchie Faktory, Mater. Simp.* **2**, 102–105 (*Chem. Abstr.* **83**, 162786b).

SHIEH T. R. and WARE J. H. (1968). Survey of microorganisms for the production of extracellular phytase. *Appl. Microbiol.* **16**, 1348–1351.

SHIEH T. R., WODZINSKI R. J. and WARE J. H. (1969). Regulation of the formation of acid phosphatases by inorganic phosphate in *Aspergillus ficuum*. *J. Bact.* **100**, 1161–1165.

SIMONYAN B. N. and GALSTYAN A. SH. (1974a). Determination of the degree of erosion of soils by enzyme activity. *Dokl. Akad. Nauk Arm. SSR.* **58**, 44–47 (*Chem. Abstr.* **81**, 2750r).

SIMONYAN B. N. and GALSTYAN A. SH. (1974b). Enzymic activity of eroded soils. *Biol. Zh. Arm.* (4), 60–67 (*Chem. Abstr.* **81**, 76824q).

SKUJIŅŠ J. J. (1967). Enzymes in soil. In "Soil Biochemistry" (A. D. McLaren and G. H. Peterson, Eds) pp. 371–414. Marcel-Dekker Inc., New York.

SKUJIŅŠ J. J. and MCLAREN A. D. (1967). Persistence of enzymic activities in stored and geologically preserved soils. *Enzymologia* **34**, 213–225.

SKUJIŅŠ, J. J., BRAAL L. and MCLAREN A. D. (1962). Characterization of phosphatase in a terrestrial soil sterilized with an electron beam. *Enzymologia* **25**, 125–133.

SOEMINTO B. (1972). Effect of added inorganic sulphate on the soil organic sulphur fractions. *Majalah BATAN* (3), 36–43 (*Chem. Abstr.* **80**, 81340n).

SOREANU I. (1972). Soil enzymic activity in an intensive apple-tree plantation and raspberry during the vegetative period. *Bul. Stiint., Inst. Pedogog., Baia-Mare, Ser. B.* **4**, 113–119 (*Chem. Abstr.* **83**, 42142s).

SPEIR T. W. (1976). Studies on a climosequence of soils in tussock grasslands 8. Urease, phosphatase and sulphatase activities of tussock plant materials and of soil. *N. Z. Jl Sci.* **19**, 383–387.

SPEIR T. W. (1977a). Studies on a climosequence of soils in tussock grasslands 10. Distribution of urease, phosphatase and sulphatase activities in soil fractions. *N. Z. Jl Sci.* **20**, 151–157.

SPEIR T. W. (1977b). Studies on a climosequence of soils in tussock grasslands 11. Urease, phosphatase, and sulphatase activities of topsoils and their relationships with other properties including plant available sulphur. *N. Z. Jl Sci.* **20**, 159–166.

SPEIR T. W. and ROSS D. J. (1975). Effects of storage on the activities of protease, urease, phosphatase and sulphatase in three soils under pasture. *N. Z. Jl Sci.* **18**, 231–237.

SPEIR T. W. and ROSS D. J. (1976). Studies on a climosequence of soils in tussock grasslands 9. Influence of age of *Chionochloa rigida* leaves on enzyme activities. *N. Z. Jl Sci.* **19**, 389–396.

STEBAKOVA V. N. (1972). Biological activity of several soils of the Tselinograd Region and effect of mineral fertilizers on it. *Trudy Tselinograd. Sel'skokhoz. Inst.* (4), 50–58 (*Chem. Abstr.* **80**, 107159n).

STEFAN A. (1963). Exosmosis of phosphates from grains and of acid phosphatase from wheat roots under conditions of stimulation with KBr and glucose. *Stud. Univ. Babeş-Bolyai, Ser. Chem.* **8**, 369–373 (*Chem. Abstr.* **61**, 11026a).

STEFANIC G. (1971). Total soil phosphatasic capacity. *Biol. Sol.* **14**, 10–11.

STEFANIC G., BOERIU I. and DUMITRU L. (1971). Effect of fertilizers and of liming range on total microflora and enzyme activity in clay-illuvial podzolic soil. *Ştiinţa Sol.* (1), 45–54.

STONE E. L. and THORP L. (1971). Lycopodium fairy rings: Effect on soil nutrient release. *Soil Sci. Soc. Am. Proc.* **35**, 991–997.

STULIN A. F. (1975). Phosphatase activity of roots of some winter wheat varieties during the use of fertilizers. *Agrokhimiya* (10), 32–36 (*Chem. Abstr.* **84**, 29762v).

SZEMBER A. (1960a). The action of soil microorganisms in making P from its organic compounds available to plants. I. The ability of soil microorganisms to mineralize organic P compounds. *Annls Univ. Mariae Curie–Skłodowska, Sect. E* **15**, 133–143.

SZEMBER A. (1960b). Influence on plant growth of the breakdown of organic phosphorus compounds by microorganisms. *Pl. Soil* **13**, 147–158.

TABATABAI M. A. and BREMNER J. M. (1969). Use of p-nitrophenyl phosphate for assay of soil phosphatase activity. *Soil Biol. Biochem.* **1**, 301–307.

TABATABAI M. A. and BREMNER J. M. (1970a). Arylsulphatase activity of soils. *Soil Sci. Soc. Am. Proc.* **34**, 225–229.

TABATABAI M. A. and BREMNER J. M. (1970b). Factors affecting soil arylsulphatase activity. *Soil Sci. Soc. Am. Proc.* **34**, 427–429.

TABATABAI M. A. and BREMNER J. M. (1971). Michaelis constants of soil enzymes. *Soil Biol. Biochem.* **3**, 317–323.

TABATABAI M. A. and BREMNER J. M. (1972a). Forms of sulphur and carbon, nitrogen and sulphur relationships, in Iowa soils. *Soil Sci.* **114**, 380–386.

TABATABAI M. A. and BREMNER J. M. (1972b). Distribution of total and available sulphur in selected soils and soil profiles. *Agron. J.* **64**, 40–46.

TESLINOVA N. A. and PERVUSHINA-GROSHEVA A. N. (1973). Enzymic activities of some soil microorganisms as a function of the place they occupy in irrigated typical sierozem. *Uzbek. biol. Zh.* (4), 24–25 (*Chem. Abstr.* **80**, 105641q).

THEODOROU C. (1968). Inositol phosphates in needles of *Pinus radiata* D. Don and the phytase activity of mycorrhizal fungi. *Trans. 9th int. Congr. Soil Sci.* **3**, 483–490.

THOMPSON E. J. and BLACK C. A. (1970a). Changes in extractable organic phosphorus in soil in the presence and absence of plants II. Soil in a simulated rhizosphere. *Pl. Soil* **32**, 161–168.

THOMPSON E. J. and BLACK C. A. (1970b). Changes in extractable organic phosphorus in soil in the presence and absence of plants III. Phosphatase effects. *Pl. Soil* **32**, 335–348.

THOMPSON L. M. and BLACK C. A. (1947). The effect of temperature on the mineralization of soil organic phosphorus. *Soil Sci. Soc. Am. Proc.* **12**, 323–326.

THOMPSON L. M., BLACK C. A. and ZOELLNER J. A. (1954). Occurrence and mineralization or organic phosphorus in soils, with particular reference to associations with nitrogen, carbon, and pH. *Soil Sci.* **77**, 185–196.

THORNTON J. I. and MCLAREN A. D. (1975). Enzymatic characterization of soil evidence. *J. Forensic Sci.* **20**, 674–692.

TOTARSKAYA-MERENOVA V. I. (1956). The direct utilization by higher plants of organic phosphorus compounds. *Pochvovedenie* (12), 17–24 (*Soils Fertil.* **20**, 981).

TYLER G. (1974). Heavy metal pollution and soil enzymatic activity. *Pl. Soil* **41**, 303–311.

TYLER G. (1976a). Influence of vanadium on soil phosphatase activity. *J. Envir. Qual.* **5**, 216–217.

TYLER G. (1976b). Heavy metal pollution, phosphatase activity, and mineralization of organic phosphorus in forest soils. *Soil Biol. Biochem.* **8**, 327–332.

TYUNYAYEVA G. N., MINENKO A. K. and PEN'KOV L. A. (1974). Effect of trifluralin on the biological properties of soil. *Soviet Soil Sci.* (6), 320–324.

VASYUK L. F. (1971). Use of a method for determining enzymic activity in soil-microbiological studies. *In* "Mikrobiol. Osn. Povysh. Plodorodiya Pochv" (L. M. Dorosinskii, Ed.) pp. 151–159. Vses-Nauch.-Issled. Inst. Sel'skokhoz. Mikrobiol., Leningrad (*Chem. Abstr.* **78**, 56909f).

VERSTRAETE W. and VOETS J. P. (1974). Impact in sugarbeet crops of some important pesticide treatment systems on the microbial and enzymatic constitution of the soil. *Meded. Fac. Landbouwwet., Rijksuniv. Gent* **39**, 1263–1267.

VERSTRAETE W., CLAES M., DE BACKERE R. and VOETS J. P. (1974). The influence of Crotodur, Isodur and Nitroform on some soil microbiological populations and processes. *Z. Pflanzenernähr. Düng. Bodenkd.* **135**, 258–266.

VLASYUK P. A. and LISOVAL A. P. (1964). Effect of fertilizers on the phosphatase activity of soil. *Vest. Sel'skokhoz. Nauki, Mosk.* (7), 52–55.

VLASYUK P. A. and LISOVAL A. P. (1969). Activity of some soil enzymes dependent on conditions of plant growth and crop rotation. *Nauk. Pr., Ukr. Sil'.gospod. Akad.* (14), 195–199 (*Chem. Abstr.* **73**, 55187d).

VLASYUK P. A., DOBROTVORSKAYA K. M. and GORDIENKO S. A. (1957). The intensity of enzyme activity in the rhizosphere of certain agricultural plants. *Dokl. Vses. Akad. Sel'skokhoz. Nauk* **3**, 14–19 (*Chem. Abstr.* **51**, 14028c).

VOETS J. P. and DEDEKEN M. (1966). Observations on the microflora and enzymes in the rhizosphere. *Annls Inst. Pasteur, Paris Suppl.* No. 3, pp. 197–207.

VOETS J. P. and VANDAMME E. (1970). The influence of 2-(thiocyanomethylthio) benzothiazole on the soil microflora and enzymes. *Meded. Fac. Landbouwwet., Rijksuniv. Gent.* **35**, 563–580.

VOETS J. P., DEDEKEN M. and BESSEMS E. (1965). The behaviour of some amino acids in gamma irradiated soils. *Naturwissenschaften* **52**, 476.

VOETS J. P., MEERSCHMAN M. and VERSTRAETE W. (1973). Microbiological and biochemical effects of the application of bitumenous emulsions as soil conditioners. *Pl. Soil* **39**, 433–436.

VOETS J. P., MEERSCHMAN P. and VERSTRAETE W. (1974). Soil microbiological and biochemical effects of long-term atrazine applications. *Soil Biol. Biochem.* **6**, 149–152.

VUKHRER E. G. and SHAMSHIYEVA K. T. (1968). Activity of some enzymes in Central Tien-Shan soils. *Soviet Soil Sci.* (3), 390–395.

WHITEHEAD J. E. M., MORRISON A. R. and YOUNG L. (1952). Bacterial arylsulphatase. *Biochem. J.* **51**, 585–594.

WILLIAMS C. H. (1975). The chemical nature of sulphur compounds in soil. *In* "Sulphur in Australasian Agriculture" (K. D. McLachlan, Ed.) pp. 21–30. Sydney University Press, Sydney.

WILLIAMSON B. and ALEXANDER I. J. (1975). Acid phosphatase localized in the sheath of beech mycorrhiza. *Soil Biol. Biochem.* **7**, 195–198.

YAROSHEVICH I. V. (1966). Effect of 50-year application of fertilizers in rotation on the biological activity of chernozem. *Agrokhimiya* **6**, 14–19 (*Chem. Abstr.* **65**, 9685b).

ZAYATA A. N. and KOCHKINA V. K. (1971). Effect of irrigating with industrial waste waters of the May Day Chemical Plant on the activity of the main soil enzymes and on the level and activity of sodium cations. *Pochv. Agrokhim.* pp. 60–62 (*Chem. Abstr.* **79**, 104286v).

ZUBETS T. P. (1973). Residual action of simizine and atrazine on the microflora and enzyme activity in sod-podzolic soil. *Nauch. Trudy, Sev.-Zapadn. Nauchno-issled. Inst. Sel'skokhoz.* **24**, 103–109 (*Chem. Abstr.* **82**, 12090k).

References added in proof

BROWMAN M. G. and TABATABAI M. A. (1977). Phosphodiesterase activity of soils. *Agron, Abstr.* **1977**, 146.

EIVAZI F. and TABATABAI M. A. (1977). Phosphatases in soils. *Soil Biol. Biochem.* **9**, 167–172.

FLOYD M. and SOMMERS L. E. (1976). Properties and stability of phosphatase enzymes in soils irrigated with waste water. *Agron. Abstr.* **1976**, 136.

GOULD W. D., COLEMAN D. C. and RUBINK A. J. (1977). Soil phosphatase activity. *Agron. Abstr.* 1977, 148.

GREENWOOD A. J. and LEWIS D. H. (1977). Phosphatases and the utilisation of inositol hexaphosphate by soil yeasts of the genus *Cryptococcus. Soil Biol. Biochem.* **9**, 161–166.

HAYANO K. (1977). Extraction and properties of phosphodiesterase from a forest soil. *Soil Biol. Biochem.* **9**, 221–223.

JUMA N. J. and TABATABAI M. A. (1977). Effects of trace elements on phosphatase activity in soils. *Soil Sci. Soc. Am.J.* **41**, 343–346.

KHAZIEV F. KH. (1976). Michaelis constants of soil enzymes. *Pochvovedenie* (8), 150–156.

PETTIT N. M., GREGORY L. J., FREEDMAN R. B. and BURNS R. G. (1977). Differential stabilities of soil enzymes Assay and properties of phosphatase and arylsulphatase. *Biochim. Biophys. Acta* **485**, 357–366.

VERSTRAETE W. and VOETS J. P. (1977). Soil microbial and biochemical characteristics in relation to soil management and fertility. *Soil Biol. Biochem.* **9**, 253–258.

VOROB'EVA E. A. and GORCHARUK L. M. (1975). Enzymatic activity of soils in the vertical series typical of the Caucasus State Reserve. *Moscow Univ. Soil Sci. Bull.* **30** (6), 41–43.

7

Interactions Between Agrochemicals and Soil Enzymes

S. CERVELLI, P. NANNIPIERI and P. SEQUI

CNR, Laboratory for Soil Chemistry, Pisa, Italy

I. Introduction

The definitions of "soil enzyme" and "agrochemical" are numerous, ambiguous and far from universally understood. A clarification of the terms employed is therefore imperative, before the interactions between agrochemicals and soil enzymes, can be considered.

The concept of "soil enzyme" has been discussed in other sections of this book. Strictly speaking soil enzymes are extracellular enzymes, somehow stabilized against adverse conditions and partially independent of changes occurring in the living organisms. For instance, acetylcholine esterase or the enzymes of the Calvin cycle cannot be considered as soil enzymes, although they are well represented in soil animals and algae. However, the unequivocal determination of soil extracellular enzyme activities is difficult and sometimes impossible. Therefore, to describe accumulated enzymes as soil enzymes can often prove more satisfying than to use the term extracellular enzymes (Kiss et al., 1972, 1975a; Sequi, 1974; Skujiņš, 1976). Accumulated enzymes include the activity of non-proliferating microbial cells and could even be thought of as a basic soil property, because starvation seems to be a normal state of soil microorganisms whilst proliferation may be considered an exceptional condition (Babiuk and Paul, 1970; Gray and Williams, 1971). A soil enzyme is a concept which can henceforth vary according to the development of methodologies. In this chapter soil enzymes will be synonymous with accumulated enzymes, although the distinction between extra- and intracellular enzymes will be effected whenever possible.

"Agrochemical" is almost a neologism and the current meaning of this word is relatively recent. Etymologically, any chemical product applied to the soil in agronomic practice is an agrochemical. A section of the recent meeting on "Agrochemicals in Soils" of the International Soil Science Society which took place in Jerusalem from 13 to 18 June 1976 was devoted to the use of sewage and sludge effluents in crop production. Since all existing substances are of a chemical nature, it seems evident that almost any compound could be considered as an agrochemical if it is applied to the soil.

In this review it may be convenient to include among agrochemicals all substances derived from an industrial process which have a more or less definite chemical composition and do not contain a significant microbial population per se. However, such a definition is not entirely satisfactory. First of all organic wastes, from domestic or industrial sources, may have varied and unpredictable chemical compositions and contain a high number of microorganisms when compared to organic residues of agricultural origin, such as a farmyard manure, poultry manure, cattle or pig slurries. Secondly many organic fertilizers (hoof, horn, bone meal, dried blood, leather waste, fish meal and so on) and every organic manure (such as sawdust) have a definite chemical composition and yet contain very little microbial life. Thirdly some inorganic (e.g. sodium nitrate) and organic agrochemicals (e.g. nicotine) can arise from either industrial processes or natural sources.

II. Transformation of Agrochemicals by Soil Enzymes

Two main difficulties hinder the writing of an up-to-date survey of the range of agrochemicals transformed by soil enzymes. First, as emphasized in the introduction, it is often difficult to recognize the true nature of the soil enzyme activity: that is whether it is purely extracellular, accumulated or induced by the proliferation of microorganisms. Secondly, the history of this aspect of soil enzyme research is recent. The first paper concerning the interactions between agrochemicals and soil enzymes, except for a few scattered reports (e.g. Rotini, 1935a; Conrad, 1940), appeared at the end of the 1950s (Kiss, 1958), and even if the topic is now expanding rapidly, only a very few interactions are clearly understood.

A main object of this section is to emphasize that substrates of soil enzymes include chemical products widely employed in agronomic practice. Since, for some transformations, it is possible to describe the biochemistry, only these will be considered as general models. In fact, an enumeration of all the agrochemicals involved in enzyme catalysed reactions would be both extensive and unnecessary and will therefore be avoided.

1. Fertilizers

1.1. Nitrogen fertilizers

Calcium cyanamide was the first nitrogen fertilizer produced from an industrial process and its use in agriculture is now decreasing. Owing to the toxic properties of CN_2^{2-}, calcium cyanamide is also effective as a pesticide (see the review of Cornforth, 1971).

The metabolic degradation of calcium cyanamide in soil involves its transformation to urea; the nature of which is still rather controversial. According to Schmalfuss (1938) it was a question of mainly enzymatic reactions, while Rotini (1940) and, more recently Rotini et al. (1967, 1971), maintained that only inorganic catalysts were responsible for calcium cyanamide degradation to urea, because of the high activation energy of the reaction. The transformation occurred even in soil samples previously sterilized by steaming or autoclaving, and was correlated both with manganese dioxide content in soil and, to a lesser extent, with ferric and aluminium hydroxides. Inorganic catalysts have also been reported as the only factors responsible for the transformation by Ulpiani (1910), Stutzer and Reis (1910), Kappen (1910), Cowie (1919, 1920), Pratolongo et al. (1934), again by Rotini (1935b, 1935c, 1939) and by Temme (1948) and Lotti (1955).

Ernst (1967) followed decomposition of ^{15}N-labelled calcium cyanamide in untreated soils and in soil samples sterilized either by γ-irradiation or by alternate autoclave and ethylene oxide treatments. In γ-irradiated soils calcium cyanamide transformation was attributed both to inorganic catalysts and to accumulated enzymes, while in autoclaved or ethylene oxide treated soil, reaction was only imputed to inorganic catalysis. Contribution of the enzymes present in proliferating microorganisms ranged from 50 to 80% of the total transformation, whilst accumulated enzymes contributed from 20 to 50%. Only 8% of the transformation was ascribed to inorganic catalysts.

Urea is a common nitrogen fertilizer. It is hydrolysed to ammonia and carbon dioxide by soil urease. Much evidence suggests that urease is present in soil mainly as accumulated enzyme. Several extraction procedures (Briggs and Segal, 1963; Burns et al., 1972a, b; Nannipieri et al., 1974, 1975) have been successfully employed in the extraction of urease-active fractions from soil. Paulson and Kurtz (1969) measured the changes in urease activity and in numbers of ureolytic microorganisms following addition of nutrients to the soil; their findings show that an extracellular background of enzyme activity exists in soil. Urease has been defined "an enzyme ubiquitous in soil" (McLaren and Pukite, 1975); literature concerning soil urease has been extensively reviewed in another section of this book.

Nitrification and denitrification are processes too complex to depend only on the presence of enzymes accumulated in soil. After treating soils with heavy γ-irradiation, the rapid oxidation of NH_3 to NO_3^- has been attributed to non-proliferating cells of the nitrifying bacteria rather than increased proliferation of survivors (Cawse 1968). The enzymes participating in nitrification remained active in the cells which lost their viability following irradiation.

Some conclusions were suggested to explain reduction of $^{15}NO_3^-$ to $^{15}NO_2^-$ in fresh soils following treatment with radiation (Cawse and Cornfield, 1969). No increase in the amount of nitrite has been observed if the irradiation was preceded by autoclaving (Cawse and Cornfield, 1972). This was explained by the fact that heat denatured the nitrate reductase. However if the γ-ray treatment is very heavy, autoclaving before irradiation fails to stop nitrite formation, which almost certainly occurs by γ-radiolysis of nitrate (Cawse and Cornfield, 1972).

1.2. Phosphate fertilizers

No doubt phosphatase is the most studied among the enzymes of the phosphorus cycle. Speir and Ross deal with this topic in Chapter 6 of this book. Rotini (1951) and Rotini and Carloni (1953) investigated hydrolysis of metaphosphates in soil. Metaphosphates were transformed to orthophosphates both by inorganic catalysts and by soil enzymes; inorganic catalysts contributed about two thirds of the transformation. The occurrence of both the enzymatic and the inorganic catalysis has been confirmed also for hydrolysis of pyrophosphates to orthophosphates by Gilliam and Sample (1968) and by Sutton et al. (1966). In flooded soils, Hossner and Philips (1971) showed that the activation energy was 18900 J.mol^{-1} for the pyrophosphatase hydrolysis and 105000 J mol^{-1} for inorganic catalysis. Khaziev (1972) and Racz and Savant (1972) found no activity in autoclaved soil samples. Analytical methods for the assay of soil pyrophosphatase activity have been recently criticized by Douglas et al. (1976).

2. Herbicides

In spite of the many studies on herbicides in soil, knowledge of the action of soil enzymes on this group of pesticides is scarce. Some information is available on the biochemical transformations of halogen-substituted anilines, which have been related to the presence of soil enzymes. Chloroanilines are released in soil during biological decay of some herbicides, such as phenylacylamides, phenylcarbamates and substituted ureas (Geissbühler et al., 1963; Kaufman and Kearney, 1965; Dalton et al., 1966; Bartha and Pramer, 1967). The compound 3, 4-dichloropropionanilide (propanil) is commonly applied as a post-emergence herbicide and selectively controls barnyard grass and other annual weeds in rice (Smith, 1965). Bartha and Pramer (1967) found that propanil was hydrolysed to 3, 4-dichloroaniline (DCA) and to 3, 3′, 4, 4′-tetrachloroazobenzene (TCAB) in two different soils (Fig. 1).

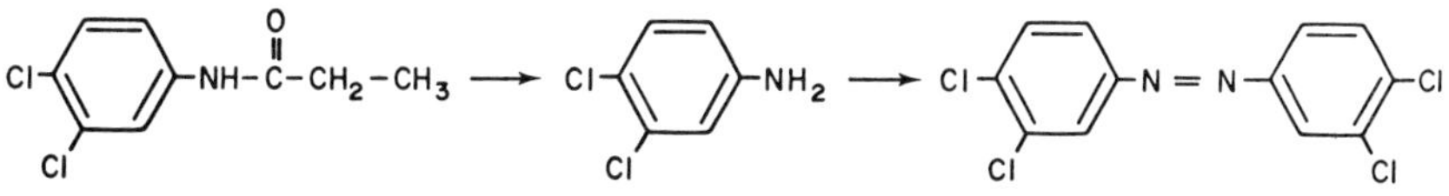

Fig. 1. Transformation of propanil in soil.

In contrast to other reports (Chisaka and Kearney, 1970), Burge (1972) found that soil texture and organic matter content were unrelated to propanil decomposition and to amounts of DCA and TCAB formed in five different soils. Burge (1972) investigated the role of soil arylacylamidase in hydrolysing the amide bond of propanil to DCA. Hydrolytic activity increased after sonification, which may suggest that an endocellular enzyme is responsible for degradation; no data are available on persistence of the enzyme outside the cells. On the other hand, the fact that no relation was found between the most probable number of propanil-decomposing bacteria and percentage of DCA produced from propanil, suggests the presence of some amounts of free enzyme in soil.

Recently Hoagland (1974) has found that arylacylamidase activity extracted from dandelion roots was inhibited some 73% by catechol. Since catechol or other aromatic compounds are present especially in organic soils, it is possible that if an extracellular arylacylamidase exists it is in part inhibited.

Saunders *et al.* (1964) have reported transformations of substituted anilines catalysed by horseradish peroxidase; Bartha *et al.*, (1968) and Bartha and Bordeleau (1969a, b) showed that soil peroxidases transform chloroanilines to chloroazobenzenes. Using 0·05 M phosphate buffer at pH 6·0 they were able to extract peroxidase from soils. Unfortunately the authors neither compared peroxidase activity in whole soils with the extracted activity nor gave the yield of extraction. They found that extracted peroxidase activity was correlated with the capacity of whole soils to convert chloroaniline to chloroazobenzene. These studies disclosed a way for further investigations aimed at revealing the possible chemical pathway of the transformation.

Sprott and Corke (1971) showed that the aerobic transformation of chloroanilines occurs biologically; at the same time they found that, following irradiation or addition of sodium azide, TCAB is initially inclined to accumulate and then to disappear. They suggested that the disappearance of TCAB could be ascribed either to residual accumulated enzyme activity or to a chemical reaction occurring through formation of free radicals. It is probable that TCAB disappearance was due to microbial activity after the effect of sodium azide declined. Recently, Kelley and Rodriguez-Kabana (1975) found reduced populations of soil bacteria two weeks after its treatment with sodium azide. However, five weeks after azide application, bacterial populations were higher in treated than untreated soils.

Bordeleau and Bartha (1972a, b) identified, in a sandy loam, a number of microbial activities responsible for the transformation of DCA to TCAB, and investigated, in detail, a strain of the soil fungus *Geotrichum candidum*. Extracellular peroxidase and aniline oxidase were isolated from the fungus cultures and their behaviour towards temperature, pH and kinetic parameters were reported.

Burge (1973) did not find any relationship between the amount of TCAB produced and the numbers of fungi and bacteria able to synthesize peroxidase. Among the five soils examined an apparently cell-free peroxidase was found in only one sample, where a very limited ability to produce TCAB was ascertained. Burge (1973) suggested that a cell-bound peroxidase was present in addition to a cell-free peroxidase; cell-free peroxidase could occur in the soil after the lysis of the cells containing peroxidase. After adding energy sources to soil, peroxidase activity increased without a parallel increase in the TCAB production. Not all soil peroxidases were suggested as able to catalyse the conversion of DCA to TCAB. On the other hand, Lay and Ilnicki (1974) found a correlation between peroxidase activity and TCAB produced after incubating the soil in the presence of both carbon and nitrogen sources. Kearney *et al.* (1970) showed that at high propanil concentrations DCA may undergo several transformations with the formation of TCAB, carbon dioxide, chlorides and coloured products more polar than TCAB. Kearney and Plimmer (1972) identified 3, 4- dichloroformanilide among the DCA transformation products.

Formation of higher polymers of DCA during propanil decay has been shown by Plimmer *et al.* (1970). Among products, 1, 3-bis (3, 4-dichlorophenyl) triazene was identified. The low amounts of TCAB recovered have been suggested by Burge (1973) as a consequence of the formation of polymers of high molecular weights. Rosen *et al.* (1970) and Linke (1970) showed the formation of the asymmetric 4-(3, 4-dichloroanilino)-3, 3′, 4′-trichloroazobenzene. Bartha (1975) has summarized the pathway of microbial transformation of propanil in soil.

Bartha (1971b) demonstrated that air-drying reduces the transformation of DCA to TCAB but not that of propanil to DCA. For these reasons he claimed the necessity to carry out biodegradation studies in biochemically intact soil samples. Evidence suggests (Bartha, 1971a) that only a small percentage of ^{14}C was lost from propanil and 4-chloroaniline. The bulk of the immobilized residues consists of intact chloroanilines that are chemically bound to humic substances. Such complexing may extend the residual life of chloroanilines in soil to as much as ten years. Two different types of chemical bonds between chloroanilines and humic substances have been suggested (Hsu and Bartha, 1974). The occurrence of such complexes, as well as the formation of higher polymers of DCA, could explain why the uneven formation of TCAB from DCA is reported not to take place (Sokolov *et al.*, 1974; Sprott and Corke, 1971; Hughes and Corke, 1974) and the lack of correlation between soil peroxidase and TCAB formed (Burge, 1973).

Concerning the mechanism of DCA transformation, Bordeleau and Bartha (1970) proposed the formation of an intermediate compound on the basis of biological oxidation processes. Electron removal in natural processes follows different steps and therefore the four-electron transfer required for the conversion of two DCA molecules to one of TCAB must involve the formation of an intermediate having an oxidation state higher than DCA and lower than TCAB. Of the various proposed intermediates, phenylhydroxylamines seem to be most likely present, because of their capacity to react with DCA to form azocompounds. The suggested mechanism (Bordeleau and Bartha, 1970) should involve two steps: a peroxidase catalysed oxidation of DCA to chlorophenylhydroxylamine and a nonenzymatic reaction between DCA and phenylhydroxylamine. Bordeleau *et al.* (1972) proposed a modified pathway which is presented in Fig. 2. It involves the presence of more intermediates some of them not yet

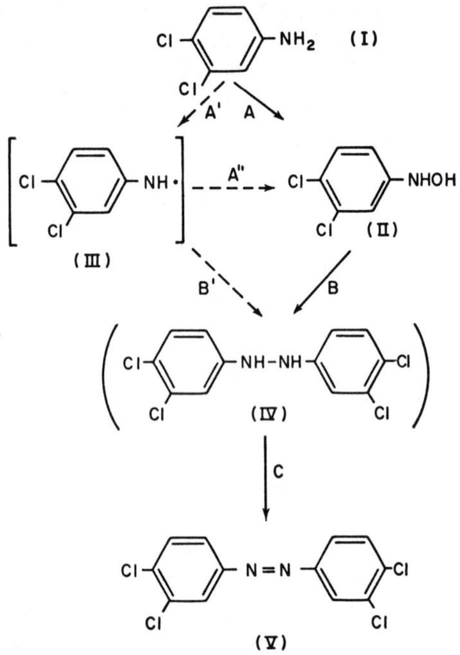

Fig. 2. Proposed pathway of TCAB formation in soil. Redrawn from Bordeleau *et al.* (1972).

identified experimentally. The compounds I, II and V have been proved to exist. Reactions A and A′ are catalysed by peroxidase while reactions A″, B, B′ and C are of a non-enzymatic nature.

3. Other pesticides

Degradation in soil of a number of insecticides (malathion, ciodrin, dichlorvos, mevinphos, methyl parathion, parathion, dimethoate, zinophos, dursban and GS 13005) has been studied by Getzin and Rosefield (1968). The hydrolytic activity of γ-irradiated soils was found lower with respect to untreated samples but higher than in soils sterilized by autoclaving. Among the different insecticides, parathion, dimethoate and GS 13005 were not hydrolysed; methyl parathion, zinophos and dursban were only partially resistant; whilst malathion, ciodrin, dichlorvos and mevinphos were easily hydrolysed.

Getzin and Rosefield (1968) attempted to extract the hydrolytic enzyme by treating the soil with a range of solutions normally employed for the extraction of the organic matter. Only 0·2 N NaOH was effective as an extractant, yielding an enzyme activity that hydrolysed malathion but not the other insecticides. The other esterase activities were therefore non-extractable or inactivated by weak alkali treatment.

An enzyme preparation active in the degradation of malathion had already been extracted from soil organisms by Matsumura and Boush (1966). It seems reasonable to suggest that the malathion esterase obtained by Getzin and Rosefield (1968) could be a soil extracellular enzyme unrelated to immediate microbial growth; γ-ray treatment inhibits growth of microorganisms but does not affect the activity of several enzymes (McLaren et al., 1957; Peterson, 1962). On the other hand, enzymes are usually rapidly inactivated by treatment with 0·2 N NaOH, unlike the malathion esterase extracted by Getzin and Rosefield (1968) which was reported to maintain its activity in 0·2 N NaOH at room temperature for more than one week.

About 85% of the heat-labile esterase activity was preserved for three months in irradiated soils, demonstrating it was a permanent soil constituent. Further evidence of the presence in soil of a cell-free enzyme which degrades malathion was again provided by Getzin and Rosefield (1971). A partially purified enzyme preparation was added to a steam sterilized soil sample; the sample was then incubated after mixing with non-sterile soil. The enzyme activity was strongly adsorbed by the soil constituents (so that extraction with Tris–HCl buffer at pH 7·0 was unsuccessful) and appeared markedly stable: active fractions were recovered for the duration of the experiment (eight weeks).

Burns and Gibson (1976) showed that malathion was rapidly degraded in both non-sterile and irradiated soil, but not in autoclaved soil. Breakdown was rapid in separated organic and organo–mineral fractions, somewhat slower in sand and silt and non-existent in all autoclaved fractions. On the basis of soil urease properties (Burns et al. 1972b; Pettit et al., 1976), Burns and Gibson (1976) suggested that the breakdown in sand, silt and clay fractions, deprived of organic matter, implies microbial and non-biological metabolism, whilst the exoenzyme malathion esterase is responsible for the rapid hydrolysis of the pesticide in organic and organo–mineral fractions.

Soil malathion esterase, a carboxyl esterase, is surprisingly resistant to proteolytic enzymes and comparatively insensitive to inhibitors like heavy metals. After incubation in the presence of hyaluronidase its activity doubles, although its stability decreases. Chemical characteristics and reaction response to hyaluronidase justify its classification as a glycoenzyme; its uncommon stability and persistence in soil as an extracellular enzyme have been ascribed to the properties of the protein–carbohydrate complex. Also, Mayaudon et al. (1975) have found that the resistance of soil protease to proteolysis was due to glycides bound to the enzyme. Malathion is hydrolysed to a monoester according to typical Michaelis–Menten kinetics (Satyanarayana and Getzin, 1973).

4. Other substances

A range of very different agrochemicals, far removed in character from fertilizers and pesticides, can reach the soil. One important class of compounds includes ionic surfactants containing sulphur; data on soil sulphatases have been reviewed by Freney and Swaby (1975) and in Chapter 6 of this book. Other compounds, such as hormones and several soil conditioners, are likely to be transformed by the enzymes. It seems possible that, although the range of agrochemicals is increasing and many microorganisms can adapt to transform them, only a limited number of relevant soil enzymes exist in an accumulated or in a truly extracellular state. However, this field of research is open and warrants further study.

III. Influence of Agrochemicals on Soil Enzymes

1. Current status of literature

Evidence for the direct influence of agrochemicals on soil enzymes is mostly circumstantial, for example, due to inhibition of the active site and current literature about direct action of agrochemicals is scarce. Many authors have studied indirect effects, like variations of selected enzyme activities in soil following application of agrochemicals, but in spite of the abundant bibliography, a critical survey of the literature concerning their effects on soil enzymes is a rather complicated undertaking.

As a matter of fact, enzyme activities are often determined only as indexes of soil fertility. Actually, a correct assessment of soil fertility based on measured values of selected enzyme activities is disputable, and this approach gives little information on the nature of the enzymes involved. In fact, many reports only show positive or negative effects on soil enzyme activities due to agrochemicals applied, and discussion concerning the nature and properties of the selected enzyme activities is minimal. References on the effects of the main classes of agrochemicals on soil enzyme activities are shown in Tables 1, 2 and 3.

TABLE 1

References concerning the influence of fertilizers on soil enzymes.

AUTHORS	Invertase	Phosphatase	Protease	Urease	Catalase	Dehydrogenase	Other Enzymes
Ambrož (1973)							Cellulase
Blagoveshchenskaya and Danchenko (1974)	x	x	x	x	x	x	Cellulase, Asparaginase.
Chunderova and Zubets (1969a)	x						
Eliade et al. (1972)	x				x		
Gamzikova (1968)	x						
Gavrilova et al. (1973)	x						
Goian (1968, 1970, 1972)	x						
Jaggi (1954)			x		x	x	
Khan (1970)	x	x		x	x	x	
Markert (1974)			x				
Muresanu and Goian (1969)	x						
Namdeo and Dube (1973a, b)			x	x			
Papacostea et al.[a]				x	x		
Papacostea et al. (1972)					x		
Rankov and Dimitrov (1971)	x						
Rosa et al. (1970)	x						
Rysavy and Macura (1972)							β-Glucosidase
Sipos et al. (1972)	x				x		
Soreanu (1972)	x						
Stefanic[a]			x	x			
Stefanic (1970)	x	x	x	x			
Stefanic et al. (1971)	x	x		x		x	Gelatinase
Stefanic et al. (1972)					x	x	
Stratonovic and Evdokimova (1975)					x	x	
Zahan and Soreanu (1969)	x						

[a] Unpublished data. Quoted by Kiss et al. (1975b).

TABLE 2

References concerning the influence of herbicides on soil enzymes.

AUTHORS	Invertase	Phosphatase	Protease	Urease	Catalase	Dehydrogenase	Other enzymes
Balasubramanian et al. (1973)	x						Amylase
,, and Siddaramappa (1974)	x						Amylase
Beck (1970)					x		
Beck (1973)				x	x	x	Amylase
Beckmann (1970)[a]					x		
Beckmann and Pestemer (1975)					x		
Berthold (1970)				x	x		
Bliyev (1973)	x				x		Amylase
Cervelli et al. (1976, 1977)				x			
Chulakov and Zharasov (1973)	x	x	x				
Chunderova and Zubets (1969)b	x	x	x	x			Amylase
Chunderova and Zubets (1971)	x	x	x	x			
Chunderova et al. (1971)	x	x	x	x			
Gamzikova (1968)	x						
Gamzikova and Syvatskaya (1968, 1970)	x		x				
Geshtovt et al. (1974)	x						
Ghinea (1964)	x	x					
Ghinea and Stefanic (1972)					x		
Giardina et al. (1970)				x	x		Cellulase
Goian (1969)		x					
Gruzdev et al. (1973)	x	x		x	x		
Hauke-Pacewiczowa (1971)					x		
Hülsenberg (1966)[a]					x		
Ishizawa et al. (1961)				x			
Karki et al. (1973a, b)					x		
Keller (1961)	x						β-Glucosidase
Kiss (1958)	x						
Klein et al. (1971)					x		
Krezel and Musial (1969)	x			x	x	x	Asparaginase, Gelatinase
Krezel and Musial[b]				x			
Krokhalev et al. (1973)	x			x	x		
Kruglov and Bei-Bienko (1969)	x	x	x	x			
Kruglov et al. (1973)	x			x	x		
Kulinska (1967)	x			x	x		
Latypova et al. (1968)				x			
Lay and Ilnicki (1974)							Peroxidase
Lenhard (1959)					x		
Leusheva and Mel'nick (1969)				x			

AUTHORS	Invertase	Phosphatase	Protease	Urease	Catalase	Dehydrogenase	Other Enzymes
Livens et al. (1973)	x	x		x		x	
Manorik and Malichenko (1969)	x	x					
Markert and Kundler (1975)		x					
Mayaudon et al. (1973)							Diphenoloiydase
Mereshko (1969)					x		
Namdeo and Dube (1973a, b)			x	x			
Naumann (1970a, b)						x	
Nikitin and Svechkov (1973)	x			x	x		
Odu and Horsfall (1971)						x	
Padenov and Molchan (1975)			x	x			
Peeples and Curl (1970)	x						
Pel'tser (1972)			x				
Pestemer and Beckman (1975)						x	
Protasov (1968, 1970)						x	
Putintseva (1970)						x	
Quilt (1972)			x				
Rankov (1970)							Cellulase
Soldatov (1968)					x		
Soldatov et al. (1971)					x		
Soreanu (1972)	x						
Spiridonov and Spiridonova (1973)			x		x	x	Peroxidase
Spiridonov et al. (1970, 1973)					x	x	Peroxidase
Svyatskaya (1972)	x		x				
Tyunyayeva et al. (1974)	x		x				
Ulasevich and Drach (1971)						x	
Verstrate and Voets (1974)	x	x		x			β-Glucosidase
Voets et al. (1974)	x	x		x			β-Glucosidase
Voinova et al. (1969)							Cellulase
Walter (1970)					x	x	
Walter (1971)	x	x		x		x	β-Glucosidase
Walter and Bastgen (1971)[a]	x	x		x		x	β-Glucosidase
Yeates et al. (1976)						x	
Zinchenko and Osinskaya (1969)	x			x	x		
Zinchenko et al. (1969)	x			x	x		
Zubets (1967)	x	x	x	x	x		Amylase
Zubets (1968a, 1973a)	x	x	x	x			
Zubets (1968b)	x	x	x	x			Amylase

[a] Cited by Kiss et al. (1975a).
[b] Unpublished data. Quoted by Grossbard and Davies (1976).

TABLE 3

References concerning the influence of non-herbicidal pesticides on soil enzymes

AUTHORS	Amylase	Invertase	Phosphatase	Urease	Catalase	Dehydrogenase	Other Enzymes
Abdel'Yussif et al. (1976)		x	x	x	x		
Ampova and Stefanov (1976)		x		x	x		
Balasubramanian and Patil (1968)[a]	x	x					
Balasubramanian et al. (1970)	x	x					
Balasubramanian et al. (1973)	x	x					
Bhavanandan and Fernando (1970)				x			
Dommergues (1959)	x						
Goian (1969)		x					
Hofer et al. (1971)	x				x	x	
Ishizawa et al. (1961)				x			
Karanth and Vasantharajan (1973)						x	
Karanth et al. (1975)		x		x		x	Tryptophanase, Phytase
Ladd et al. (1976)							Protease
Lethbridge and Burns (1976)				x			
Livens et al. (1973)	x	x	x		x		
Markert (1974)				x			
Markert and Kundler (1975)				x			
Naumann (1970a, c, 1972)						x	
Pel'tser (1972)				x			
Teuber and Poschenrieder (1964)	x	x					
Tsirkov (1970)					x	x	Protease
Van Faassen (1973, 1974)						x	
Verstraete and Voets (1974)	x	x	x				β-Glucosidase
Voets and Vandamme (1970)	x	x	x				β-Glucosidase

[a] Cited by Kiss et al. (1975a).

2. Direct effects of agrochemicals on soil enzymes

Urease and its inhibition in soil have been considered by Bremner and Mulvaney in chapter 5 of this book. Only one example of the direct effect of herbicides on soil urease will be reported here. Substituted ureas (Fig. 3), a class of commonly used herbicides, can inhibit jack bean urease differently as a function of both the halogen substitution on the phenyl ring and the steric hindrance of molecular substituents (Cervelli et al., 1975c). It was

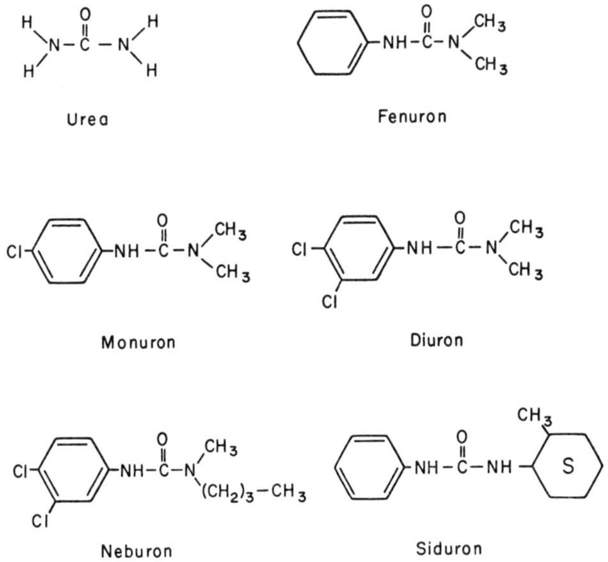

Urea

Fenuron

Monuron

Diuron

Neburon

Siduron

Fig. 3. Urea and some substituted ureas.

proposed that the enzyme molecule reacts with inhibitors by means of the oxygen atom of the carboxyl group in the substituted ureas forming a complex which can be stabilized by resonance (Fig. 4). Such herbicides may even provide an excellent tool to study the mechanism of action of the urease itself. Similar effects have been reported for substituted ureas in soil (Cervelli *et al.*, 1976, 1977) and it is reasonable to suggest a direct effect of these herbicides on free extracellular and accumulated soil urease.

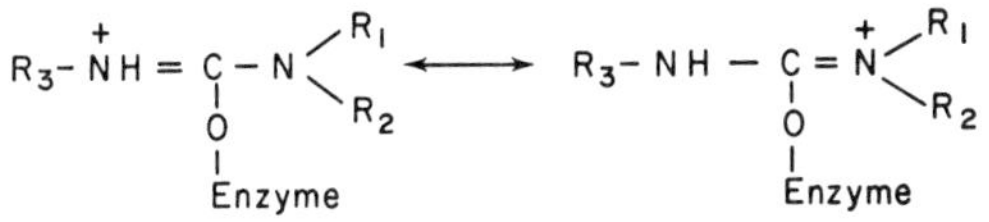

Fig. 4. Proposed resonance formulae of urease-inhibitors complex. From Cervelli *et al.* (1975c).

Recently Corbett (1975), in discussing the biochemical design of pesticides, pointed out that pesticides can be built up on the basis of their supposed interaction with target enzymes. Once the enzyme has been selected, an inhibitor could be designed for the active site that will bind either covalently or non-covalently. Inhibitors that bind non-covalently are usually close analogues of the substrate that compete with it for the site and with which they are in reversible equilibrium. In general, the occurrence of direct effects is foreseeable whenever the molecular shape of the agrochemical can fit, at least partially, either the active site or an allosteric site of a soil enzyme. Unfortunately few studies on this matter are available at present.

The soil matrix can strongly influence the behaviour of agrochemicals, as will be discussed hereafter. In particular, inorganic agrochemicals can be involved in a variety of reactions and their effect on soil enzymes can only be postulated from experimental results obtained with purified soil extracts. In this sense, soil enzymes are generally reported to be surprisingly indifferent to several factors which cause activation or inhibition of enzymes extracted from microbes or plant and animal tissues. For instance, malathion esterase, partially purified from soil extracts, is unusually stable in the presence of metal ions and most of the common enzyme inhibitors: it is partially inhibited only by the heavy metals Ag^+, Hg^{2+} and Pb^{2+} at high concentrations (Satyanarayana and Getzin, 1973). Therefore a high content of heavy metals in fertilizers could play a secondary role in influencing the activity of some soil enzymes.

3. Indirect effects of agrochemicals on soil enzymes

3.1. Methodological problems

Any indirect effect of agrochemicals on soil enzymes will involve the action of an agrochemical on soil organisms which, in time, contribute to accumulated enzyme activity. The study of these indirect effects may give an insight into the genesis of soil enzymes.

3.2. Alteration of life functions of soil organisms

Soil enzymes are produced by living organisms. It seems obvious therefore that any action altering the life functions of soil organisms could indirectly affect soil enzyme activities. This will be true both if enzymes are excreted by or are accumulated in microorganisms, also if they are derived from plant residues produced by the actions of bacteria, fungi or even earthworms. The level of each enzyme activity should vary in relation to its specific turnover in soil.

Agrochemicals are often a source of nutrients and consequently stimulate growth of soil populations. Fertilization, in general, promotes growth of soil organisms, although some fertilizers can inhibit specific groups; for instance, nitrogen fertilization decreases nitrogen fixing activities in soil and earthworms are reduced by acidifying fertilizers. On the other hand, application of pesticides enhances those bacteria which are able to degrade and utilize them as a source of nutrients.

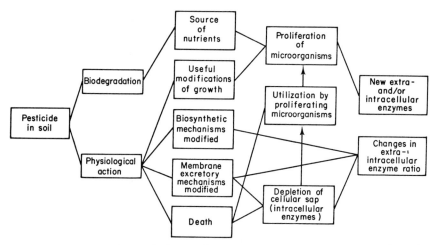

Fig. 5. An outline of some indirect effects of pesticides on soil enzymes through their action on soil microorganisms.

A pesticide, if not immediately degraded in soil, may exert a physiological effect on living organisms (Fig. 5). If the organisms are not able to survive, their death may involve cell lysis and a temporary increase of the extracellular fraction of soil enzymes. Such an effect could be a consequence of necrosis of plant root cells also, although it may be difficult to think about a change of extracellular enzymes produced in soil in this way. More generally, dead tissues and cells will be utilized by other soil organisms; a similar effect could be induced by modifications produced by any agrochemical on the composition of the waste products of soil animals, or of the earthworm casts. If the organisms are able to survive the application of the agrochemical, other effects may become apparent. Inhibition of some enzymes may occur inside the cells. Also, inhibition of the biosynthesis of some enzymes may occur.

A model of the action of some herbicides and pesticides on soil organisms is shown in Table 4. It includes the effects on most organisms: earthworms, algae, fungi and bacteria; uncertainty about genesis of soil enzymes can justify the tentative generalization of the scheme. The effect of plant roots is not included in this scheme but they may also be thought of as soil organisms involved in enzyme genesis; their response to pesticides may resemble that of algae.

Some specific, indirect effects of agrochemicals on soil enzymes will be considered briefly in the following paragraphs.

3.2(a). Modifications of biosynthetic mechanisms
Many of the modifications induced by agrochemicals occur at the level of protein biosynthesis. Distinction between endo- and extracellular enzymes

is essential. If the action of the agrochemical is of importance to the biosynthesis of merely endocellular enzymes, the effect on the pool of extracellular enzymes will be revealed only after the death of the cell. However, the effect should be immediately evident if inhibition occurs at the level of biosynthesis of enzymes normally excreted by the cell. This could be the case of chlordane and aminotriazole, reported to inhibit synthesis of endopeptidase and aminopeptidase in *Aeromonas proteolytica* (Lichfield and Huben, 1973). Unfortunately, our knowledge of the mode of action of supposed inhibitors of biosynthetic reactions is rather scanty and more study with cell-free systems is required to confirm postulated mechanisms (Ashton and Crafts, 1973; Corbett, 1974; Cherry, 1976).

TABLE 4

A scheme of the likely action of selected herbicides and pesticides on soil organisms

Herbicide or pesticide	Life function altered	Primary process involved	Notes
Bipyridylium herbicides, thiolcarbamates, petroleum oils	Structural organization	Disruption of membranes	
Polyoxin D	Structural organization	Inhibition of chitin synthesis	Only arthropods and fungi
Bipyridylium herbicides, ureas, triazines, acylamides, uracils, etc	Energy supply	Diversion and inhibition of photosynthetic electron transport	Only photosynthetic organisms (algae, etc.)
Arsenicals, dialkyl dithiocarbamates	Energy supply	Inhibition of pyruvate dehydrogenase	
Antibiotics, benzimidazoles, N-phenylcarbamates, dinitroanilines	Growth and reproduction	Inhibition of DNA, RNA or protein biosynthesis and cell or nuclear division	
Organophosphorus and carbamate insecticides	Nervous system	Inhibition of acetyl-cholinesterase	Only organisms with nervous system (earthworms, etc.)

3.2(b). Regulation of protein biosynthesis

Induction or repression of the synthesis of specific enzymes could be the most important consequence of the application of an agrochemical. Many enzymes are reported to be strongly influenced by soil treatments. Urease activity, for instance, is non-specifically inducible in soil both by nitrogen compounds (such as ammonium sulphate or carbonate) and by urea itself (Paulson and Kurtz, 1969; Zantua and Bremner, 1976). In contrast, soil cellulase activity is specifically inducible: it increases after incubation with cellulose and decreases after addition of glucose; a cellulase (C_1) is also stimulated by the presence of nitrogen, whilst another cellulase (C_X) increases after addition of phosphates (Ambrož, 1973).

Fertilization can markedly influence soil phosphatase activity (Rankov and Dimitrov, 1971); Muresanu and Goian (1969) showed that phosphatase activity increased after application of moderate amounts of fertilizers but decreased if higher amounts were applied. The level of soil phosphatase has been related to the lack of inorganic phosphates (Skujiņš, 1967). Actually, in pure cultures phosphatase is a repressible enzyme decreasing in content when microorganisms are transferred from deficient to normal phosphate medium (Mills and Campbell, 1974; Ihlenfeldt and Gibson, 1975). Recently Nannipieri *et al.* (1978) showed that in a sandy loam a marked increase of phosphatase activity occurred after addition of glucose and sodium nitrate. The increase was related to the rise of bacteria populations, as determined by the direct count. In the presence of inorganic phosphate, phosphatase activity did not increase and remained fairly constant, as did the activity of control soil. Since soil phosphatase does not decrease after addition of inorganic phosphate, unlike the phosphatase of microorganisms in pure culture, it is possible that a substantial fraction of the activity could be stabilized as extracellular enzyme.

3.2(c). Effects on membranes

The action of agrochemicals on membranes may vary from the disorganization of the physical structure to the modification of the transport or the excretion processes. The consequences of these changes may include shifts in the intra/extracellular enzyme ratio, or biochemical changes in the activity of enzymes bound to the cells.

Our current ideas of biological membrane structure have developed from the continuous bilayer theory (Danielli and Davson, 1935), the unit membrane concept (Robertson, 1957) and the particulate lipoprotein model (Vanderkooi and Green, 1970). It is now generally acknowledged that membrane proteins are at least partially submerged in this phase and in some cases may span the whole membrane. In this way, both ionic and apolar interactions between the proteins and the phospholipids play an important part in membrane stability. This type of structure seems to be more capable of fulfilling the physical and biochemical requirements of a membrane than some of their earlier models. A complete disorganization of membrane structure can occur whenever an agrochemical is able to dissociate the stabilizing polar or hydrophobic bonds between the component proteins and phospholipids. For instance, dodine and guazatine can damage cell membranes; these fungicides have a long alkylic chain linked to a guanidine group, and their molecular structure may disorganize array of membranes (Brown and Sisler, 1960; Pressman, 1963; Somers and Pring, 1966). Examples of both types of reactions are cited in the literature but no detailed studies have been made to elucidate the molecular events involved.

Disarray of the membrane structure can also depend on non-active substances present in the commercial products. For instance, pesticides are frequently formulated as emulsifiable concentrates, which are solutions of the pesticides in oil, or in some other convenient organic solvents, together with emulsifying agent(s). Petroleum oils consist largely of aliphatic hydrocarbons, both saturated and unsaturated. Currier (1951) showed that short exposures of barley, carrot or tomato plants to the vapour of benzene, toluene, xylene or trimethyl benzene produced a rapid darkening of the leaf tips, which he thought indicated the leakage of cell sap. When the hydrocarbon dissolves in the membrane, the fatty acid chains of the phospholipids will be forced apart. Initially, this may only increase the permeability, but as the concentration of hydrocarbon increases in the membrane, the bilayer configuration could be completely dissociated. Besides, affecting lipid–lipid interactions, the hydrocarbons probably disrupt hydrophobic bonding between lipids and the apolar regions of membrane proteins (Dallyn and Sweet, 1951; Currier and Peoples, 1954).

Damage to cell membranes may also occur as a result of the interaction between lipids and free radicals, as in the case of application of bipyridylium herbicides. Harris and Dodge (1972) suggest that this effect may be due to an attack on the cell membrane lipids by peroxy and hydroxyl free radicals themselves derived from hydrogen peroxide formed after reaction of oxygen with bipyridylium free radicals (Calderbank, 1968). Thiolcarbamate, sulphallate and chlorinated aliphatic acid herbicides cause wax reduction in plants (Gentner, 1966; Mann and Pu, 1968; Martin and Juniper, 1970; Still et al., 1970; Wilkinson and Hardcastle, 1970). Such reduction can be caused by action on lipid synthesis, which in turn can alter cell excretory processes.

Mechanisms of enzyme excretion by cells are not clarified, although, recently, great attention has been paid to their elucidation (Allison and Davies, 1974; Lampen, 1974; Satir, 1974; Palade, 1975). However, it seems evident that any action of the agrochemicals on cell compartments and cell walls as well as on the transport processes, may be of primary importance. The loss of compartmentalization would release hydrolytic enzymes from the vacuole into the cytoplasm, so producing a secondary effect (Harris, 1970). A similar mechanism of action has been reported for diquat in Chlorella vulgaris (Stokes et al., 1970). The effect of DDT on transport processes in synthetic membranes has been studied by O'Brien

(1975). The action of herbicides on ion transport through cell membranes has been reviewed by Morrod (1976). Changes in phosphatase activity as a consequence of the cell wall abnormality have been shown in *Chlamydomonas reinhardii* (Loppes and Deltour, 1975); it seems reasonable to foresee a possible action of agrochemicals both at the level of the enzyme–cell wall complexes and of the mechanisms of secretion and release of the enzyme from whole cells, as shown for the alkaline phosphatase of *Pseudomonas aeruginosa* (Ingram *et al.*, 1973; Day and Ingram, 1975).

3.2(d). Other physiological effects

Some pesticides act like plant growth regulators, influencing transport of indoleacetic acid, gibberellin synthesis or ethylene levels. Other compounds may disrupt co-ordination and energy supply processes. However, the mode of action of many agrochemicals is not well known. Many quinones, for instance, are used as fungicides (Webb, 1966; Rich, 1969); dichlone hinders germination of spores of many fungi (McCallan and Miller, 1958) and according to Owens and Novotny (1958) acts simultaneously on a variety of metabolic processes. Many classes of pesticides (fluoride, those containing mercury, alkylene bis-dithiocarbamate fungicides etc.) show a non-specific mode of action. Not even an outline knowledge exists concerning the mode of action of about 100 pesticides (Corbett, 1975).

3.3. Influence on dynamics of soil populations

Except for unusual or very drastic treatments, agrochemicals rarely have a permanent impact on total soil populations. Nevertheless an agrochemical can modify the interrelationships between particular groups of soil organisms and so influence the amount and type of enzyme produced. In a broad sense, agrochemicals can stimulate growth of one class of soil organisms by inhibiting other classes. For example, the number of springtails in soil increases after application of DDT, due to a reduction of predatory mites (Edwards and Thompson, 1973). A more intriguing situation occurs in the behaviour of the organisms living in obligate symbiosis or mutual co-operation in soil. This is of especial importance in agriculturally-valuable associations such as those between rhizobia and legumes and most rhizosphere effects.

It is essential to recall that in soil, owing to the relatively large biomass with respect to available substrate for growth, almost all microorganisms are living under starvation conditions. Three agrochemical-induced events can be imagined as capable of disrupting this condition:

(1) the death of sensitive organisms with the consequent utilization of the organic residues by the surviving populations;

(2) the direct utilization of agrochemicals by the organisms which are able to

degrade or to metabolize them—in this sense the behaviour of organic pesticides as a source of carbon may be the best example of agrochemicals influencing dynamics of soil populations;
(3) the development of microbial populations which depend on secondary nutrient sources, e.g. metabolites produced from the decomposition of the agrochemical or excreted by the proliferating microflora.

Some theoretical influences of these events on soil microorganisms have been represented in Fig. 6. All herbicides are assumed to have a lethal action on one group of living organisms; all other microbial populations are assumed to be completely insensitive and can proliferate by utilizing the residue of the killed cells. Herbicide A is degradable, after a lag period, by induced enzymes. Its effect as a source of nutrients on the proliferation of microorganisms has been simplified by separating it from the influence of the dead organic matter arising from sensitive organisms. Herbicide B is degraded immediately and the effects are additive. The impact of herbicide C is similar to that of herbicide A, but a secondary microflora can multiply by utilizing a metabolite of the herbicide, together with the products of cell lysis and other substances excreted by the primary population.

The utilization of dead organic matter and the subsequent proliferation of microorganisms is not, of course, restricted to microbial interrelationships. It includes the effects of dead soil animals and plant residues on microorganisms, as well as the utilization of this organic material by other living organisms, such as earthworms. Also, modifications induced by the agrochemicals on the composition of plant root exudates, the excreta of soil animals (including the casts of the earthworms) can be of importance in the evolution of soil enzymes.

A variety of situations can arise from the effects of agrochemicals on less sensitive or completely insensitive soil organisms. For instance, in an excellent review by Cullimore (1971), the inhibition of insensitive microorganisms by herbicides is presented as the most common case: that is, resistant microbes neither utilize dead cells nor multiply during the period of enzyme induction. Separation of the two effects, as shown in Fig. 6, seems a more intuitive approach.

The evolution of microbial communities in soil is of considerable interest and it can sometimes play a major role in the elimination of organic agrochemicals. Research on this subject has generally been aimed at isolating single strains of microorganisms which are able to degrade specific pesticides, independent of the presence of other organisms. An example of enzyme evolution in a microbial community growing on the herbicide Dalapon (2, 2'-dichloropropionic acid) has recently been described by Senior et al. (1976). Dalapon undergoes microbial degradation (Kearney et al., 1964), but is considered a partially recalcitrant herbicide. It can

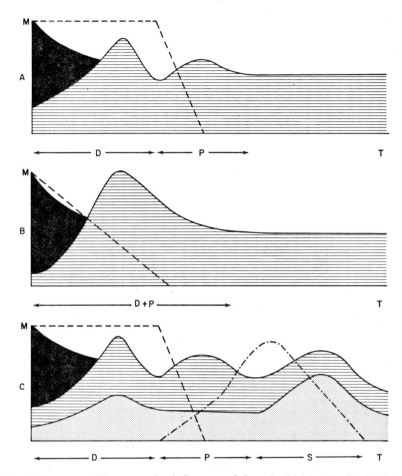

Fig. 6. Some possible composite influences of three herbicides (A, B, C; dashed lines) on soil microbial populations (M) over a period of time (T). Single effects are the death (D) of sensitive organisms (in black) and their utilization by other organisms: the utilization of the herbicides by a group of microbial populations (P, lined shade); the development of a secondary community of microorganisms (S, dotted), which are not able to degrade the pesticide, but utilize metabolites and pesticide intermediate products (dotted and lined strokes) excreted by the primary populations. Herbicide B is degraded by enzymes present in soil; herbicides A and C are degraded by enzymes induced after a lag period.

however be attacked by a community of microbes isolated by a continuous-flow culture enrichment procedure. The microbial community includes primary and secondary utilizers, which can grow on metabolites produced directly from the catabolism of the herbicide, on metabolites excreted by the primary utilizers, or on cell lysis products. Senior *et al.* (1976) also showed the occurrence of a mutation in one member of the secondary community, identified as *Pseudomonas putida*, which had acquired the ability to grow on the herbicide as its sole source of carbon and energy through the evolution of a dehalogenase. Much of the research into the degradation of organic agrochemicals in soils has ignored secondary populations, which can obviously be important for the stability of the whole microbial community. On the other hand, some metabolites produced from the primary catabolism of the organic agrochemicals could be toxic even for primary utilizers, as discussed by Cullimore (1971). Modification of soil enzymes through food chains and the succession of microbial populations represents one of the more promising trends in current research.

IV. Influence of Soil Environment

The interactions between agrochemicals and soil enzymes include both decay of the agrochemicals by the enzymes and the direct or indirect impact of agrochemicals on enzyme activities. However, all interactions depend themselves on the influence of the soil physical and chemical environment. If the given agrochemical is not adsorbed onto the soil matrix, its fate will depend on other environmental conditions such as pH, soil water content, temperature etc. In contrast, if the agrochemical is strongly adsorbed onto soil particles, its concentration will be reduced until no enzyme-catalysed reaction is possible. Both the enzymes adsorbed onto soil colloidal particles and those localized within humus particles behave differently in respect to free enzymes in solution (McLaren and Packer, 1970; Ladd and Butler, 1975). Unfortunately environmental effects, except in some cases (Cervelli *et al.*, 1973, 1975b, 1977; Benesi and McLaren, 1975; Irving and Cosgrove, 1976), have been neglected in the determination of kinetic constants.

The influence of the soil matrix on agrochemical fate can be represented by means of the competition of two effects, that is *"matrix effect"* and *"enzyme effect"*. The first one will take place when the binding between soil particles and agrochemical does not permit the substrate to react with the enzyme to form the product (Fig. 7). A striking example of this is the hydrolysis of some organophosphorus insecticides by phosphatases; Heuer *et al.* (1976) added acid and alkaline phosphatases to a loamy loessial sierozem sterilized by irradiation and investigated the persistence of pyrimiphos-methyl, parathion and guthion in both soil and soil extracts. Pesticide degradation occurred in soil extracts or in solutions of commercial phosphatases. No effect was observed when pesticides were added to soil.

The lack of activity was supposed to be due to adsorption of both enzymes and pesticides onto clays which were present at a concentration of 13% in the examined soil.

When the enzyme effect overcomes the matrix effect, the agrochemical is only lightly bound to soil particles and can react with the enzyme to form the product. This is the case of urea hydrolysis by urease. When urea is added to soil containing urease its hydrolysis is slower than in solution but, as showed by Durand (1965), this is due to pH variation near soil particles rather than to the matrix.

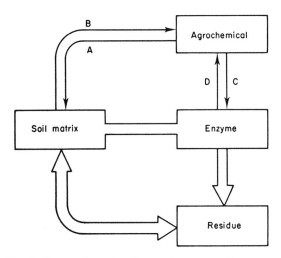

Fig. 7. A simplified scheme of matrix effect and enzyme effect on agrochemicals in soil. If A > > B and A > > C, matrix effect predominates and enzyme reaction can become impossible. If C > > A, matrix has no effect on the action of enzymes. If B ~ A, matrix and enzyme effects can be comparable also in the case that A > > C.

When matrix and enzyme effects are comparable we have an intermediate and more complex case. This is perhaps very common and, of course, the limiting factors in these instances are derived from both the concentration of agrochemicals in the soil solution (or the activity in the case of ions) and the ability of the soil system to replenish the agrochemical when its concentration is decreased owing to the transformation by soil enzymes. Thus each soil can have a different set of parameters that define the intensity of A in Fig. 7, depending on reactivity of its constituents.

As previously discussed, contrasting findings concerning DCA transformation to TCAB can be rationalized because of the influence of adsorption. DCA persistence in soil in fact is only partially explicable by the formation of TCAB and other polymeric residues. Kearney and Plimmer (1972) investigated the influence of DCA concentration on TCAB formation. Their data suggested that TCAB formation was limited by DCA availability. If all available binding sites are saturated by DCA, the excess DCA can be available to peroxidase and TCAB can be formed. At low concentrations the bulk of DCA is tightly bound to the soil and cannot be extracted by organic solvents and salt solutions (Bartha, 1971a; Chisaka and Kearney, 1970). Hsu and Bartha (1974, 1976) investigated the adsorption of DCA on soil organic matter and suggested a model mechanism for binding of DCA. Hughes and Corke (1974) investigated the influence of pH on TCAB formation from propanil and DCA at recommended field rates and found that transformation occurred only between pH 4 and 5·5. TCAB formation was negligible above pH 6. The soil matrix can also affect the behaviour of inhibitors of soil enzymes. Recently Cervelli *et al.* (1977) investigated the effect of the substituted urea herbicides fenuron, monuron, diuron and linuron on soil urease. In all soils examined herbicides behaved as mixed type inhibitors. Similar results were previously obtained for jack bean urease (Cervelli *et al.*, 1975c).

The mixed type inhibition is described by equation (1)

$$\frac{1}{V} = \frac{1}{V_m}\left[1 + \frac{I_s}{aK_i}\right] + \frac{K_s}{V_m}\left[1 + \frac{I_s}{K_i}\right]\frac{1}{S} \tag{1}$$

where I_s is the inhibitor concentration in solution and a represents the change in affinity induced by the inhibitor $(1 < a < \infty)$.

The soil matrix affects the inhibitor concentration; in fact the substituted urea herbicides are adsorbed onto soil and therefore their concentration I_s depends upon the adsorption equilibrium. Since the adsorption of the urea herbicides follows the Freundlich law (Cervelli *et al.*, 1975a) a new equation (2) must be taken into account,

$$I_a \rightleftharpoons I_s \qquad\qquad I_a = K\,I_s^{1/n} \tag{2}$$

where I_a correspond to x/m and I_s to C in the Freundlich isotherm. From equations 1 and 2:

$$\frac{1}{V} = \frac{1}{V_m}\left[1 + \frac{K_s}{S}\right] + \frac{1}{V_m K_i K^n}\left[\frac{1}{a} + \frac{K_s}{S}\right]I_a^n \tag{3}$$

For two substrate concentrations (S_1 and S_2) the straight lines described by equation (3) will intersect each other. The abscissa value of the intersection point will be $K_i K_n$, and K_i will be obtained simply by knowing K and n of the Freundlich equation.

In such a way Cervelli *et al.* (1977) were able to calculate the inhibition constants. K_i values in soil were greater than those obtained for jack bean urease. Besides the different sources of soil urease the observed differences in K_i could be ascribed to differences in diffusion phenomena.

Other evidence for the effects of microenvironment on enzyme–agrochemical relationships is circumstantial. First charge–charge interactions between soil matrix and agrochemicals can occur. Recently Ladd and Butler (1975) have compared humus–enzyme to synthetic polyanionic-enzyme systems. Concentrations of agrochemicals with an opposite charge to that of the matrix are greater in the microenvironment surrounding the enzymes than in soil solution. In such a way smaller amounts of agrochemicals will be sufficient to influence the soil enzyme activity. On the contrary, enzymes will be influenced only by higher amounts of agrochemicals with a charge equal to that of matrix surrounding the protein. Secondly, steric restrictions can occur in soil depending on the surrounding microenvironment, either as a result of restricted movement of agrochemicals to the enzyme, or because of the orientation of enzyme in relation to the adsorbant surface (the manner of attachment of the enzyme involving functional groups being so located that interaction with agrochemicals is prevented). Finally the rate of diffusion of the agrochemicals to the enzyme can be influenced by microenvironment.

Enzyme–agrochemical relationships can be affected indirectly through the influence of the soil matrix on microbes. The matrix can influence the ecology of soil microorganisms and consequently the origin of soil enzymes. An outstanding example of the importance of sorptive interactions between microorganisms and soil particles is the correlation between the lack of montmorillonite-type clays in soil and the rapid spread of *Fusarium oxysporum* and *Histoplasma capsulatum* (Stotzky, 1972). The so-called "long-life" soils of Central America, in which *Fusarium oxysporum* failed to spread, all contained montmorillonite-type clay.

Stotzky (1972) reported that microhabitats differing in clay mineral composition influence the ecology of microorganisms through their buffering capacity. In conditions of sporadic addition of substrate into the micro-habitats an initial growth of asporogenous bacteria is induced. Subsequently a sequence of bacteria, actinomycetes and then fungi appears. The fungi are the slow colonizers because, in addition to their slow germination rates, they may require growth factors provided by previous populations. In microhabitats devoid of montmorillonite clay the buffering capacity of the environment against accumulation of acidic metabolites is absent and the development of acid sensitive populations is restricted. In such a way when fungal spores germinate they find a low level of competition and, owing to their tolerances to acidic pH, they proliferate by using the remaining substrate. Fungi can thus invade adjacent environments, using their original microhabitats as a food base. In contrast, in the presence of montmorillonte clays an exchange of cations from the clay surface for H^+ produced during metabolism occurs and the development of acid conditions is prevented. Bacteria and actinomycetes can develop to a much higher level than in the absence of clay. If no inhibitors are produced by prior populations, when the spore germinates it must compete with other populations for the remaining available oxygen and substrate. In addition, the high activity of prior populations increases carbon dioxide levels. The combination of all these factors reduces fungal proliferation, preventing their extension to adjacent microhabitats.

The addition of pesticides that inhibit the growth of fungi or of primary and secondary populations can completely change any ecological sequence, as previously seen. The situation is even more complicated if interactions of agrochemicals with clays can occur. In fact, if adsorption occurs onto the same surfaces which form the microhabitat of the specific soil organisms involved, smaller quantities of agrochemicals will be sufficient to influence the ecological sequence. On the other hand, if the movement of agrochemicals to microhabitats is prevented (owing to opposite charge or steric hindrance) interaction with microbes is unlikely. Finally, if no interaction at all occurs between agrochemical and soil matrix, the amount of agrochemical in the soil solution is higher and the possibility of an effect upon a microbial sequence is greater.

V. Conclusions

Soil enzymes are involved in transformations of agrochemicals commonly employed in agriculture. Several agrochemicals are not directly transformed by soil enzymes but most (if not all) of them have some influence on soil organisms and hence on soil enzymes.

A schematic representation of interactions between agrochemicals and soil enzymes is given in Fig. 8. The direct effects of agrochemicals on soil enzymes are represented in the dashed, squared area. Almost all other reactions, however, influence the interactions of the agrochemicals with the soil enzymes, i.e., withdrawing intermediate products from the soil solution.

In order to draw any practical implications from the interactions between agrochemicals and soil enzymes it is necessary to bear in mind two important points:

(1) Agrochemicals can be directly applied to the soil at very heavy rates (hundreds or even thousands of kilograms per hectare) or in comparatively limited amounts (hundreds of grams per hectare). Other agrochemicals, applied to the aerial portions of plants, reach the soil incidentally. Their interaction with soil enzymes depends both on the amount and on the kind of chemical applied. However, realistic rates of the agrochemical concerned are sometimes insufficiently considered in laboratory experiments.

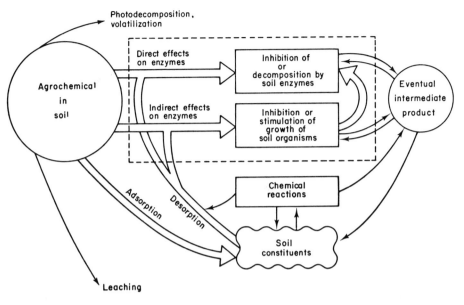

Fig. 8. Fate of agrochemicals in soil.

(2) Combinations of different agrochemicals are often used in agronomic practice. A very important development in fertilizers in recent years involves their use as carriers of pesticides (Petty *et al.* 1971). The economic advantages of such mixtures are evident. Technical difficulties such as compound stability, physical problems with mixing, metal corrosion of pumps and so on, are generally taken into account during the formulation of agrochemical combinations. However, their influence on soil enzymes has been completely disregarded. It seems possible that the application of urea, together with an inhibitor of urease, could aggravate the leaching of nitrogen fertilizer by rainwater.

Agrochemicals are of critical importance in achieving improvements in agricultural production and the use of agrochemicals in developed countries has undergone tremendous growth in the last few decades. The need for substantial increases in agricultural products is an urgent problem, particularly in the less-developed areas of the world and further increases in agrochemical use can be foreseen. It seems safe to say that the impact of agrochemicals on soil life in general as well as interaction with soil enzymes requires much more research. This will depend upon the careful design of experiments, critical interpretation of the results and hence modelling of the behaviour of agrochemicals in the field.

References

ABDEL'YUSSIF R. M., ZINCHENKO V. A. and GRUZDEV G. S. (1976). The effect of nematicides on the biological activity of soil. *Izv. Timiryazev. Sel'skokhoz. Akad.* (1), 206–214.

ALLISON A. C. and DAVIES P. (1974). Mechanisms of endocytosis and exocytosis. *Symp. Soc. Exptl. Biol.* 28, 419–446.

AMBROŽ. (1973). Study of the cellulase complex in soil. *Rostl. Vyroba* 19, 207–212.

AMPOVA G. and STEFANOV D. (1969). Microbiological processes in soil fumigated with nematicides. *Pochvozn. Agrokhim.* (6), 117–123.

ASHTON F. M. and CRAFTS A. S. (1973). "Mode of action of herbicides". John Wiley, New York.

BABIUK L. A. and PAUL E. A. (1970). The use of fluorescein isothiocyanate in the determination of the bacterial biomass of grassland soil. *Can. J. Microbiol.* 16, 57–62.

BALASUBRAMANIAN A. and SIDDARAMAPPA R. (1974). Effect of simazine application on certain microbiological and chemical properties of a red sandy loam soil. *Mysore J. Agric. Sci.* 8, 214–219.

BALASUBRAMANIAN A., BAGYARAJ D. J. and RANGASWAMI G. (1970). Studies on the influence of foliar application of chemicals on the microflora and certain enzyme activities in the rhizospher of *Eleusine coracana* Gaertn. *Pl. Soil* 32, 198–206.

BALASUBRAMANIAN A., SIDDARAMAPPA R. and OBLISAMI G. (1973). Effect of biocide on microbiological and chemical properties of soil. II. Effect of simazine and dithane M-45 on soil microflora and certain soil enzymes. *Pesticides* 7, 13.

BARTHA R. (1971a). Fate of herbicide derived chloroanilines in soil. *J. Agric. Fd Chem.* 19, 385–387.

BARTHA R. (1971b). Altered propanil degradation in temporarily air-dried soil. *J. Agric. Fd Chem.* 19, 394–395.

BARTHA R. (1975). Microbial transformations and environmental fate of some phenylamide herbicides. *Rocz. Glebozn.* 26, 17–24.

BARTHA R. and BORDELEAU L. (1969a). Cell-free peroxidases in soil. *Soil Biol. Biochem.* 1, 139–143.

BARTHA R. and BORDELEAU L. (1969b). Transformation of herbicides-derived chloroanilines by cell free peroxidases in soil. *Bact. Proc.* 4, A26.

BARTHA R. and PRAMER D. (1967). Pesticide transformation to aniline and azo-compounds in soil. *Science* 156, 1617–1618.

BARTHA R., LINKE H. A. B. and PRAMER D. (1968). Pesticides transformations: production of chloroazobenzenes from chloroanilines. *Science* 161, 582–583.

BECK T. (1970). Der mikrobielle Abbau von Herbiziden und ihr Einfluss auf die Mikroflora des Bodens. *Zentralbl. Bacteriol., Parasitenkd., Infectionskr. Hyg., Abt.* II 124, 304–313.

BECK T. (1973). Über die Eignung von Modellversuchen bei der Messung der biologischen Activität. *Bayer. Landw. Jahrb.* 50, 270–88.

BECKMANN E. O. and PESTEMER W. (1975). Wirkung von unterschiedlicher Humusversorgung auf Herbicidabbau und biologische Aktivität des Bodens. *Landwirt. Forschg.* 28, 24–33.

BENESI A. and MCLAREN A. D. (1975). Redox dependence of papain enzyme activity on charged kaolinite surfaces and in solution. *Soil Biol. Biochem.* **7**, 379–381.

BERTHOLD W. (1970). Investigations on the enzyme activities of herbicide-treated vineyard soils. *Summs. Paps. 7th Int. Congr. Pl. Prot.*, Paris, pp. 295–297.

BHAVANADAN V. P. and FERNANDO V. (1970). Studies on the use of urea as a fertilizer for tea in Ceylon. 2. Urease activity in tea soils. *Tea Quart.* **41**, 94–106.

BLAGOVESHCHENSKAYA Z. K. and DANCHENKO N. A. (1974). Activity of soil enzymes after prolonged application of fertilizer to a corn monoculture and crops in rotation. *Soviet Soil Sci.* **6**, 569–575.

BLIYEV YU. K. (1973). Effect of herbicides on the biological activity of soils. *Soviet Soil Sci.* **5**, 423–429.

BORDELEAU L. M. and BARTHA R. (1970). Azobenzene residues from aniline-based herbicides; evidence for labile intermediates. *Bull. Environ. Contam. Toxicol.* **5**, 34–37.

BORDELEAU L. M. and BARTHA R. (1972a). Biochemical transformations of herbicide-derived anilines in culture medium and in soil. *Can. J. Microbiol.* **18**, 1857–1864.

BORDELEAU L. M. and BARTHA R. (1972b). Biochemical transformations of herbicide-derived anilines: purification and characterization of causative enzymes. *Can. J. Microbiol.* **18**, 1865–1871.

BORDELEAU L. M., ROSEN J. D. and BARTHA R. (1972). Herbicide-derived chloroazobenzene residues: pathway of formation. *J. Agric. Fd Chem.* **20**, 573–578.

BRIGGS M. H. and SEGAL J. (1963). Preparation and properties of free soil enzyme. *Life Sci.* **1**, 69–72.

BROWN I. F. and SISLER H. D. (1960). Mechanisms of fungitoxic action of N-dodecyl-guanidine acetate. *Phytopathology* **50**, 830–839.

BURGE W. D. (1972). Microbial populations hydrolysing propanil and accumulation of 3, 4-dichloroaniline and 3, 3′, 4, 4′-tetrachloroazobenzene in soils. *Soil Biol. Biochem.* **4**, 379–386.

BURGE W. D. (1973). Transformation of propanil-derived 3, 4-dichloroaniline in soil to 3, 3′-4, 4′-tetrachloroazobenzene as related to soil peroxidase activity. *Soil Sci. Soc. Amer. Proc.* **37**, 392–395.

BURNS R. G. and GIBSON W. P. (1976). The disappearance of 2, 4-D, diallate and malathion from soil and soil components. *Abstr. I.S.S.S. Conference "Agrochemicals in Soils"*. Jerusalem, 13–18 June.

BURNS R. G., EL SAYED M. H. and MCLAREN A. D. (1972a). Extraction of an urease-active organo-complex from soil. *Soil Biol. Biochem.* **4**, 107–108.

BURNS R. G., PUKITE A. M. and MCLAREN A. D. (1972b). Concerning the location and persistence of soil urease. *Soil Sci. Soc. Am. Proc.* **36**, 308–311.

CALDERBANK A. (1968). The bipyridylium herbicides. *Adv. Pest. Control Res.* **8**, 127–235.

CAWSE P. A. (1968). Effects of gamma radiation on accumulation of mineral nitrogen in fresh soils. *J. Sci. Fd Agric.* **19**, 395–398.

CAWSE P. A. and CORNFIELD A. H. (1969). The reduction of ^{15}N-labelled nitrate to nitrite by fresh soils following treatment with gamma radiation. *Soil Biol. Biochem.* **1**, 267–274.

CAWSE P. A. and CORNFIELD A. H. (1972). Biological and chemical reduction of nitrate to nitrite in γ-irradiated soils and factors leading to eventual loss of nitrite. *Soil Biol. Biochem* **4**, 497–511.

CERVELLI S., NANNIPIERI P., CECCANTI B. and SEQUI P. (1973). Michaelis constant of soil acid phosphatase. *Soil Biol. Biochem.* **5**, 841–845.

CERVELLI S., NANNIPIERI P., GIOVANNINI G. and CIARDI C. (1975a). La sostanza organica e l'assorbimento degli erbicidi ureici nel terreno. *Agr. Ital.* **104**, 291–303.

CERVELLI S., NANNIPIERI P., GIOVANNINI G. and PERNA A. (1975b). Concerning the distribution of enzymes in soil organic matter. Studies about Humus, *Trans. Int. Symp. Humus et Planta* VI, Prague, pp. 291–296.

CERVELLI S., NANNIPIERI P., GIOVANNINI G. and PERNA A. (1975c). Jack bean urease inhibition by substituted ureas. *Pest. Biochem. Physiol.* **5**, 221–225.

CERVELLI S., NANNIPIERI P., GIOVANNINI G. and PERNA A. (1976). Relationships between substituted urea herbicides and soil urease activity. *Weed Res.* **16**, 365–368.

CERVELLI S., NANNIPIERI P., GIOVANNINI G. and PERNA A. (1977). Effects of soil on urease inhibition by substituted urea herbicides. *Soil Biol. Biochem.* **9**, 393–396.

CHERRY J. H. (1976). Action on nucleic acid and protein metabolism. *In* "Herbicides physiology, biochemistry, ecology" (L. J. Audus, Ed.) vol. 1, pp. 525–546, Academic Press, New York.

CHISAKA H. and KEARNEY P. C. (1970). Metabolism of propanil in soils. *J. Agric. Fd Chem.* **18**, 854–858.

CHULAKOV SH. A. and ZHARASOV SH. V. (1973). Biological activity of southern soils of Kazakhstan during herbicide use. *Izv. Akad. Nauk Kaz. SSR, Ser. Biol.* **11**, 7–13.

CHUNDEROVA A. I. and ZUBETS T. P. (1969a). Phosphatase activity in derno-podzolic soils. *Pochvovedenie* (11), pp. 47–53.

CHUNDEROVA A. I. and ZUBETS T. P. (1969b). Effect of herbicides on the biological processes in sod-podzolic soil cropped with maize without tilling between rows. *Byull. Vses. Nauch.-Issled. Inst. Sel'-skokhoz. Mikrobiol.* (14), 52–59.

CHUNDEROVA A. I. and ZUBETS T. P. (1971). Mathematical reliability of the results of soil microbiological and biochemical analyses. *Mikrobiol. Biokhim. Issled. Pochvy, Mater. Nauch. Konf.* **1969**, 30–34.

CHUNDEROVA A. I., ZUBETS T. P. and SOFINSKII A. M. (1971). The effect of herbicides on the soil microflora with their systematic use in the crop rotation. *Khim. Sel. Khoz.* **9**, 527–530.

CONRAD J. P. (1940). Hydrolysis of urea in soil by thermolable catalysis. *Soil Sci.* **49**, 253–263.

CORBETT J. R. (1974). "The biochemical mode of action of pesticides". Academic Press, New York.

CORBETT J. R. (1975). Biochemical design of pesticides. *Proc. 8th British Insect. Fung. Conf.* 981–993.

CORNFORTH I. S. (1971). Calcium cyanamide in agriculture. *Soil Fert.* **34**, 463–470.

COWIE G. A. (1919). Decomposition of cyanamide and dicyanodiamide in the soil. *J. Agric. Sci.* **9**, 113–137.

COWIE G. A. (1920). The mechanics of the decomposition of cyanamide in the soil. *J. Agric. Sci.* **10**, 163–176.

CULLIMORE D. R. (1971). Interaction between herbicides and soil microorganisms. *Residue Rev.* **35**, 65–80.

CURRIER H. B. (1951). Herbicidal properties of benzene and certain methyl derivatives. *Hilgardia* **20**, 383–406.

CURRIER H. B. and PEOPLES S. A. (1954). Phytotoxicity of hydrocarbons. *Hilgardia* **23**, 155–173.

DALLYN S. L. and SWEET R. D. (1951). Theories on the herbicidal action of petroleum hydrocarbons. *Proc. Am. Soc. Hort. Sci.* **57**, 347–354.

DALTON R. L., EVANS A. W. and RHODES R. C. (1966). Disappearance of diuron from cotton field soils. *Weeds* **14**, 31–33.

DANIELLI F. J. and DAVSON H. (1935). The permeability of thin films. *J. Cellular Comp. Physiol.* **5**, 495–508.

DAY D. F. and INGRAM J. M. (1975). In vitro studies of an alkaline phosphatase-cell wall complex from *Pseudomonas aeruginosa*. *Can. J. Microbiol.* **21**, 9–16.

DOMMERGUES Y. (1959). The effects of nematocides upon the biological activity of soil. *Fruits* **14**, 177–181.

DOUGLAS L. A., RIAZI-HAMADANI A. and FIELD J. F. B. (1976). Assay of pyrophosphatase activity in soil. *Soil Biol. Biochem.* **8**, 391–393.

DURAND G. (1965). Les enzymes dans le sol. *Rev. Ecol. Biol. Sol* **2**, 141–205.

EDWARDS C. A. and THOMPSON A. R. (1973). Pesticides and the soil fauna. *Residue Rev.* **45**, 1–79.

ELIADE C., GHIRITA V., PETRESCU R., STEFANIC G. and CICOTTI M. (1972). Aspects of biological decomposition of vegetable matter in soil. *Third Symp. Soil Biol. Bucharest*, pp. 90–101.

ERNST D. (1967). Transformation of cyanamide in arable soils. *Z. Pflanzenernähr. Düng. Bodenkd.* **116**, 34–44.

FRENEY J. R. and SWABY R. G. (1975). Sulphur transformations in soils. *In* "Sulphur in Australasian Agriculture". (K. E. McLachlan Ed.) pp. 31–39.

GAMZIKOVA O. I. (1968). Sucrase activity in soils of the Omsk region and the effect of some agricultural procedure on it. *Sb. Nauch. Rab., Sib. Nauch.-Issed. Inst. Sel'-skokhoz.* **14**, 60–63.

GAMZIKOVA O. I. and SVYATSKAYA L. N. (1968). Action of the herbicides 2, 3, 6-trichlorobenzoic acid and 2, 4-dichlorophenoxyacetic acid on the urease and saccharase activity of soil. *Sb. Dokl. Simp. Ferment. Pochvy* **1967**, 333–339.

GAMZIKOVA O. I. and SVYATSKAYA L. N. (1970). Herbicides and biological activity of soil. *Sb. Nauch, Rab., Sib. Nauch.-Issled. Inst. Sel'-skokhoz.* (15), 38–42.

GAVRILOVA A. N., SHIMKO N. A. and SAVCHENKO V. F. (1973). Dynamics of organic phosphorus compounds and phosphatase activity in pale yellow sod-podzolic soil. *Soviet Soil Sci.* **5**, 320–328.

GEISSBÜHLER H., HASELBACH C., AEBI H. and EBNER L. (1963). The fate of N-(4-chlorophenoxy)-phenyl-N, N-dimethylurea in soil and plants. *Weed Res.* **3**, 277–297.

GENTNER W. A. (1966). The influence of EPTC on external foliage wax deposition. *Weeds* **14**, 27–31.

GESHTOVT YU. N. ZHARASOV RH. U., BAINAZAROVA KH. T. and KOSTUTINOV E. K. (1974). Use of herbicides in fallow-cereal crop-rotation under dry conditions. *Khim. Sel. Khoz.* (5), 53–57.

GETZIN L. W. and ROSEFIELD I. (1968). Organophosphorous insecticide degradation by heat-labile substances in soil. *J. Agric. Fd Chem.* **16**, 598–601.

GETZIN L. W. and ROSEFIELD I. (1971). Partial purification and properties of a soil enzyme that degrades the insecticide malathion. *Biochim. Biophys. Acta.* **235**, 442–453.

GHINEA L. (1964). L'influence des aminotriazines sur l'activité microbiologique du sol. *Trans. 8th Int. Cong. Soil Sci., Bucharest* **3**, 857–869.

GHINEA L. and STEFANIC G. (1972). Influence of herbicides on soil biological activity. I. Dehydrogenase activity. *Third Symp. Soil Biol., Bucharest*, 139–145.

GIARDINA M. C., TOMATI U. and PIETROSANTI W. (1970). Hydrolytic activities of the soil treated with paraquat. *Meded. Fac. Landbouw-Wet., Rijksuniv. Gent* **35**, 615–626.

GILLIAM J. W. and SAMPLE E. G. (1968). Hydrolysis of pyrophosphate in soil pH and biological effects. *Soil Sci.* **106**, 352–357.

GOIAN M. (1968). Cercetari privind activitatea fosfatazica din soluri. *Lucr Stiint. Inst. Agron. Timisoara, Ser. Agron.* **11**, 139–152.

GOIAN M. (1969). Cercetari privind activitatea fosfatazica din sol (II). *Lucr. Stiint. Inst Agron. Timisoara, Ser. Agron.* **12**, 489-496.

GOIAN M. (1970). Activitatea fosfatazica in solurile saraturate. *Stiinta Solului* **8**, 13–17.

GOIAN M. (1972). Contributii la stabilizea posibilitatii de apreciere a unor activitatii biochemice din sol prin cercetarea activitatii fosfatazei alcaline. *Cerc. Biol. in partea de vest a Romaniei* **2**, 508–516.

GRAY T. R. G. and WILLIAMS S. T. (1971). Microbial productivity in soil. *Symp. Soc. Gen. Microbiol.* **21**, 255–287.

GROSSBARD E. and DAVIES H. A. (1976). Specific microbial responses to herbicides. *Weed Res.* **16**, 163–169.

GRUZDEV G. S., MOZGOVOI A. F. and KUZHMINA I. V. (1973). Effect of teflan on biological activity of soil under sunflower. *Izv. Timiryazev. Sel'skokhoz. Akad.* (4), 136–141.

HARRIS N. (1970). Studies on the mode of action of paraquat and diquat. Ph.D. Thesis, University of Bath.

HARRIS M. and DODGE A. D. (1972). Effect of paraquat on flax cotyledon leaves. Physiological and biochemical changes. *Planta* **104**, 210–219.

HAUKE-PACEWICZOWA T. (1971). Influence of herbicides on the activity of the soil microflora. *Pamiet. Pulawski* **46**, 5–48.

HEUER B., BIRK I. and YARON B. (1976). Effect of phosphatases on the persistence of organophosphorus insecticides in soil and water. *J. Agric. Fd Chem.* **24**, 611–614.

HOAGLAND R. E. (1974). Hydrolysis of 3′, 4′-dichloropropionalinide by an aryl acylamidase from *Taraxacum officinale*. *Phytochemistry* **14**, 383–386.

HOFER I., BECK T. and WALLNÖFER P. (1971). Effect of fungicide benomyl on the microflora of soil. *Z. Pflanzenkr. (Pflanzenpathol.) Pflanzenschutz* **78**, 398–405.

HOSSNER L. R. and PHILLIPS D. P. (1971). Pyrophosphatase hydrolysis in flooded soil. *Soil Sci. Soc. Am. Proc.* **35**, 379–383.

HSU T.-S. and BARTHA R. (1974). Interaction of pesticide-derived chloroaniline residues with soil organic matter. *Soil Sci.* **116**, 444–452.

HSU T.-S. and BARTHA R. (1976). Hydrolysable and nonhydrolysable 3, 4-dichloroaniline-humus complexes and their respective rates of bio-degradation. *J. Agric. Fd Chem.* **24**, 118–122.

HUGHES F. and CORKE C. T. (1974). Formation of tetrachloroazobenzene in some

Canadian soils treated with propanil and 3, 4-dichloroaniline. *Can. J. Microbiol.* **20**, 35–39.

IHLENFELDT M. J. A. and GIBSON J. (1975). Phosphate utilization and alkaline phosphatase activity in *Anacystis nidulans. Arch. Microbiol.* **102**, 23–28.

INGRAM J. M., CHENG K. J. and COSTERTON J. W. (1973). Alkaline phosphatase of *Pseudomonas aeruginosa:* the mechanisms of secretion and release of the enzyme from whole cells. *Can. J. Microbiol.* **19**, 1407–1415.

IRVING G. C. J. and COSGROVE D. J. (1976). The kinetics of soil acid phosphatase. *Soil Biol. Biochem.* **8**, 335–340.

ISHIZAWA S., TANABE I. and MATSUGUHI T. (1961). Effect of DD, EDB and PCP upon microorganisms and their activities in soil. Part II. Effects on microbial activity. *Soil Pl. Fd* **6**, 156–163.

JAGGI W. (1954). Soil microbiological studies in a fertilizer trial. *Schweiz. Landwirt. Forsch.* **13**, 531–554.

KAPPEN H. (1910). The decomposition of cyanamide by mineral constituents of the soil. *Fuehlings Landwirt. Ztg.* **59**, 657–679.

KARANTH N. G. K. and VASANTHARAJAN V. N. (1973). Persistence and effect of dexon on soil respiration. *Soil. Biol. Biochem.* **5**, 679–684.

KARANTH N. G. K., CHITRA C. and VASANTHARAJAN V. N. (1975). Effect of fungicide, dexon on the activities of some soil enzymes. *Indian J. Exp. Biol.* **13**, 52–54.

KARKI A. B., COUPIN L., KAISER P. and MOUSSIN M. (1973a). Effects of sodium chlorate on soil microorganisms, their respiration and enzymatic activity. I. Ecological study in the field. *Rev. Ecol. Biol. Sol* **10**, 3–11.

KARKI A. B., COUPIN L., KAISER P. and MOUSSIN M. (1973b). Effects du chlorate de sodium sur les microorganismes du sol, leur respiration et leur activité enzymatique. *Weed Res.* **13**, 133–139.

KAUFMAN D. D. and KEARNEY P. C. (1965). Microbial degradation of isopropyl-N-3-chlorophenyl carbamate and 2-chloroethyl-N-3-chlorophenyl carbamate. *Appl. Microbiol.* **13**, 443–446.

KEARNEY P. C. and PLIMMER J. R. (1972). Metabolism of 3, 4-dichloroaniline in soils. *J. Agric. Fd Chem.* **20**, 584–585.

KEARNEY P. C., KAUFMAN D. D. and BEALL M. L. (1964). Enzymatic dehalogenation of 2, 2-dichloropropinate. *Biochem. Biophys. Res. Commun.* **14**, 29–33.

KEARNEY P. C., SMITH R. J., Jr., PLIMMER J. R. and GUARDIA F. S. (1970). Propanil and TCAB residues in rice soils. *Weed Sci.* **18**, 464–466.

KELLER T. (1961). The effect of some herbicides on the microbiological activity of the soil of a chestnut grove. *Mitt. Schweig. Anst. Forstl. Versuchsw.* **37**, 401–418.

KELLEY W. D. and RODRIGUEZ-KABANA R. (1975). Effects of potassium azide on soil microbial populations and soil enzymatic activities. *Can. J. Microbiol.* **21**, 565–570.

KHAN S. V. (1970). Enzymatic activity in a grey wooded soil as influenced by cropping systems and fertilizers. *Soil Biol. Biochem.* **2**, 137–139.

KHAZIEV F. Kh. (1972). Determination of the pyrophosphatase activity in soil. *Pochvovedenie* (2), 67–73.

KISS S. (1958). Talajenzimec. *In* "Talajtan" (M. J. Csapo, Ed.) pp. 491–622, Agro-Silvica, Bucharest.

KISS S., DRĂGAN-BULARDA M. and RADULESCU D. (1972). Biological significance of the enzymes accumulated in soil. *3rd Symp. Soil Biol., Bucharest*, 19–78.

KISS S., DRĂGAN-BULARDA M. and RADULESCU D. (1975a). Biological significance of enzymes accumulated in soil. *Adv. Agron.* **27**, 25–87.

KISS S., STEFANIC G. and DRĂGAN-BULARDA M. (1975b). Soil enzymology in Romania. (Part II). *Contrib. Bot., Cluj*, 197–207.

KLEIN D. A., LOH T. C. and GOULDING R. L. (1971). A rapid procedure to evaluate the dehydrogenase activity of soils low in organic matter. *Soil Biol. Biochem.* **3**, 385–87.

KREZEL Z. and MUSIAL M. (1969). The effect of herbicides on soil microflora. 2. The effect of herbicides on enzymatic activity of the soil. *Acta Microbiol. Polon. (Ser. B)* **18**, 93–97.

KROKHALEV A. K., NIKITIN E. S. and SVECHKOV V. I. (1973). Effect of herbicides on the biological activity of soil in nurseries of Sakhalin. *In* "Khimicheskii Ukhod za Lesom" (V. P. Bel'kov, Ed.), pp. 175–177. Lenizdat, Pskov. Otd., Pskov.

KRUGLOV YU. V. and BEI-BIENKO N. V. (1969). Methods for studying the residual action of herbicides on the microbiological processes in soil. *Mikrobiol. Biokhim. Issled. Pochvy, Mater. Nauch. Konf.* **1969**, 143–148.

KRUGLOV YU. V., GERSH N. B. and BEI-BIENKO N. V. (1973). Effect of meturin on biological soil activity. *Khim. Sel. Khoz.* **11**, 294–296.

KULINSKA D. (1967). The effect of simazine on soil microorganisms. *Roczn. Nauk. Roln.* **93A**, 229–262.

LADD J. N. and BUTLER J. H. A. (1975). Humus-enzyme systems and synthetic, organic polymer-enzyme analogs. *In* "Soil Biochemistry" (E. A. Paul and A. D. McLaren, Eds) Vol. 4, pp. 143–194. Marcel Dekker, New York.

LADD J. N., BRISBANE P. G., BUTLER J. H. A. and AMATO M. (1976). Studies on soil fumigation. III. Effects on enzyme activities, bacterial numbers and extractable ninhydrin reactive compounds. *Soil Biol. Biochem.* **8**, 255–260.

LAMPEN O. G. (1974). Movement of extracellular enzymes across cell membrane in transport at the cellular level. *Symp. Soc. Exp. Biol.* **28**, 351–374.

LATYPOVA R. M., NOVITSKII S. A., MASLOVA L. G. and SHAPOVAL'SKAYA L. A. (1968). Effect of herbicides on the biological activity of the soil. *Sb. Nauch. Tr. Beloruss. Sel'skokhoz. Akad.* **57**, 61–69.

LAY M. M. and ILNICKI R. D. (1974). Peroxidase acitivity and propanil degradation in soil. *Weed Res.* **14**, 111–113.

LENHARD G. (1959). The effect of 2, 4-D on certain physiological aspects of soil microorganisms. *S. Afr. J. Agric. Sci.* **2**, 487–497.

LETHBRIDGE G. and BURNS R. G. (1976). Inhibition of soil urease by organophosphorus insecticides. *Soil Biol. Biochem.* **8**, 99–102.

LEUSHEVA M. I. and MEL'NIK N. M. (1969). Biological activity of orchard soil tilled around the tree trunks by various methods. *In* "Rol'Mikroorganizmov v Pitanii Rastenii i Plodorodii Pochvy", pp. 214–219, Nauka i Tekhnika, Minsk.

LICHFIELD C. D. and HUBEN R. P. (1973). Effect of selected pollutants on the specific growth rate and extracellular enzyme synthesis of *Aeromonas proteolytica*. *Bull. Ecol. Res. Comm.* (Stockholm) **17**, 464–466.

LINKE H. A. B. (1970). 3, 3′, 4′-trichloro-4-(3, 4-dichloroanilino) azobenzene, a degradation product of the herbicide propanil in soil. *Naturwissenschaften* **57**, 307–308.

LIVENS J., MAYAUDON J., SAIVE R., VOETS J. P. and VERSTRAETE W. (1973). Etude preliminaire de l'effet des pesticides utilisés en culture betteraviere sur l'activité biologique du sol. *Inst. Belge Amelior. Betterave Pubbl.* (4), 135–179.

LOPPES R. and DELTOUR R. (1975). Changes in phosphatase activity associated with cell wall defects in *Chlamidomonas reinhardii. Arch. Microbiol.* **103**, 247–250.

LOTTI G. (1955). La trasformazione della calciocianamide nei terreni calcarei. *Ann. Fac. Agric. Univ., Pisa* **16**, 35–46.

MANN J. D. and PU M. (1968). Inhibition of lipid synthesis by certain herbicides. *Weed Sci.* **16**, 197–198.

MANORIK A. V. and MALICHENKO S. M. (1969). The effect of symmetrical triazines on phosphatase and urease activity in the soil. *Fiziol. Biokhim. Kul't. Rast.* **1**, 173–178.

MARKERT S. (1974). Über die Möglichkeit zur Stenerung von Harnstoffumsetzungen in Boden durch einige Handelsübliche Pestizide. *Trans. 10th Int. Cong. Soil Sci., Moscow* **9**, 113–140.

MARKERT S. and KUNDLER P. (1975). Model trials to test the effect of commercial pesticides on nitrogen decomposition in the soil. *Arch. Acker-u. Pflanzenbau u. Bodenkd., Berlin* **19**, 487–497.

MARTIN J. T. and JUNIPER B. E. (1970) "The cuticle of plants". Edward Arnold, London.

MATSUMURA F. and BOUSH G. M. (1966). Malathion degradation by *Trichoderma viride* and a *Pseudomonas* species. *Science* **153**, 1277–1280.

MAYAUDON J., BATISTIC L. and SARKAR J. M. (1975). Properties des activities proteolitiques extraites des sols frais. *Soil Biol. Biochem.* **7**, 281–286.

MAYAUDON J., EL-HALFAWI M. and CHALVIGNAC M. A. (1973). Properties des diphenol oxydases extraites des sols. *Soil Biol. Biochem.* **5**, 369–383.

MCCALLAN S. E. A. and MILLER L. P. (1958). Innate toxicity of fungicides. *Adv. Pest Control Res.* **2**, 107–134.

MCLAREN A. D. and PACKER O. L. (1970). Some aspects of enzyme reactions in heterogeneous systems. *Adv. Enzymol.* **33**, 245–308.

MCLAREN A. D. and PUKITE A. (1975). Ubiquity of some soil enzymes and isolation of soil organic matter with urease activity. *In* "Humic substances. Their structure and functions in the biosphere". (D. Povoledo and M. L. Golterman, Eds), pp. 187–193. Pudoc, Wageningen.

MCLAREN A. D., RESHETKO L. and HUBER W. (1957). Sterilization of soil by irradiation with an electron beam and some observation on soil enzyme activity. *Soil Sci.* **83**, 497–501.

MERESHKO M. YA. (1969). The effect of herbicides on the biological activity of the microflora of chernozem soils. *Mykrobiol. Zh.* **31**, 525–529.

MILLS C. and CAMPBELL J. N. (1974). Production and control of extracellular enzymes in *Micrococcus sodonensis. Can. J. Microbiol.* **20**, 81–90.

MORROD R. S. (1976). Effects on plant cell membrane structure and functions. *In* "Herbicides" (L. J. Audus, Ed.) pp. 292–295. Academic Press, London and New York.

MURESANU P. L. and GOIAN M. (1969). Investigation of the phosphomonoesterase activity in fertilized and limed soils. *Agrokem. Talajt.* **18**, 102–106.

NAMDEO K. N. and DUBE J. N. (1973a). Proteinase enzyme as influenced by urea and herbicides applied to grassland oxisol. *Indian J. Exp. Biol.* **11**, 117–119.

NAMDEO K. N. and DUBE J. N. (1973b). Residual effect of urea and herbicides on hexosamine content and urease and proteinase activities in a grassland soil. *Soil Biol. Biochem.* **5,** 855–859.

NANNIPIERI P., CERVELLI S. and PEDRAZZINI F. (1975). Concerning the extraction of enzymatically active organic matter from soil. *Experientia* **31,** 513–514.

NANNIPIERI P., JOHNSON R. L. and PAUL E. A. (1978). Criteria to measure microbial growth and activity in soil. *Soil Biol. Biochem.* In press.

NANNIPIERI P., CECCANTI B., CERVELLI S. and SEQUI P. (1974). Use of 0·1 M pyrophosphate to extract urease from a podzol. *Soil Biol. Biochem.* **6,** 359–362.

NAUMANN K. (1970a). Dynamics of the soil microflora after adding plant protective agents. VII. Effect of some disinfectants on soil microorganisms. *Zentralbl. Bacteriol., Parasitenkd., Infektionskr. Hyg., Abt.* II **125,** 478–491.

NAUMANN K. (1970b). Dynamics of the soil microflora following the application of plant protective agents. IX. Experiments with the herbicides, calcium cyanide, simazine and sodium trichloroacetate. *Kuehn-Arch.* **81,** 329–350.

NAUMANN K. (1970c). Soil microflora after application of the fungicides alpisan (trichlorodinitrobenzene), captain and thiuram. *Arch. Pflanzenschutz* **6,** 383–398.

NAUMANN K. (1972). The effect of some environmental factors on the reaction of the soil microflora to pesticides. *Zentralbl. Bakteriol., Parasitenkd., Infektionskr. Hyg., Abt.* II **127,** 379–396.

NIKITIN E. S. and SVECHKOV V. I. (1973). Effect of dalapon on the biological activity of the mountain forest brown soil. *In* "Povyshenie Produktivnosti Lesov Dal'nego Vostoka", pp. 228–229. Lesnaya Promyshlennost', Moscow.

O'BRIEN R. D. (1975). Nonenzymic effects of pesticides on membranes. *In* "Environmental dynamics of pesticides" (R. Haque and V. H. Freed, Eds), pp. 331–342. Plenum Press, New York and London.

ODU C. T. I. and HORSFALL M. A. (1971). Effect of chloroxuron on some microbial activities in soil. *Pest. Sci.* **2,** 122–125.

OWENS R. G. and NOVOTNY H. M. (1958). Mechanisms of action of the fungicide dichlone (2, 3-dichloro-1, 4-naphthoquinone). *Contrib. Boyce Thompson Inst.* **19,** 462–482.

PADENOV K. P. and MOLCHAUN A. P. (1975). The effect of prometryne activity of the soil. *Vest. Akad. Navuk Belarus. SSR, Ser. Sel'skokhoz. Navuk* (2), 49–51.

PALADE G. (1975). Intracellular aspects of process of protein secretion. *Science* **189,** 347–358.

PAPACOSTEA P., PETRE N., DANCAN H. and CICCOTTI M. (1972). Preliminary research concerning the influence of some plant and soil stimulating preparations. *Third Symp. Soil Biol., Bucharest,* pp. 79–85.

PAULSON K. N. and KURTZ L. T. (1969). Locus of urease activity. *Soil Sci. Soc. Am. Proc.* **33,** 897–901.

PEEPLES J. L. and CURL E. A. (1970). Effect of the herbicide EPTC on growth and enzymatic activity of *Sclerotium rolfsii* and *Trichoderma viride*. *Phytopathology* **60,** 586.

PEL'TSER A. S. (1972). Kinetics of C^{14}-urea decomposition in soil depending on soil moisture, addition of toxic chemicals, liming and soil pH. *Soviet Soil Sci.* **4,** 571–576.

PESTEMER W. and BECKMAN E. O. (1975). The effect of linuron in a soil containing various amounts of humus. *Z. Pflanzenkrank. Pflanzenschutz* **82**, 109–122.

PETERSON G. H. (1962). Respiration of soil sterilized by ionizing radiations. *Soil Sci.* **94**, 71–74.

PETTIT N. M., SMITH A. R. J., FREEDMAN R. B. and BURNS R. G. (1976). Soil urease: activity, stability and kinetic properties. *Soil Biol. Biochem.* **8**, 479–484.

PETTY H. B., BURNSIDE O. C. and BRYANT J. P. (1971). Fertilizer combinations with herbicides or insecticides. *In* "Fertilizer Technology and Use" (R. A. Olson, T. J. Army, J. J. Hanway and V. J. Kilmer, Eds) 2nd edn, pp. 494–515. Soil Sci. Soc. Am., Madison, Wisconsin.

PLIMMER J. R., KEARNEY P. C., CHISAKA H., YOUNT J. B. and KLINGEBIEL U. I. (1970). 1, 3-bis(3, 4-dichlorophenyl)triazene from propanil in soils. *J. Agric. Fd Chem.* **18**, 859–861.

PRATOLONGO U., ROTINI O. T. and JOB R. (1934). La transformazione della calcio-cianamide nel terreno. *Ital. Agric.* **71**, 3–12.

PRESSMAN B. C. (1963). The effects of Guanidine and Alkylguanidines on the Energy Transfer Reactions of Mitochondria. *J. Biol. Chem.* **238**, 401–409.

PROTASOV N. I. (1968). Use of herbicides to control weeds in fodder lupine. *Sb. Nauch. Tr., Beloruss. Sel'skokhoz. Akad.* **57**, 84–89.

PROTASOV N. I. (1970). Action of herbicides on the biological activity of the soil. *Sb. Nauch. Tr., Beloruss. Sel'skokhoz. Akad.* **64**, 151–155.

PUTINTSEVA L. A. (1970). Effect of atrazine and simazine on the biological activity of soil. *Biol. Nauki (Stravopol')* PART 1, 38–46.

QUILT P. (1972). The effect of carbyne on soil microorganisms. Ph.D. Thesis, University of Bath.

RACZ G. J. and SAVANT N. K. (1972). Pyrophosphate hydrolysis in soil as influenced by flooding and fixation. *Soil Sci. Soc. Am. Proc.* **36**, 678–682.

RANKOV V. (1970). The effect of certain herbicides on soil cellulase-decomposing activity. *Pochvoz. Agrokhim.* **5**, 73–80.

RANKOV V. and DIMITROV G. (1971). Soil phosphatase activity resulting from the manuring of the tomato and cabbage. *Pochvozn. Agrokhim.* **6**, 93–98.

RICH S. (1969). Quinones (Fungicides). *In* "Fungicides, an advanced treatise" (D. C. Torgeson, Ed.) Vol. II, pp. 447–475. Academic Press, New York.

ROBERTSON J. D. (1957). The ultrastructure of cell membranes and their derivatives. *Biochem. J.* **65**, 43 P.

ROSA V., DOMOCOS T., ROGOBETE G. and FAUR A. (1970). Cercetari asupra activitatii zaharazice a solului smolnita (Gataia-Timis). *Microbiologia* **1**, 487–492.

ROSEN J. D., SIWIERSKY M. and WINNET G. (1970). FMN-Sensitized photolyses of chloroanilines. *J. Agric. Fd Chem.* **18**, 495–496.

ROTINI O. T. (1935a). La trasformazione enzimatica dell'urea nel terreno. *Ann. Labor. Ric. Ferm. L. Spallanzani, Milano* **3**, 179–192.

ROTINI O. T. (1935a). La trasformazione catalitica della cianamide in urea. *Chim. Ind.* **17**, 14–30.

ROTINI O. T. (1935c). La trasformazione della calciocianamide del terreno. *Ann. Lab. Chim. Agr. Univ., Milano* **1**, 101–149.

ROTINI O. T. (1939). Sopra la pretesa trasformazione microbiologica della calcio-cianamide nel terreno agrario. *Ann. Tecn. Agr.* **12**, 210–228.

ROTINI O. T. (1940). Gli agenti della trasformazione della calciocianamide nel terreno. *Chim. Ind.* **22,** 7–14.

ROTINI O. T. (1951). Rapporti dei fertilizzanti con il terreno agrario e la loro trasformazione. *Ann. Fac. Agr. Univ., Pisa* **12,** 159–178.

ROTINI O. T. and CARLONI L. (1953). The transformation of metaphosphates into orthophosphates promoted by agricultural soil. *Ann. Sper. Agrar. Pisa* **7,** 1789–1799.

ROTINI O. T., GALOPPINI C. and BELLI P. G. (1967). Trasformazione della cianamide in terreno sterile. *Chim. Ind.* **49,** 374–379.

ROTINI O. T., SOLDATINI G. F. and GIOVANNINI G. (1971). La trasformazione della calciocianamide nei terreni sommersi. *Agrochimica* **15,** 523–530.

RYSAVY P. and MACURA J. (1972). The formation of β-galactosidase in soil. *Folia Microbiol.* **17,** 375–381.

SATIR B. (1974). Membrane events during the secretory process. *Symp. Soc. Exp. Biol.* **28,** 399–418.

SATYANARAYANA T. and GETZIN L. W. (1973). Properties of a stable cell-free esterase from soil. *Biochemistry* **12,** 1566–1572.

SAUNDERS B. C., HOLME-SIEDLE A. G. and STARK B. P. (1964). "Peroxidase". Butterworth, Washington D.C. and London.

SCHMALFUSS K. (1938). Der Abbau des Zyanamids. *Z. Pflanzenernähr. Dung. Bodenkd.* **9,** 273–305.

SENIOR E., BULL A. T. and SLATER J. M. (1976). Enzyme evolution in a microbial community growing on the herbicide dalapon. *Nature* **263,** 476–479.

SEQUI P. (1974). Gli enzimi del terreno. *Ital. Agric.* **111,** 91–109.

SIPOS G., STEFANIC G., ELIADE G., MOGA R. and PETRESCU R. (1972). Modificari chimice provacate in sol si planta prin efectul cumulat al ingrasamintelor cu NPK si organice aplicate la grin si porumb. *Analele I.C.C.P.T., Fundulea, Ser. B* **38,** 151–162.

SKUJIŅŠ J. J. (1967). Enzymes in soil. *In* "Soil Biochemistry" (A. D. McLaren and G. H. Peterson, Eds) Vol. 1, pp. 371–414. Marcel Dekker, New York.

SKUJIŅŠ J. J. (1976). Extracellular enzymes in soil. *CRC Critical Rev. Microbiol.* **6,** 383–421.

SMITH R. G., Jr. (1965). Propanil and mixtures with propanil for weed control in rice. *Weeds* **13,** 236–238.

SOKOLOV M. S., KNYR L. L., STREKOZOV B. P., AGARKOV V. D. and KRYZHKO B. A. (1974). Behaviour of some herbicides under rice irrigation. *Agrokhimiya* **3,** 95–106.

SOLDATOV A. B. (1968). Effect of some herbicides on the biological soil processes and the yield beets. *Sb. Nauch. Tr., Beloruss. Sel'skokhoz. Akad.* **57,** 70–78.

SOLDATOV A. B., BARBUT'KO N. and LATUSHKIN L. (1971). Effect of herbicides on the biological activity of the soil. *Sb. Nauch. Tr., Beloruss. Sel'skokhoz. Akad.* **76,** 131–137.

SOMERS E. and PRING R. J. (1966). Uptake and binding of dodine acetate by fungal spores. *Ann. Appl. Biol.* **58,** 457–466.

SOREANU I. (1972). Dinamica activitatii zaharazice a unui sol brun de padure dintr-o plantatic intensiva de mar si zmeur. *Contrib. Bot., Cluj,* pp. 371–376.

SPIRIDONOV YU. YA. and SPIRIDONOVA G. S. (1973). Effect of long-term use of sym triazines on the biological activity of the soil. *Soviet Soil Sci.* **5,** 162–171.

SPIRIDONOV YU. YA., SKHILADZE V. SH. and SPIRIDONOVA G. S. (1970). The effects of diuron and monuron in a red soil of the moist subtropics of Adizhariya. *Subtropicheskie Kul'tury*, pp. 152–160.

SPIRIDONOV YU. YA., SKHILADZE V. SH., SPIRIDONOVA G. S. and MEPARISHVILI U. KH. (1973). Effect of diuron and monuron on the biological activity of subtropical soils. *In* "Povedenie, Prevrashchenie i Analiz Pestitsi dov i Ikh Metabolitov v Pochve" pp. 116–118. Nauch. Tsentr Biol. Issled., Pushchino-na-Oke.

SPROTT G. D. and CORKE C. T. (1971). Formation of 3, 3′, 4, 4′-tetrachloroazobenzene from 3, 4-dichloroaniline in Ontario soils. *Can. J. Microbiol.* **17**, 235–240.

STEFANIC G. (1970). Effect of pea crops and of basal fertilizers on the microflora and on the enzymatic activity in a podzol. *Lucr. Conf. Nat. Stiinta solului,* Eforie **1967**, 241–249.

STEFANIC G., BOERIU I. and DIMITRIU L. (1971). Influentia ingrasamintelor si a nivelului de calcarizare asupra microflorei totale si a activitati enzimelor dintr-un sol podzolic argilo-iluvial pseudogleizat. *Stiinta Solului* **9**, 45–54.

STEFANIC G., ELIADE G., PETRESCU R. and PICU I. (1976). Influence of ploughing depth and of fertilizers on soil biology changes in the dehydrogenase activity and total carbon and available phosphorus content. *3rd Symb. Soil Biol.,* Bucharest, pp. 129–138.

STILL G. G., DAVIS D. G. and ZANDER G. L. (1970). Plant epicutilar lipids. Alteration by herbicidal carbamates. *Pl. Physiol.* **46**, 307–314.

STOKES D. M., TURNER J. S. and MARKUS H. (1970). Effects of the dipyridyl diquat on the metabolism of *Chlorella vulgaris* II. Effects of diquat in the light of chlorophyll bleaching and plastid structure. *Aust. J. Sci.* **23**, 265–274.

STOTZKY G. (1972). Activity, ecology and population dynamics of microorganisms in soil. *CRC Crit. Rev. Microbiol.* **2**, 49–137.

STRATONOVICH M. U. and EVDOKIMOVA N.V. (1976). The effect of fertilizers on some indices of the biological activity of a derno-podzolic soil. *Izv. Tim. Sel'skokhoz. Akad.* (2), 140–149.

STUTZER A. and REIS F. (1910). Untersuchungen über Kalkstickstoff und einhge seiner Umsetzungs Produkte. *J. Landwirt.* **58**, 65–76.

SUTTON C. D., GUNARY D. and LARSEN S. (1966). Pyrophosphate as a source of phosphorus for plants. II. Hydrolysis and initial uptake by a barley crop. *Soil. Sci.* **101**, 199–204.

SVYATSKAYA L. N. (1972). Biological activity of weakly leached chernozem following use of herbicide mixture. *Nauch. Tr., Sib. Nauch.-Issled. Inst. Sel'skokhoz.* (3), 83–85.

TEMME J. (1948). Transformation of calcium cyanamide in the soil. *Pl. Soil* **1**, 145–166.

TEUBER M. and POSCHENRIEDER H. (1964). Investigation on the effect on the microflora of cultivated fen soil nematocides that contain mustard oil. *Bayer. Landwirt. Jahrb.* (3), 1–8.

TSIRKOV Y. (1970). Effect of the organic chlorine insecticides hexachlorane, heptachlor, lindane and dieldrin on activity of some soil enzymes. *Pochv. Agrokhim.* **4**, 85–88.

TYUNYAYEVA G. N., MINENKO A. K. and PENAKOV L. A. (1974). Effect of trifluralin on the biological properties of soil. *Soviet Soil Sci.* **6**, 320–324.

ULASEVICH E. I. and DRACH YU. O. (1971). Effect of atrazine on the nitrogen-fixing bacteria in grey podzolized soil and maize rizosphere. *Mikrobiol. Zh., Kiev* **33**, 735–736.

ULPIANI C. (1910). Sulla trasformazione della calciocianamide nel terreno agrario. *Gazz. Chim. Ital.* **40**, 613–666.

VANDERKOOI G. and GREEN D. E. (1970). Biological membrane structure. I. The protein crystal model for membranes. *Proc. Nat. Acad. Sci. US* **66**, 615–621.

VAN FAASSEN H. G. (1973). Effects of mercury compounds on soil microbes. *Pl. Soil* **38**, 485–487.

VAN FAASSEN H. G. (1974). Effect of the fungicide benomyl on some metabolic processes, and on numbers of bacteria and actinomycetes in the soil. *Soil Biol. Biochem.* **6**, 131–133.

VERSTRAETE W. and VOETS J. P. (1974). Impact in sugar beet crops of some important pesticides treatment systems on the microbial and enzymatic constitution of the soil. *Meded. Fac. Landbouwwett. Rijksuniv. Gent* **39**, 1263–1277.

VOETS J. P. and VANDAMME E. (1970). Influence of 2-(thiocyanomethylthio)benzo-thiazole on the microflora and enzymes of the soil. *Meded. Fac. Landbouwwett. Rijksuniv. Gent* **35**, 563–580.

VOETS J. P., MEERSCHMAN P. and VERSTRAETE W. (1974). Soil microbiological and biochemical effects on long-term atrazine application. *Soil Biol. Biochem.* **6**, 149–152.

VOINOVA zh., ULAKHOVA S. and PETROVA P. (1969). Soil microbiology with the use of certain herbicides. *Pochv. Agrokhimiya* **4**, 103–110.

WALTER B. (1970). The influence of various herbicides on structure and micro-biology of the soil. *Z. Pflanzenkr. Pflanzenschutz 77, Sonderheft* **5**, 29–31.

WALTER B. (1971). Investigations on CO_2 production and the enzymatic activity of a vineyard soil after treatments with herbicides. *C. R. 6th Conf. du Comité-Francais de Lutte coutres les Mauvaises Herbes., Columa,* 79–82.

WEBB, J. L. (1966). "Enzyme and metabolic inhibitors" vol. III. Academic Press, New York.

WILKINSON R. E. and HARDCASTLE W. S. (1970). EPTC effects on total fatty acids and hydrocarbons. *Weed Sci.* **18**, 125–128.

YEATES G. W., STOUT J. D., ROSS D. I., DUTCH M.E. and THOMAS R. F. (1976). Long-term effects paraquat-diquat and additional weed control treatments on some physical, biological and respiratory properties of a soil previously under grass. *New Zealand Jl. Agric. Res.* **19**, 51–61.

ZAHAN P. and SOREANU I. (1969). Influentia unui complex de factori asupra sporirii productiei de mazare in conditiile solurilor acidedin *Nord-vestul Transilvaniei. Bul. Stiint. Inst. Pedag. Baia Mare, Ser B* **1**, 167–175.

ZANTUA M. I. and BREMNER J. M. (1976). Production and persistence of urease activity in soil. *Soil Biol. Biochem.* **8**, 369–374.

ZINCHENKO V. A. and OSINSKAYA T. V. (1969). Changes in the biological activity of soil incubated with herbicides. *Agrokhimiya* **6**, 94–101.

ZINCHENKO V. A., OSINSKAYA T. V. and PROKUDINA N. A. (1969). The effect of herbicides on the biological activity of the soil. *Khimiya Sel'skokhoz.* **7**, 850–853.

ZUBETS T. A. (1967). Effect of herbicides on the biological activity of soil. *In* "Issledovanya Molodykh Uchenykh", pp. 105–113. Lenizdat, Leningrad.

ZUBETS T. P. (1968a). Activity of hydrolytic enzymes in sod-podzolic soil following use of herbicides. *Sb. Dokl. Simp. Ferment. Pochvy* **1967**, 280–288.

ZUBETS T. P. (1968b). Effect of herbicides on the processes of mineralization of organic matter in peat soil. *Nauch. Tr., Sev.-Zapad. Nauch.-Issled. Inst. Sel'skokhoz.* **12,** 46–50.

ZUBETS T. P. (1973a). Residual action of simazine and atrazine on the microflora and enzyme activity in sod-podzolic soil. *Nauch. Tr., Sev.-Zapad. Nauch.-Issled. Inst. Sel'skokhoz.* **24,** 103–109.

ZUBETS T. P. (1973b). Microbiological and biochemical activity of soil as index of the presence of herbicides and their metabolites in soil. *In* "Povedenie, Prevrashchenie i Analiz Pestitsidov i ikh Metabolitov v Pochve" pp. 82–89. Nauch. Tsentr Biol. Issled., Pushchino-na-Oke.

8

Enzyme Activity in Soil: Some Theoretical and Practical Considerations

R. G. BURNS

Biological Laboratory, University of Kent, Canterbury, Kent, England.

I. The Soil Environment

1. Introduction

In recent years there have been a number of attempts to envisage the actual environment where soil microorganisms reside (see reviews by Alexander, 1971; Stotzky, 1972; Hattori and Hattori, 1976; Burns, 1978b; attempts which have been hindered by the paucity of relevant data due, in turn, to an indaequate methodology. As a consequence, it has frequently proven more rewarding (and certainly easier) to isolate microorganisms from soil and study their properties in axenic culture. It is doubtful, however, whether this can still be described as *soil* microbiology, although a generation of pesticide scientists would probably dispute that statement. Certainly the extrapolation of in-vitro observations to the natural environment is an unreliable and frequently misleading exercise. Even the interpretation of in-vitro evidence (e.g. from undisturbed soil cores) must be viewed with caution due to probable climatic, botanical, zoological and manipulative changes. Paradoxically however, the dissection of a soil's interacting components is often the sole experimentally-acceptable approach to unravelling the mysteries of its ecology.

Another disturbing aspect of our fragile understanding of microbial interactions in soil is that a good deal of it is founded upon plant and animal ecology. This is far from totally acceptable even though some of the basic tenets, such as the dispersal–colonization–succession–climax sequence and the categories of harmful and beneficial associations between organisms, are broadly applicable to the microbial world. Furthermore, macro-ecological principles are often based upon observational rather than experimental data; an intrinsically weak foundation in the minds of formally-trained microbiologists and biochemists.

A pessimistic conclusion from this preamble might be that an understanding of the interrelated heterogeneity of the soil is quite beyond the reach of mortal man. Indeed, the range of scientific disciplines (chemistry, physics, hydrology, microbiology, biochemistry, mineralogy, enzymology, etc.) requiring mastery and the inherent complexity of the problem are alarming in the extreme. Notwithstanding, the last dozen years or so have seen the emergence of a veritable constellation of techniques (Pochon *et al.*, 1969; Roswell, 1973) which have permitted a closer, and possibly more realistic look at the environment of the soil microbe. These include transmission and scanning electron microscopy (Gray, 1967; Hagen *et al.*, 1968) and the electron microprobe (Todd *et al.*, 1973), fluorescent staining (Eren and Pramer, 1966; Schmidt, 1973; Anderson and Slinger, 1975), soil sectioning (Jones and Griffiths, 1964; Bascomb and Bullock, 1974) and the image analyser (Murphy *et al.*, 1977), microcalorimetry (Wadso, 1973), ATP analysis (Ausmus, 1973), infra-red photography (Casida, 1968), the pedoscope (Aristovskaya, 1973; Wagner, 1974), diffusion gradients (Caldwell and Hirsch, 1973) and the use of radioisotopes (Brock, 1971; Smith *et al.*, 1973). Finally, the impetus given by research into the fundamental problems of soil–pesticide interactions (Guenzi, 1974; Burns, 1975), pollution microbiology in general (Higgins and Burns, 1975) and, within the present context, soil enzyme activities (Kiss *et al.*, 1975; Skujiņš, 1976; Burns, 1977) has had a significant effect on our comprehension of the soil microenvironment.

2. The bulk soil environment

It is now generally recognized that the gross physical, chemical and biological characteristics of soil give a severely distorted picture of the soil
microhabitat. Thus the spacial variabilities we, as soil microbiologists,
should be concerned with understanding and integrating involves metres
and centimetres (bulk soil), millimetres (soil aggregates) and micrometres
(colloidal clays, organic matter and individual microorganisms).

On a macroscopic scale the surface horizons of an undisturbed soil often
exhibit an homogeneity of microbial and enzymic activity, both in time and
space. The reproducibility of enzyme determinations has even been suggested as a valuable clue in confirming whether or not a suspect was at the
site of a crime (Thornton and McLaren, 1975). In our experience the
urease activity in five separate collections of soil (air-dried and sieved)
made over a period of three years, and sampled within five metres of a
central point, has varied by only 5·9% and is independant of season; a
continuity generally in harmony with other reports (McGarity and Myers,
1967). Ramírez-Martínez and McLaren (1966) described phosphatase as
being stable throughout the year although Cortez *et al.* (1972) recorded
significant fluctuations in invertase and amylase activities. Soils subject to
extreme climatic variations (Cooper, 1972) and those still undergoing morphogenesis, be it due to weathering or leaching of mineral and organic
components, will, in all probability, exhibit fluctuations in enzyme levels.

Cultivated soils must be expected to express biochemical variability.
This may be caused by such events as crop rotation, ploughing, pH adjust-
ment (e.g. liming), pesticide spraying and fertilizer amendment. The
influence of the last two have been discussed in detail in Chapter 7. The
rhizosphere effect (Brown, 1975) has long been recognized although in-
completely understood yet it is quite obvious that the biological events in
the plant root region are markedly different from those in the non-rhizosphere soil. Furthermore, these events vary according to species, age and
general well-being of the plant. In addition, the death, harvesting and
ploughing-in of plant residues will have a significant influence on microbial
activities. Ross (1966, 1968) compared the ratio of invertase to amylase
activity and found it higher in soil under the grasses *Dactylis glomerata*
(cocksfoot) and *Bromus* sp. (brome) than that under red (*Trifolium pra-
tense*) and white (*Trifolium repens*) clover. Neal (1973) and Saunders and
Metson (1971), studying a prairie soil, found that higher phosphatase activities were associated with the invader species than with the dominant
indigenous flora. Possibly the invaders' greater potential to solubilize and
absorb phosphate was responsible for their success. Pancholy and Rice
(1973a, b) observed that amylase, cellulase and invertase activities declined

during succession from prairie to oak/pine forest. The opposite trend occurred with urease and dehydrogenase. It is, of course, sometimes difficult to ascertain the proportion of enzyme produced by the root *per se* and that accounted for by the rhizosphere effect. A number of others (e.g. Berger-Landefeldt, 1965; Franz, 1973; Ross, 1976) have also studied the influence of variable vegetative cover on soil enzyme activities.

Enzyme activity, like microbial numbers, declines with depth (Myers and McGarity, 1968) and is inevitably related to soil type. The horizon of high activity may descend to some 20 cm below the surface in a cultivated soil (Franz, 1973) or only a few centimetres in a desert soil (Skujiņš, 1973).

3. The soil aggregate environment

The primary inorganic fragments of soil tend to be clumped together to form crumbs or aggregates. These are of variable size ($\approx$ 1–10 mm) and their formation and stability is irrevocably linked with soil fertility. It has long been known, for example, that well-aggregated soils drain efficiently, stimulate the diffusion of gases and solutes, aid root anchorage and growth, and encourage an active microbial population. Poorly-aggregated soils, on the other hand, may become waterlogged, anaerobic and inhibit microbial proliferation. In the light of this knowledge it is not surprising that there has been much research into the formation and properties of soil aggregates during the past quarter of a century (Low, 1954; Griffiths, 1965; Harris *et al.*, 1966; Allison, 1968; Baver, 1968; Aspiras *et al.*, 1971b; Cheshire, 1977). This has shown that a multitude of factors contribute to the development of aggregates and their subsequent stabilization. They include, possibly in order of occurrence, the flocculation of clay particles: due to a combination of electrostatic attraction and a variety of other inorganic and organic mechanisms (Santoro and Stotzky, 1967). These microaggregates then become associated with each other due to their physical entanglement by plant roots and fungal hyphae (Swaby, 1949; Aspiras *et al.*, 1971a). Subsequently, microbial proliferation contributes extracellular polysaccharides which have a cementing effect (Mehta *et al.*, 1960; Acton *et al.*, 1963; Hepper, 1975) whilst aromatic polymers may be, at least in part, responsible for the consolidation of newly-formed aggregates (Griffiths and Burns, 1972). Climatic changes, agricultural practices and the activities of soil animals are also important.

Because of the many microbiological contributions made towards aggregate formation it is probable that their organic composition is different from that found in non-aggregated soil. This variation may be further exaggerated by the subsequent metabolism of microbes located within the crumb. Thus, in quantitative terms the biochemistry of the aggregated fraction may be different from that of the bulk soil, as it will also be qualitatively different from that of the unconsolidated particulates. Whether these variations will be reflected in the activity of accumulated enzymes is uncertain (see Cerná, 1966 and 1970 for indications), but if differences are observed they could be attributed either to greater microbial proliferation (during and after crumb formation) or more efficient protection of the extracellular enzyme (by a concentration of humic polymers). Whatsoever, these possibilities indicate a need for caution when separating soil for experiments as there may be differences other than those attributable to fraction size alone. In passing, it is worth recording that Drăgan-Bularda and Kiss (1972 and Chapter 4) have suggested the involvement of dextranase in soil aggregate degeneration.

Within a soil aggregate there exists a plethora of microsites each presenting a discrete niche for microbial colonization. Consequently microbial discontinuity is the rule rather than the exception in this habitat (Tyagny-Ryadno, 1968; Nishio, 1970). Both aerobic and anoxic environments abound and their relative proportions and positions are determined by such diverse factors as water content, the size of pores between the particulate components of the aggregate, surface area and microbial consumption of oxygen within the crumb. As a general rule anaerobic bacteria are concentrated towards the core (Greenwood, 1968), whereas fungi and other obligate aerobes tend to reside in peripheral regions. Moisture gradients will exist, especially during periods of drought, when external surfaces rapidly become dehydrated. Notwithstanding, microbes located in the surface layers will be the first to benefit from re-wetting and the appearance of soluble nutrients. Microbes living deep within the aggregate may be protected from predators (especially protozoa) but may be physically prevented from extensively colonizing their microenvironment. It has been suggested that most of the internal microorganisms (or at least their ancestors) were trapped during aggregate formation (Allison, 1973)

4. The soil colloid environment

4.1. Components

To a large extent microbial activity occurs at interfaces; in the soil environment these interfaces are essentially solid–liquid. Thus it is not surprising that the smallest soil particles, the clays and the organic colloids, with their high unit surface areas assume a prominent role in soil microbial ecology. Clay colloids are aluminosilicates with, at least in relation to biological activity, two crucial properties: one structural and the other physico–chemical (Table 1). Some clays (e.g. montmorillonite) are constructed of silicon and aluminium plates which are only held together by weak bonds.

As a result they tend to expand upon hydration. Others (e.g. kaolinite) have more tightly bound components and non-expanding lattices. As a consequence, clays not only have an external adsorption surface but sometimes a significant internal one as well. The second property, shared by all clays, is that they are predominantly negatively charged. This arises, during morphogenesis, from the isomorphous replacement of the Al^{4+} and Si^{3+} by ions of a lower valency (e.g. Mg^{2+}, Fe^{2+}) imparting a charge imbalance. The clay particle will attempt to neutralize its abundance of negative charges by adsorbing cations from solution which will, in turn, be exchanged for other positively charged ions, molecules and indeed microorganisms.

TABLE 1

Characteristics of soil colloidal particles

Soil constituent	Surface area (m^2 g^{-1})	Cation exchange capacity (mEq. $100g^{-1}$)
Kaolinite[a]	25–50	2–10
Illite[a]	75–125	15–40
Vermiculite[a]	500–700	120–200
Montmorillonite[a]	700–750	80–120
Organic matter[b]	500–800	200–400

[a] Data from Weber (1972).
[b] Data from Bailey and White (1970).

The nature of soil organic colloids is extremely complex and has occupied a number of able minds for many years (Kononova, 1966; Felbeck, 1971; Schnitzer and Khan, 1972; Flaig, 1975; Flaig et al., 1975; Haider et al., 1975). Like the expanding lattice clays they have both internal and external surface areas due to the amorphous polymeric network of which they are composed. Humic matter may also carry an abundance of negative charge due, not to isomorphous substitution, but to the ionization of its constituents. Therefore the charge of a soil organic colloid is related to the soil pH and the dissociation constants of its constituent molecules. Incidentally the charge of microbial cells results from the same mechanisms. It is apparent, therefore, that microbe–substrate and enzyme–substrate interactions predominate at surfaces and are governed by the physical and chemical factors which characterize these zones.

4.2. Adsorption

Adsorptive and other methods by which microbes, substrates, products, enzymes and inorganic ions become immobilized on soil surfaces have been discussed elsewhere and in a variety of contexts (Bailey and White, 1970; Marshall, 1971; Zaborsky, 1973; Ponec *et al.*, 1974). They are merely summarized here (Table 2). It suffices to say that, due to the numerous physical, chemical and biological mechanisms available, association with surfaces is not solely restricted to those cationic moieties which can become involved in ion-exchange. The consequential environmental changes of immobilization and their potential effect on biochemical activity have been commented upon by many (McLaren and Packer, 1970; Katchalski *et al.*, 1971; Filip, 1973) particularly in comparison to reactions in homogenous systems. It is worth recalling that the starting-point for a sizeable proportion of these reviews is the sub-cellular environment described by Danielli (1937) and that the comparison of soil to a living tissue (Skujiņš, 1967) is a rewarding and pertinent exercise.

TABLE 2

Adherence of microbes, enzymes, substrates, products and inorganic ions to soil surfaces

A. Sorptive mechanisms:	cation exchange
	anion exchange
	protonation
	van der Waals forces
	hydrogen bonding
	co-ordination complexes
	ligand exchange
B. Microbial structures:	exopolysaccharides
	prostheca
	flagella
	pili
C. Miscellaneous:	chemical incorporation into humic polymers
	physical entrapment within organic colloids
	lipophilic association with organic matter

Adsorbed reactants will find themselves in a very different environment to that previously encountered in the ambient menstruum. For instance, it is well known that hydrogen ions are concentrated at the negatively charged surfaces of clays. This phenomenon may give rise to a diffuse double layer (DDL) micro-environment between one and three pH units more acid than the surrounding soil aqueous phase (Hartley and Roe, 1940; McLaren, 1960; Mortland, 1968). Other adsorbed organic and inorganic metabolites may also cause a pH shift ($\triangle$ pH). As a result one would expect to record changes in the pH-activity profiles of enzymes when monitored in the presence of colloidal particles. This has been noted on numerous occasions and is presumably due to a combination of pH-mediated events (e.g. protonation, ionization, denaturation, etc.) affecting the enzyme and/or its substrate.

Microorganisms liberate and consume H^+ which results in significant pH gradients across membranes (see, for example, Riebeling et al., 1975). This has profound effects on permeability, transport, kinetic behaviour and the formation of ATP (Mitchell, 1966). It is possible that any pH difference between colloid surfaces and the cytoplasmic membrane has a similar effect on the fundamental functions of microorganisms residing in close proximity. Hydrogen ions concentrate at the cell wall of the microbe itself and may, at the least, induce changes in shape and porosity (Ou and Marquis, 1970). Conversely alkaline micro-environments, induced by ammonium adsorption, may be responsible for the success of actinomycetes in acid soils (Williams and Mayfield, 1971).

Another effect of ion (and molecule) concentration at surfaces is to offer high levels of nutrients (NH_4^+, K^+, Ca^{2+}, Mg^{2+}, etc.) and thus competitive advantages to adsorbed cells (and indeed adjacent plant root hairs). This frequent assumption needs qualification, however, as microbes may only be able to benefit fully from adsorbed nutrients if they can extract them directly from the surface. If the nutrients are desorbed and enter into solution prior to uptake then the concentration available to the microbe will depend upon the extent of the DDL and volume of the immediate water film together with the microbe's own anionic abilities. Microbial and enzymic activities will, themselves, induce nutrient gradients.

Many enzymes, given the opportunity, become adsorbed between the clay lattices where both their accessibility to bacterial proteases and their catalytic activity are generally, although not exclusively, reduced. These interlammelar proteins virtually act as molecular calipers. In an elegant experiment McLaren and Peterson (1961) found that the basal spacing of bentonite increased from 1·1 to 4·7 nm on addition of protein and then decreased after enzymic digestion. In contrast, catalase (Harter and Stotzky 1973) tends to be adsorbed mainly on the external surfaces of clay and yet is completely unavailable as a bacterial carbon source. Furthermore its breakdown of hydrogen peroxide is enhanced (Stotzky, 1972). Another report by the same authors (Harter and Stotzky, 1971) revealed that the adsorption of a variety of proteins to homoionic clay was more closely related to the molecular weight, the number of active sites which were available to react with the clay surface, and the valency of the clay cation, rather than to the pH or the isoelectric point of the enzyme. With reference to this final observation, it is generally true that clays strongly adsorb enzymes at pH values below their isoelectric point (Zaborsky, 1973). The sorption of alkaline proteases to clays is substantially stronger than that of neutral proteases

(Ambrož, 1970). Adsorption which led to a decrease in activity and stimulated denaturation.

The consequences of enzyme association with organic matter have been discussed at length by Ladd and Butler (1975). They summarized research in this area and concluded that synthetic enzyme–organic matter derivatives were more resistant to denaturation and biological degradation. This resistance may be due to a combination of physical and chemical mechanisms—such as viscosity, covalent bonding, hydrogen bonding and other ionic linkages. The bacteriostatic properties of the aromatic components of humus may also make a significant contribution to enzyme longevity.

Examination of the relationship between specific elements of soil organic matter and enzyme activity has revealed a variety of effects. For instance, tannins reduced the activities of a large number of enzymes (Lyr, 1961; Williams, 1963; Goldstein and Swain, 1965; Lewis and Starkey, 1968; Gnittke and Kunze, 1975; Fernando and Roberts, 1976) although it is not always clear whether this inhibition was due to an enzyme–tannin or a substrate–tannin association. Humic acids, the most abundant constituents of recalcitrant soil humus, commonly inhibit enzyme activity (Mato et al., 1971). They appear to be aromatic polymers complexed with (amongst other organic moieties) amino acids and proteins. This certainly imparts stability against microbial and enzyme attack although it may also retard substrate turnover. Verma et al. (1975) have suggested that proteins are linked to the phenolic units of humic matter through the ϵ-amino group of lysine or the –SH group of cysteine. Ladd and Butler (1969) observed humic acid inhibition of the non-specific proteolytic enzyme pronase, and noted that it was more dramatic with some substrates than others. This implied that humic acid was selectively blocking catalysis. They later showed (Ladd and Butler, 1970) that the inhibitory effect of humic acids on carboxypeptidase, chymotrypsin and trypsin could be reversed by the addition of monovalent and divalent cations. The conclusion drawn from this work is that the cations displace the bound and partially inhibited enzyme at the surface of the anionic humic acid. Humic acids generally have a more pronounced influence on enzymes than fulvic acids possibly due to a combination of their molecular weight, aromatic content and the condensed state of the molecule (Butler and Ladd, 1971). Humates may also influence biological activity by supplying organic and inorganic nutrients and adsorbing inhibitory substances (Martin et al., 1976).

Lipids are common components of biological membranes and, as such, make a significant contribution ($\approx 5\%$) to soil organic matter. Cellular studies (Fleischer et al., 1962), concerned with the activity of enzymes in relation to membranes, have implied an involvement of lipids with enzyme orientation, the formation of a permeability barrier for substrates and enzymes and the protection of the enzyme from harmful solutes. Some subcellular enzymes actually require lipids for activity. It is possible that these influences exist in the lipid containing organic gel where many extracellular soil enzymes are located. Other naturally-occurring organic carriers of enzymes (see Zaborsky, 1973) include chitin, lignin, cellulose and starch.

Additional factors which have been observed to modulate enzyme activity and which may have relevance to the soil microenvironment include: polyelectrolytes (Goldstein et al., 1970; Goldman et al., 1971; Elbein, 1974) such as histones, lysozyme and the teichoic acids and peptidoglycans of bacteria; calcium ions (Heinen and Lauwers, 1975) and any stearic and electrostatic effects induced by adsorbates occupying the same domain (Ackerman, 1962).

Upon association with a soil surface the enzyme may undergo unfolding (Morgan and Corke, 1976) or coagulation (see James and Augenstein, 1966) which can change its affinity for the substrate, in addition a number of active sites may become directly involved with, or masked by adsorption and thus unavailable for reaction with the substrate, or the enzyme may become aligned in an advantageous way. More radical changes concerning the actual specificity of the enzyme have been suggested. Oparin (1937) observed that subcellular soluble invertase hydrolysed sucrose to glucose and fructose, whereas it performed a synthetic role when in the bound state. Whether this type of functional change can occur in soil extracellular bound enzymes is unknown. However, Ladd and Butler (1975) have described how proteolytic enzymes adsorbed to an organic matrix will, due to stearic effects, become more efficient at hydrolysing low molecular weight peptides than high molecular weight proteins.

4.3. Location of exoenzymes

Any assessment of the involvement of the mechanisms discussed above in the stabilization of enzymes in soil can only be intuitive. Notwithstanding, a general method of enzyme entrapment has been proposed (Burns et al., 1972a, b) which suggests that enzymes are physically and chemically immobilized within the discontinuous organic colloidal material which is itself associated with soil clay particles (the "organo–mineral complex"); an immobilization that occurs during humic matter genesis when exoenzymes, and endoenzymes from lysed cells, become trapped. The location does not impede diffusion of substrate to, and product away from the enzyme but does offer protection against macromolecules such as pronase (Burns et al., 1972b; Aliev and Zviagintsev, 1974; Nannipieri et al., 1974; Cacco and Maggioni, 1976). This model was originally postulated for urease (and inspired by the seminal research of Conrad performed some thirty-two years earlier (Conrad, 1940) and, within that limited context, is still somewhat appealing. However not all accumulated enzymes have low molecular weight soluble substrates and it is difficult to imagine the initial depolymerization of substrates such as cellulose and starch (especially when complexed as macro-organic matter) occurring within these organic matter–enzyme films. This shortcoming suggests an extension of the model (Fig. 1) whereby some enzymes are located on the surface of the preformed humic material, possibly attached through lipophilic moieties and with their critical denaturation sites screened from attack.

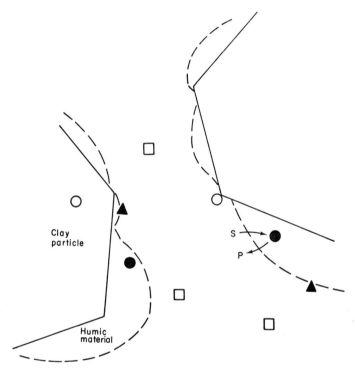

Fig. 1. Suggested location of exoenzymes in the soil microenvironment.
● enzyme trapped within colloidal humus of organo-mineral complex;
○ enzyme associated with clay—either on the surface or in the interlamellar spaces;
▲ enzyme attached to the surface of organic colloid film;
□ ephemeral enzyme free in soil solution.
S = substrate, P = product.

It appears that the exoenzymes (e.g. urease) are distributed between most of the humus fractions rather than associated with just one (McLaren *et al.*, 1975) and that their overall stabilization may involve polysaccharides (Martin and Haider, 1971; Mayaudon *et al.*, 1975) as well as the aromatic polymers.

II. Enzyme-Substrate Interactions

1. Introduction

Microorganisms in soil are concerned with an host of biogeochemical transformations involving macro- and micro-nutrients crucial to the success of all flora and fauna. These are cyclic changes in which bacteria and fungi have both a catabolic and an anabolic role. Additionally, earthworms and a myriad of other invertebrates have an important function in the preliminary fragmentation of organic debris. Any nutrients which are not utilized for cell construction are released for subsequent use by plants, animals and indeed other microorganisms. Needless to say microorganisms are not altruistic and it is fortunate, therefore, that they are far from totally effective in assimilating substrates. If this were not so the microflora of terrestrial and aquatic environments would constitute a sink, effectively immobilizing large quantities of nitrogen, carbon, sulphur, phosphorus, etc. Wagner (1975), in discussing microbial growth, quotes figures for carbon assimilation by aerobes of 20–70% and, for the far less efficient anaerobes, of 2–5%.

In the lithosphere a proportion of these transformations is non-biological involving a variety of physical and chemical catalysts. Most, however, are mediated by enzymic mechanisms located in, or arising from microorganisms, plants and soil animals.

2. Components of soil enzymic activity

As discussed previously in this volume (Chapters 1, 2 and 7) the total enzymic capability of a soil is a composite of many different fractions, easily described but measured with some difficulty. It includes those which are intracellular, those associated to varying degrees with the external surface of the cell wall, those which have diffused from their point of origin into the soil solution and those which have become temporarily or permanently bound to the soil colloids. Some microbiologists regard an enzyme as extracellular once it has passed through the plasma membrane (Glenn, 1976) and thus this category includes those enzymes in the periplasmic space, within the wall, attached to the outer surface of the wall, and liberated into the surrounding medium. Others (Pollock, 1962) have restricted the extracellular term to those found solely in the supernatant and produced during normal cell growth (excluding lysis). The expression *soil* extracellular enzyme has come to be accepted as applying essentially to that fraction which is separated both spatially and probably temporally from its cellular origin. This ambiguous terminology has been referred to elsewhere and is obviously in need of clarification. Skujiņš (1976) has suggested adopting the adjective "abiontic" to describe soil exoenzymes.

Intracellular catalysis can occur in both extant and undamaged dead cells. Enzymes at the external surface of the cell wall may be somewhat restricted, in diffusion terms, by being partially attached to or trapped within an exopolymer coat. Numerous microbial cells secrete homo- and hetero-polysaccharides (Sutherland, 1972; Hepper, 1975) which may serve many protective and adhesive functions (Dudman, 1977) phenolic extraction of this fraction reveals the presence of a variety of proteins and amino acids. This association may be in the form of a discontinuous slime layer or, if the polymeric material is comparatively rigid, a capsule. Consequently, some exoenzymes may be retained within a viscous network which will still allow the passage of substrates and products (cf. the plant root mucigel–Jenny and Grossenbacher, 1963; Greaves and Darbyshire, 1972), a system analogous to that hypothesized for enzymes associated with humus (Burns et al., 1972a) and referred to earlier.

A logical extension, based on our knowledge of the categories of soil exoenzymes, is to conclude that at least a proportion of this significant catalytic component is caused by leakage from viable cells and rupture of moribund cells. The net result is the extracellular accumulation of intracellular, periplasmic and intramural enzymes. Subsequent stabilization of these and the true extracellular enzymes is brought about by their incorporation into humic material which arises itself from the degradation of macro-organic matter and the synthetic activities of bacteria and fungi.

The presence of this recalcitrant enzyme fraction is not in doubt. Evidence gleaned from inhibitor studies, from observing the linear consumption of the product and by the extrapolating microbial counts to zero are supportive, whilst the appearance of activity in 9000 year old permafrost peat soils from Alaska (Skujiņš and McLaren, 1968) and lake sediments (Rădulescu and Kiss, 1971) is spectacular corroboration of the phenomenon of enzyme accumulation. As a caveat to this final statement one can only speculate whether these ancient soils were free of biological imput for the duration of their natural preservation, there is the possibility that many enzymes might survive for thousands of years in a permanently frozen state. Interestingly 32 000 year old permafrost soils, contaminated in transit, showed little or no urease and phosphatase activities *despite* the presence of microorganisms.

3. Synthesis and secretion of enzymes

The principal mechanisms involved in the synthesis and release of exoenzymes by eukaryotes are well documented (Sachs, 1969; Palade, 1975;

Chrispeels, 1976). It is a sequence of events initiated by the ribosomes and involving the transport of proteins through the cisternal space of the endoplasmic reticulum to the Golgi apparatus. Here the enzymes become enclosed in vesicles which ultimately fuse with the plasma membrane, undergo exocytosis and excrete their contents.

In prokaryotes the processes, although spatially simpler and reviewed by many (Bull, 1972; Payne, 1976; Glenn, 1976), are far less well defined. Exoenzyme synthesis probably occurs on cytoplasmic membrane-bound ribosomes (Lampen, 1974), a location which has profound translational and transcriptional ramifications. Passage through the membrane must involve mechanisms which will permit the transport of a charged protein through a hydrophobic lipid barrier. A cytoplasmically-located vesicular apparatus, the mesosome, has been implicated (Lampen, 1965) yet has proved somewhat contentious (Salton, 1974; Greenawalt and Whiteside, 1975; Salton and Owen, 1976) whilst the involvement of polysaccharides, aminoterminal groups, oscillating carrier proteins, fatty acids and lipids have all been suggested. A tentative transport scheme, involving a switch from hydrophobic to hydrophilic conformations during passage through the membrane, has been postulated by Lampen (1974).

Many enzymes of Gram negative bacteria are restricted to and accumulate in the periplasmic space (Heppel, 1971; Salton, 1971; Gould et al., 1975) and are only released when the cell is converted to a spheroplast, whilst truly extracellular enzymes (more often produced by Gram positive than Gram negative bacteria, Costerton et al., 1974) diffuse out through cell wall pores with (Dancer and Lampen, 1975) or without (Beacham et al., 1976) the assistance of phospholipids. It is reasonable to suggest that these pores may also be used for substrate uptake (see later) and product excretion. Transport of large molecules through small apertures, as well as self-destructive problems may be overcome by the protein having a flexible structure (Pollock and Richmond, 1962) and only assuming its active configuration outside the cytoplasmic membrane or even the cell wall. Dimerization of inactive alkaline phosphatase subunits (zymogens) has been reported to occur in the periplasmic space (Schlesinger, 1968; Torriani, 1960) subsequent to secretion by the cytoplasmic membrane. Containment within a vesicle would also protect the parent cell from newly formed enzymes as would the presence of aspecific inhibitor. A variety of post-translational modifications (e.g. addition of carbohydrate residues, subtraction of amino- and carboxy-terminal residues) may occur on the way from the cytoplasmic membrane to the external medium and be mediated by deformylases and peptidases. Phosphatase enzymes, as they relate to the periplasmic and extracellular environment, have been studied intensively and it is clear that a number of different forms of this enzyme exist with different solubilities and pH optima (Loppes, 1976).

Alkaline phosphatase, for one, is understood to be a periplasmic enzyme although considerable leakage occurs during growth (Ingram *et al.*, 1973) possibly due to defects in the lipopolysaccharide (Lindsay *et al.*, 1973) and protein (Singh and Reithmeier, 1975) components of the outer membrane. The properties of a second group of well studied enzymes, the exo- and endopencillinases of *Bacillus licheniformis*, have been reviewed by Lampen (1974).

4. Uptake of substrates

Little is known concerning the factors which determine whether a substrate is degraded intracellularly or extracellularly although, no doubt, cell, substrate and environmental characteristics all contribute. The molecular size, solubility, configuration, charge distribution and association with other compounds are all important substrate properties. In addition, it would be a disadvantage to degrade a substrate intracellularly if the product were either inherently toxic or severely altered the pH of the cytoplasm. The porous microbial cell wall is likely to act selectively as a barrier to large molecules, behaving differently according to species (Naider *et al.*, 1974). Payne (1976), Payne and Gilvarg (1968) and Payne and Matthews (1974) have recorded a broad relationship between the size of peptides and their utilization for growth by *Escherichia coli* and suggest that an exclusion limit of approximately 530 daltons exists. The molecular sieving function of a variety of Gram negative enteric bacteria has also been examined by Decad and Nikaido (1976). They reported that di- and trisaccharides diffused through the cell wall and into the periplasmic space whilst oligosaccharides did not. This indicated a cut-off of between 550 and 650 daltons. Different limits were observed with polyethylene glycols which exhibited some penetration at sizes of 1540 daltons. The authors did not find these results incompatible and suspected that they were due to the conformational differences of the two types of substrates in solution. The work of Scherrer and Gerhard (1971) with *Bacillus megaterium* indicates that Gram positive bacteria have much larger pores with a cut-off of about 100 000 daltons. Other workers have studied cell wall uptake of certain antibiotics, dyes and β-thiogalactosides. It has been postulated that non-specific diffusion occurs through water-filled pores although this should not exclude transport of large molecular weight substrates facilated by proteins (Nakae and Nikaido, 1975). It is suggested (Epstein, 1975), at least for Gram negative bacteria, that periplasmic binding proteins act as specific transmembrane carriers releasing substrates at the plasma membrane. ATP may function as the driving force in this process. Active transport mechanisms, for subsequent transfer through the principal permeability barrier, the cytoplasmic membrane, have been described elsewhere (Kaback, 1970).

5. Role of extracellular enzymes

It is unlikely that unprotected extracellular enzymes survive for long in the hostile soil environment. Enzymes are, after all, merely proteinaceous substrates for many microorganisms. Moreover, soil chemical and physical properties will encourage denaturation. Research concerned with adding enzymes to soil (Roberge, 1970; Drozdowicz, 1971) or by stimulating their synthesis (Zantua and Bremner, 1976) has often indicated that only an ephemeral surge in activity occurs; inactivation being attributed to a combination of adsorption, degradation and denaturation. Eukaryotic exoenzymes tend to contain disulphide bridges and exist as glycoproteins (Eylar, 1965) a combination known to enhance stability to elevated temperatures, pH fluctuations and proteolytic activity (Pazur and Aronson, 1972). Prokaryotic exoenzymes, as a general rule (for exceptions see Berry *et al.*, 1970; Glew and Heath, 1971), lack these advantages (Pollock and Richmond, 1962; Kelley *et al.*, 1973). Other features of extracellular proteins are that they possess a polarity range of 39·4–54·4% (Glenn, 1976) and have molecular weights between 20 K and 40 K.

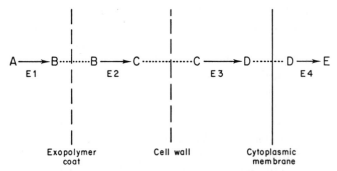

Fig. 2. A model for the sequential decay of polymeric substrates.
E1 = extracellular enzyme; E2 = exopolymer-located enzyme;
E3 = periplasmic enzyme; E4 = cytoplasmic enzyme.

The inherent lack of stability shown by many extracellular enzymes presents us with a dilemma in our understanding of their ecological and evolutionary importance. For an enzyme to be of any value to its parent microorganism it must first endure for a sufficient period to allow it to both locate and react with a substrate. Secondly the microbe must be able to utilize the product of that reaction as a nutrient and as an energy source. This, after all, is the purpose of an extracellular enzyme. Many protozoal organisms avoid these problems by engulfing their substrates in vacuoles prior to digestion. Fig. 2 suggests a scheme which would maximize the efficiency of bacterial degradative enzymes.

Very low levels of exoenzymes are continually being released by microorganisms. Some of these diffuse away from the cell whilst others remain associated, electrostatically or otherwise, with the wall. The former group, "enzyme scouts", will be denatured unless they contact a suitable substrate or adhere to the soil colloids. If they do become associated with soil colloids, the presence of a stable catalytic fraction will lessen (or negate) the microbial populations requirement for "enzyme scouts". Thus the ecological advantages of an accumulation of enzymes are emphasized. The solubilized product(s) of any successful depolymerization reaction may diffuse to the cell or, better still, the cell may exhibit a tactic response. In either case it is only at this stage that enzyme induction would be initiated by product uptake and the microbe can respond efficiently to the presence of an exogenous substrate. If the products of the initial bound enzyme catalysis were still too large or insoluble for cytoplasmic absorption then the opportunity exists for their further breakdown by enzymes associated with the cell wall, possibly its polymer coat should there be one, and within the periplasmic space. As a result (and this is a key feature of the model) a series of molecular sieves may exist, all with enzymes controlling the passage of organic substances of decreasing complexity to the next compartment. The number of compartments involved will depend on cell type (Gram positive or Gram negative) and the presence or absence of structures on the outer surface of the bacteria. The range may be from two (Gram positive) to four or more (Gram negative). The number of compartments actually required will depend on the substrate.

Because exopolysaccharide and periplasmic enzymes are somewhat protected from denaturation pressures they may be at an high enough indigenous level to obviate the delay of induction. Exopolysaccharidases may even enhance the activity of trapped enzymes (Kuwabara and Lloyd, 1971).These pools would impart a significant competitive advantage to those microorganisms with the correct sequence of enzymes. Other benefits of only one or a small number of truly extracellular steps in a sequence is the decrease in energetically-wasteful catalyses and the elimination of the spatial problems associated with multi-enzyme extracellular reactions.

To substantiate this hypothesis one requires three sorts of evidence:

(1) that microbes will respond chemotactically to the products of extracellular catalysis;
(2) that the soluble end-product of a sequence will induce the enzymes responsible for the previous depolymerization reactions;
(3) that the enzymes involved in a sequential degradation are somewhat restricted to the cytoplasm, periplasmic space, exopolysaccharide coat and medium surrounding the cell according to the reaction they catalyse.

The first piece of evidence is easy to come by for the chemotactic behaviour of microbes is well established (Chet and Mitchell, 1976). More specifically microbial migration in response to a monosaccharide product of extracellular glycosidases has been suggested for the slime mould *Physarum polycelphalum* (Carlile, 1970; Kilpatrick and Stirling, 1977). Product induction of an enzyme is a well recognized phenomenon, as are other regulatory mechanisms (see review by Glenn, 1976). A search of the literature has not revealed any direct evidence for enzyme compartmentation of the type suggested in (3) which is perhaps not surprising when one considers that our knowledge of the enzymology and biochemistry of polymer breakdown is far from complete. For example, the saccharification of the commonest carbohydrate polymer, cellulose, may involve four or more poorly-defined steps performed by the enzymes endoglucanase (C_1), exoglucanase (C_x), cellobiohydrolase and β—glucosidase (Hannij and Reese, 1969; Bailey *et al.*, 1975). It has recently been suggested (Sternberg, 1976) that the β-glucosidase responsible for cellobiose hydrolysis (cellobiase) and produced by *Trichoderma viride*, is composed of two enzymes and that there are two endoglucanases (Petterson, 1975). Regulation of cellulase activity involves both inducer and repressor systems (Stewart and Leatherwood, 1976). The topology and efficiency of cell-bound cellulases has been the subject of speculation by Hofsten (1975). Similarly complex and poorly defined sequences are described for starch, proteins and pectins. Exogenous substrates of variable composition (e.g. hemicellulose, lignin) cloud the picture even more. Furthermore, the techniques for fractionating cell components are far from satisfactory and many of the products may be artifactual. The Gram negative bacteria, with their complex cell envelopes, present especial problems.

The potential effects of pesticides on cell membranes, compartments and excretion mechanisms and their subsequent contribution to soil enzyme levels, have been discussed by Cervelli *et al.* in Chapter 7.

III. Methodology

1. Introduction

Many of the primary obstacles impeding advances in soil enzymology are associated with methodology because, as with any evolving scientific discipline, those methods have yet to stabilize and throw up a cogent series of experimental techniques. "Methods in Enzymology" (Academic Press) has now reached forty-seven volumes and only two articles deal, even superficially, with soil enzyme measurement. Not surprisingly therefore, the major research groups have either developed their own assays from first principles or radically adapted those already published. The resulting disparity is illustrated in Table 3 for one such enzyme, urease. It is patently obvious that this variability is not merely the expression of scientific independance but rather the widespread inapplicability of, and dissatisfaction with, published methods. To a significant degree this is, in itself, a reflection on the heterogeneity of the soil environment. An environment which has not only the obvious mineralogical and biological variations but also some apparently less easily defined geographical ones as well. In our experience many of the assays, which have worked well in laboratories located in other parts of the world, are unsatisfactory in our hands. Indeed, it is often difficult to comprehend how they have ever produced satisfactory results! Needless to say, the development of a soil enzyme assay is a tedious task for any research worker, especially one whose previous contacts with enzymology are based upon a tried and tested methodology. The process can take up to six months for it to reach a level from which one can begin to investigate, with any degree of confidence, specific problems associated with enzyme–soil interactions. Not surprisingly, therefore, the methodology of soil enzyme assays has often been accorded less attention than it deserved.

Classical enzymology has an established and respected pedigree. Enzymes are assayed in the presence of excess substrate dissolved in a buffer and the reaction mixture is agitated for a fixed period of time. Both the pH of the reaction and the temperature of incubation are the optimum for activity. These conditions give the *maximum potential* catalytic activity; a phrase to be borne in mind when designing an assay and interpreting the results. The concept of potential rather than realized soil microbial activity was discussed some years ago by Clark (1967). There is, of course, a world of difference between pure enzymology and the in-vivo activity of soil enzymes. These enzymes are rarely presented with excess substrate, the moisture level is extremely variable and whilst soils do have a significant buffering capacity it is certainly not up to the standard of, say 0·5 M phosphate or tris-HCl. A further point of divergence also concerns the soil pH which is rarely the optimum for activity. A word of caution is required here: these are all macro-environment (i.e. bulk soil) measurements and may not be synonymous with those in the molecular size zone where the enzyme and its substrate interact (see Section I. 4.). One is tempted to say that at least the catalyst is the same, but this is a daring assumption without the evidence of a *purified* enzyme extract; frequently attempted (Chalvignac and Mayaudon, 1971; Burns *et al.*, 1972b; Satyanarayana and Getzin, 1973; Mayaudon and Sarkar, 1974) but not yet achieved.

2. Buffers and pH

There has been considerable discussion in the literature (and elsewhere in this volume) concerning the choice of buffers for soil enzyme assays, and

TABLE 3

Variable methodology in the assay of soil urease

Soil wt (g)	Buffer	Buffer conc. (M)	pH	Reaction time (h)	Temperature (°C)	Substrate final conc. (M)	Analytical method	Reference
1	tris–maleate	t0·5	7·0	1	20	1·5	Titrimetric analysis of evolved ammonia	Pettit et al. (1976–modified)
5	tris–H$_2$SO4	0·05	9·0	2	37	0·02	Microdistillation of ammonia	Tabatabai and Bremner (1972)
3	phosphate	0·067	8·8	4	37	0·007	Microdistillation of ammonia	May and Douglas (1976)
5	K–citrate	0·5	6·8	2·5	35	0·5	Colorimetric analysis of ammonia	Thente (1970)
5	phosphate	?	6·7	24	37	0·83	Microdistillation of ammonia	Galstyan (1965)
0·2	water	—	soil pH	20	30	0·014	Colorimetric analysis of ammonia	Speir and Ross (1975)
2	phosphate	0·1	7·0	72	28	3·5% in unstated volume	Titrimetric analysis of evolved ammonia	Giardina et al. (1970)
1	K–acetate	0·05	5·5	continuous	?	0·33	$^{14}CO_2$	Skujiņš and McLaren (1969)

indeed some have suggested dispensing with their use altogether. Tabatabai and Bremner (1972) have debated, at length, the factors influencing their selection of tris–H_2SO_4 (THAM) buffer at pH 9·0 for urease assays of soils with bulk pH values of 6·1–7·5; release of NH_3 was of paramount importance. Tabatabai and Singh (1976), when designing an assay for rhodanese, based their final choice on pH-activity optima and low extraction of humic matter. Ryšavý and Macura (1972) also observed that phosphate buffer solubilized soil organic matter which interfered with subsequent spectometry. Consequenly they used tris–maleate buffer for β-galactosidase determinations Lethbridge et al. (1978) were faced with similar inaccuracies when using citrate citrate–phosphate and phosphate buffers in the assay of β–1, 3 glucanase and their final choice was sodium acid maleate (pH 5·4). Ross and McNeilly (1972) were of the opinion that neither buffered nor unbuffered media were ideal for glucose oxidase assays, but that acetate–phosphate buffer was the lesser of the two evils. In their work citrate buffer was rejected because it reacted with ferric ions to enhance glucose oxidase activity by 21%. In contrast, citrate buffer was adopted for soil pyrophosphatase assays (Douglas et al., 1976) because it separated higher amounts of orthophosphate from pyrophosphate than either tris–H_2SO_4 or tris–maleate–NaOH. Sodium maleate was selected as a buffer for p-nitrophenyl phosphatase determinations because it extracted little or no enzyme activity from the soil (Brams and McLaren, 1974; Irving and Cosgrove, 1976). These randomly selected experimental techniques are presented solely to illustrate the plethora of reasons influencing the choice of a buffer in soil enzyme determinations. Notwithstanding, the predominant reasons for selecting a particular buffer should be to poise the pH and then to stabilize it during the reaction. It is clear, therefore, that a unified approach to the selection of buffers is desperately needed—possibly based upon the following criteria:

(1) Assays should always be performed at, or close to, the optimum pH for activity. This is a basic enzymological axiom, although there are some who would argue for determinations to be made at the soil pH. It is really a question of whether the experimenter's principal objectives are fundamental or applied. The latter approach is concerned with relevance to *in situ* activity (see Appendix chapter) whilst the former concentrates on absolute measurements. Occasionally these contrasting viewpoints merge and it is a fortuitous event if the optimum pH for activity is approximately the same as the bulk soil pH; a coincidence which has occurred in our measurement of β–1, 3 glucanase (optimum pH 5·4) and arylsulphatase (optimum pH 5·8) in Hamble silt loam (pH 5·4), but not with urease (optimum pH 7·0) or phosphatase (optimum pH 6·9). It is also desirable to measure activity on the highest plateau of the pH activity curve thus ensuring that minor fluctuations in pH during the incubation period do not affect the result. Fortunately, a large number of the soil enzymes that we

and others have investigated have plateaux of high activity ($>85\%$ maximal) over a pH range of 0·5 to 2·0 units. This compares with plant, animal and microbial enzymes which tend to have sharper peaks of activity.

(2) Buffers perform either one or both of their principal functions during an assay: they adjust the pH to that required for optimum activity and maintain that pH for the duration of the reaction. With reference to the first point it is well to remember that the buffer pH is not necessarily synonymous with that of the reaction mixture. This is due to the buffering capacity of the soil itself, which tends to alleviate any change brought about by the buffer (Table 4). Thus the pH of the soil suspension should be measured during and at the completion of the incubation period and pH activity curves should be plotted as activity versus average reaction mixture pH. Concerning the second point, the pH may fluctuate during the catalysis itself due to acid or alkaline metabolites and decay of the in-hibitor (e.g. sodium azide) or other additives (e.g. pesticides). We have recorded a change of pH from 5·40 to 5·85 in an unbuffered system during the seventeen hours required for the assay of β–1, 3 glucanase in soil.

TABLE 4

Efficiency of buffers at poising and maintaining pH during the assay of β-1,3 glucanase in a soil pH 5.4

Buffer	Buffer pH	Reaction mixture pH	
		time zero	end of in-cubation period (17 h)
Sodium acid maleate	5·0	5·10	5·15
	5·6	5·55	5·60
	6·2	6·18	6·18
	7·0	6·80	6·70
Citrate phosphate	3·0	3·30	3·32
	4·0	4·20	4·28
	5·0	5·02	5·20
	6·0	6·10	6·19
	7·0	7·05	7·00
	8·0	7·75	7·65
Glycine–NaOH	8·0	6·88	6·75
	9·0	8·50	8·10
	10·0	9·40	9·18
	11·0	9·70	9·47

(3) Zantua and Bremner (1975a) have advocated the use of distilled water instead of buffer in urease assays when measuring the hydrolysis of urea fertilizer. Previous workers have also adopted this principle either by design or default. We believe, however, that this is an unacceptable

approach for a variety of reasons. In the first instance, assays can only be carried out at the soil pH, which in most cases will not be the optimum for activity. In addition, comparisons often need to be made with purified enzymes which cannot themselves be assayed in substrate–distilled water solutions. In this, laboratory-distilled water varies unpredictably between pH 6 and pH 8. Soil pH will also fluctuate with sampling time, due to such factors as plant growth, manuring and liming. As a result, one of the key parameters of the assay, pH, is a potential variable and comparisons between experiments are fraught with danger. It is suggested that activity in soil-distilled water suspensions is used solely as a guide to the final choice of buffer avoiding, whenever possible, those which stimulate or inhibit activity. In instances when the assay pH is significantly different from that of the soil it is advisable to compare a range of buffers and select from those with a common response.

Other factors influencing the choice of buffer include the avoidance of those which extract high quantities of organic matter and which would interfere with any subsequent spectrophotometric analysis.

3. Duration of assay

Most exoenzyme assay procedures involve the use of a microbial inhibitor such as toluene or sodium azide, the object being to prevent microbial proliferation and thus the catalysis of substrate and/or the further production of exoenzymes. The comparative merits and demerits of these and other antimicrobial agents (e.g. antibiotics) have been debated at length elsewhere (Drobnik, 1961; Beck and Poschenrieder, 1963; Kiss and Boaru, 1965; Ross, 1968; Kelley and Rodriguez-Kabana, 1975; Skujiņš, 1976). Suffice to say that such considerations as efficacy and specificity of growth inhibition, contact between inhibitor and soil particles, induction of cell lysis, changes in cell wall permeability and interference with enzyme activity *per se* are all of paramount importance when choosing an inhibitor. Taking these and other criteria into account, it is readily apparent that a totally satisfactory inhibitor has yet to be discovered. Ideally, one should minimize the contribution of microbial growth (and thus the *need* for an inhibitor) during an assay by choosing the shortest time necessary (< 2 h) to produce a reasonable level (cf. 10x control) of activity. In longer assays (i.e. > 5 h) the inhibitor may assume a greater degree of importance for, although doubling times in soil are normally considerably slower than in in-vitro studies (Hissett and Gray, 1976), the response to an abnormally-concentrated slug of substrate may well be a rapid surge in growth. Even then, if the contribution of proliferating microbes to total enzyme activity is insignificant it matters little whether or not an inhibitor is used.

However large differences in substrate turnover between sterile and non-sterile soil indicate the significance of microbial growth. Both these contrasting forms of response have been observed by Kiss *et al.* (1972) and have been reviewed later Kiss *et al.* (1974). The restriction of measurement to that activity ascribed to accumulated enzymes is experimentally necessary and may also be conceptually valid if the catalytic process being monitored is an exclusively extracellular event (e.g. initial depolymerizing steps in cellulose, starch and protein degradation). In these instances the cessation of microbial growth halts production of enzymes, limiting activity to that caused by previously generated exoenzymes. If, however, activity is the sum of intra- and extracellular reactions the inhibition of growth may not stop leakage of intracellular enzyme nor indeed substrate uptake and turnover in latent cells. For example, soil urease activity is not only cytoplasmic (*sensu strictu*, see Seneca *et al.*, 1962) but also occurs as a bound extracellular component largely derived from dead and lysed cells. The confirmation of purely extracellular enzymic activity is of some difficulty in other areas of microbiology. Even cell-free supernatants will contain a proportion of lysed cell material (derived naturally from dead cells or from mechanical disruption during centrifugation). To be certain that an enzyme is genuinely extracellular and is not released from cells during preparation, one must also look for marker enzymes which are known to be uniquely intracellular. Their presence in a cell-free extract would imply that some lysis had occurred and suggest caution in interpreting the origins of the actual enzyme being studied. Ideally, what is required for this type of enzyme is an inhibitor which halts microbial growth and prevents passage of substrates and enzymes through the cell wall. One such possibility is formaldehyde which is known to block the entry of ONPG (o-nitrophenyl-β-D-galactopyranoside) into intact cells, and thus prevent intracellular β-galactosidase activity (Singh, 1975), although a direct effect on extracellular enzymes may limit its value.

At any particular moment a combination of events is responsible for the measured activity. These events are broadly determined by the location of the enzyme whether solely intracellular, a combination of intra- and extracellular or solely extracellular.

4. Miscellaneous

Other methodological shortcomings, not always given the attention they deserve, range from the elementary to those which require considerable forethought if they are to be avoided. Examples of the former include: pre-incubation of reactants to bring them to assay temperature; confirmation (by kinetics and the linear relationship between quantity of soil and activity) that the substrate concentration in the reaction mixture is in excess $\rightarrow$ 5 $\times$ K_m (a potential source of error during long assays); making sure that the volume of liquid in the reaction mixture is adequate to allow free diffusion of substrate; determining that the reaction is irreversibly halted at the end of the incubation period; and using an high enough buffer concentration to ensure that there is no pH shift during the reaction. Perhaps less obvious pitfalls are caused by soil fixation and microbial assimilation (see for example: glucose oxidase and the products of cellulase activity—Ross, 1974) rendering products unavailable for subsequent analysis; the non-biological conversion of substrate or destruction of products during determination (e.g. steam distillation—see May and Douglas, 1976), and the possible concentration (and exclusion from analysis) of products in centrifuged pellets. The comparative merits of using fresh soil, stored moist soil and stored air-dried soils for enzyme assays have been argued at length (Skujiņš, 1967; Spier and Ross, 1975; Zantua and Bremner, 1975b), as have their broader applications to soil microbiological research. Adequately designed controls should eliminate other sources of error.

The relevance of the classic Michaelis–Menten model of enzyme kinetics to soil analysis (Irving and Cosgrove, 1976) and the comparative values of Lineweaver–Burk, Eadie–Hofstee and direct linear plots have been examined (Pettit et al., 1977).

5. Quantification of enzyme components

Many attempts have been made to identify and quantify the various elements of total enzyme activity, particularly the accumulated fraction. Paulson and Kurtz (1969), in a sophisticated and often quoted piece of research, used multiple regression analysis to estimate the proportional contributions of proliferating microbes and stable extracellular enzymes to total soil urease. Ureolytic activity in a soil, with a maximum microbial population induced by carbon and nitrogen amendment, was estimated as 33% extracellular and 67% directly due to microbial activity. Under steady-state conditions, however (prior to enrichment), the extracellular component was as high as 89%.

Other estimates of the percentage of accumulated urease from comparisons of soils, with or without a microbial inhibitor, include 45–90% (Anderson, 1962), 58–83% (McGarity and Myers, 1967) and 51–67% (Bhavanandan and Fernando, 1970). Dalal (1975) discovered enormous variations between different soils ($< 10\% - > 90\%$) independant of microbial activity). The discordant nature of these figures is reflected in the techniques employed and the individual author's interpretation of the experimental results, as well as the soil type.

TABLE 5

Suggested scheme for the sequential elimination of soil enzyme
components (exo/endo type)

Soil pretreatment	Microbial growth	Uptake of substrate by intact dead cells	Intracellular enzyme released from ruptured cells and adhering to cell fragments	Extracellular	
				Free	Bound to soil colloids
Fresh soil, field wetness, long assay (48 h+)	+	+	+	+	+
Air-dried, long assay (48 h+)	+	+	+	−	+
Air-dried, bacteriostat, short assay (< 4 h)	−	+	+	−	+
Store wet with bacteriostat (24 h), short assay (< 4 h)	−	+	+	−	+
Air-dried, bacteriostat, inhibitor of substrate diffusion into cell, short assay (< 4 h)	−	+	+	−	+
Irradiate (< 5 Mrad)	±	+	+	−	+
Irradiate (> 5 Mrad)	−	−	−	−	progressive decline
Autoclave	−	−	−	−	−

Others (McLaren, 1972; Nannipieri *et al.*, 1974), by extrapolating microbial numbers to zero, have confirmed the existence of a significant background level of urease unrelated to microbial numbers. Pettit *et al.* (1976), drawing upon information from irradiated soil, estimated the extracellular urease component to be 60%. The contribution of active microbes to the overall activity of phosphatase has been discussed (Ramírez-Martínez and McLaren, 1966).

It may be possible to induce the sequential destruction of the individual enzyme components and, as a consequence, assess their quantitative imput. In the instance of a true exoenzyme this would be comparatively easy (see previous statements) and be achieved by comparing the disappearance of substrate over a long period (say 48 h) in soil without microbial inhibitor to that in soil containing an inhibitor. For more complicated intra/extra-cellular enzymes, such as urease, the approach portrayed in Table 5 might clarify the issue. The value of gamma irradiation as a tool in the study of soil microbiology is readily apparent from the reviews of McLaren (1969) and, more recently, Cawse (1975). Most microorganisms are sensitive to irradiation doses of between one and five Mrads; their response depending upon species, soil type and moisture content, and the volume of soil being treated. The cessation of microbial growth does not necessarily coincide with the disappearance of metabolic activity and the persistence of enzymic catalysis in irradiation-sterilized soils is common (Cawse and Mableson, 1971; Roberge, 1971; Powlson and Jenkinson, 1975).

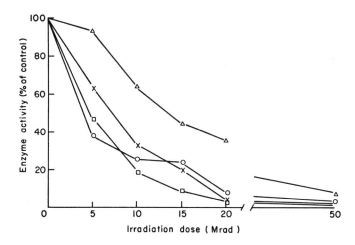

Fig. 3. The effect of irradiating air-dried soil on the activity of arylsulphatase ($\square$), phosphatase ($\triangle$), β-1, 3 glucanase ($\times$) and urease ($\bigcirc$).

Extracellular enzymes differ widely in their resistance to irradiation and are much influenced by the moisture status of the soil. Enzymes in wet soil appear more susceptible than those in dry soil possibly due to the de-naturing effect of activated water radicals (Peterson, 1962b). Surges in activity in irradiated soils may be due to increased cell wall permeability and/or cellular lysis followed by release of enzymes (McLaren et al., 1957; Skujiņš et al., 1962; Thente, 1970). We have recently compared the effect of varying levels of gamma irradiation on urease, phosphatase, arylsulphatase and β-1, 3 glucanase activities in both wet and dry soils (Burns et al. 1978). The results (Fig. 3) indicate that these enzymes do not all respond in a like manner, either because of inherent differences or due to the sundry locations of the components contributing to total activity.

The usefulness of this differential technique as an aid to the quantification of enzyme components is presently under investigation. It is suggested that extracellular unbound enzyme is rapidly destroyed during irradiation, as is any released from lysed cells. Subsequent denaturation of bound exoenzymes may be due either to the direct effect of higher levels of irradiation or to the disruption of their protective organic colloid associations. The radiation-induced depolymerization of microbial polysaccharides has been demonstrated by Griffiths and Burns (1968) whilst the solubilization of humus fractions in irradiated soils is well known (see Cawse, 1975). In more general terms, one concludes that irradiation is a far less drastic sterilization technique than autoclaving but suffers from many of the shortcomings of chemical inhibitors (cell lysis and release of endoenzymes, failure to prevent substrate uptake by dead cells etc.).

IV. Conclusions

It will be evident to the reader that a better understanding of soil enzymes could have relevance in areas where immobilized catalysts have already assumed a commercial significance. In other words, our present comprehension strongly indicates that soil-bound enzymes would offer cheap, long-lived alternatives to some of the carrier systems currently in use. Many of these systems tend to suffer from leakage, susceptibility to microbial decay and general instability. The dominant mechanisms of immobilization have been summarized by Weetall (1975) and are diagrammatically reproduced in Fig. 4. They are essentially all mechanisms which have been suggested at one time or another to explain the immobilization and protection of enzymes by soil colloids.

The importance of immobilized enzymes in the chemical, pharmaceutical and medical industries has been discussed in detail in Zaborsky's monograph (Zaborsky, 1973) and elsewhere (Sizer, 1972; Smiley and Strandberg, 1972; Mosbach, 1977), whilst applications in the production of food and drink have been reviewed recently by Weetall (1975). The journals Process Biochemistry (Morgan–Grampian Ltd.) and Biotechnology and Bioengineering (John Wiley and Sons) are important sources of review and research material in applied enzymology. Notwithstanding, it is difficult to discover any soil- or soil component-enzyme interaction studies that have been principally concerned with the feasability of their commercial application.

The expanding use of urea fertilizers has spurred research into the factors which influence urea hydrolysis and the subsequent oxidation of ammonia by soil chemolithotrophs. The number of potential environmental

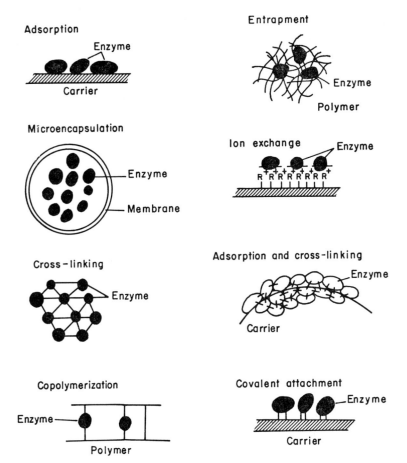

Fig. 4. Schematic representation of methods of immobilizing enzymes. From Weetall (1975) with permission.

problems associated with this aspect of the nitrogen cycle is enormous (Table 6), in addition to the economic importance of efficient fertilizer use. As a consequence, a large effort has been directed towards controlling the sequence of events commencing with urea and ending with nitrate (Prasad *et al.*, 1971). This has resulted in a number of slow-release fertilizers either produced by synthetic means (urea-form) or by physically protecting the urea (sulphur and polymer coats). Direct inhibition of the nitrification process has also been achieved with such as N-serve (2-chloro-6 trichlo-romethyl) pyridine, AM (2-amino-4 chloro-6-methylpyrimidine) and ST (2-sulphanilamidothiazole). Bremner and Mulvaney, in Chapter 5 of this book, have detailed much of this work. Interest in polyphosphate fertilizers

(Hughes and Hashimoto, 1971) has stimulated research into the relevant enzymes (Douglas *et al.*, 1976).

TABLE 6

Ecological consequences of urea-N cycle disruption

$$\text{Urea} \xrightarrow{\;1\;} \text{Ammonia} \xrightarrow{\;2\;} \text{Nitrite} \xrightarrow{\;3\;} \text{Nitrate} \xrightarrow{\;4\;} \text{Plants}$$

1 faster than 2 (NH_3 accumulation and alkalinity)
    volatilization losses of N as NH_3 from microbe-plant cycle
    inhibition of nitrifiers (especially *Nitrobacter*)
    localized air pollution
    damage to seedlings
    chemodenitrification ($NH_3 + HNO_2 \longrightarrow N_2 + 2H_2O$)
    potential accumulation of oxidation products—N_2O, NO—in atmosphere
    (catalyse breakdown of O_3)

2 faster than 3 (NO_2 accumulation)
    toxic to a wide range of soil microbes
    leaching of nitrite anion (mammalian toxicity-methaemoglobinaemia)

3 faster than 4 (NO_3 accumulation)
    leaching of nitrate anion (eutrophication)
    concentration of unassimilated NO_3 in plants
    subsequent reduction of NO_3 to NO_2 in food, water and intestine
    (mammalian toxicity)

From Burns (1977)

The interactions of pesticides with enzymes, discussed by Cervelli *et al.*, in Chapter 7 of this book, have profound implications both in the design and registering of novel pesticides and the control of subsequent persistence. Should legislation demand that soil microbiological properties be included in pesticide screening then the accuracy of enzyme determinations (cf. plate counts—Table 7) would be telling support for their use as an indicators. Prior to their adoption, however, the nonuniformity of techniques discussed earlier, would need resolution. As a cautionary note, our work with organophosphorus insecticides (Lethbridge and Burns, 1976), albeit at 50, 100 and 1000 ppm, indicates that soils should continue to be monitored for some time after the parent compound (or even its metabolites) have disappeared. The point being that the indigenous enzyme activity of a soil has accumulated over a period of time and reflects previous generations of microorganisms. Thus the inhibition or death of a group of microbes immediately after the addition of a pesticide may not make a significant difference to enzymic activities until much later when leakage

TABLE 7

The reproducibility of soil enzyme assays

Enzyme	Error (%)	Replicates
β−1, 3 glucanase	4	
Urease	4	
Phosphatase	8	4
Arylsulphatase	1	
Microbial Counts		
Total (nutrient agar)	22	
Total (glucose–soil extract agar)	37	20
Bacteria (Bunt and Rovira agar)	48	

of bound enzymes becomes apparent. Re-establishment of the exoenzyme level will probably depend upon recovery and proliferation of the relevant microbes and incorporation of the enzymes into newly-synthesized humic material. (Fig. 5.) This may be the simplest form of relationship between an event and its manifestation in soil microbial ecology. We have also discovered a significant lag between the disappearance of the herbicide 2, 4–D and the recovery of the urease activity which it inhibits (Burns 1978a).

The control of a pesticide's persistence, once it is in the soil, has been

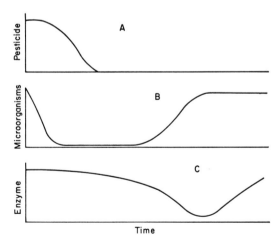

Fig. 5. Time lapse between pesticide-induced inhibition of microorganisms and its effect on soil enzyme levels.
A = decay of pesticide; B = inhibition and recovery of sensitive microbial population; C = decline and re-establishment of indigenous enzyme level.

investigated and a variety of chemical and physical methods explored (Burns, 1976). Our observations, concerning the stimulation of malathion decay in soil by the addition of esterase–organic matter extracts (Gibson and Burns, 1977, Burns and Gibson, 1978) may just possibly be worth evaluation as a control method; an idea parallel to the suggestion that immobilized enzymes could be used to detoxify pesticide-contaminated waste waters (Munnecke, 1978). The presence of accumulated extracellular enzymes capable of having a significant role in the breakdown of pesticides is almost certainly not restricted to the few reports involving organophosphates (Getzin and Rosefield, 1968; Kishk *et al.*, 1976), propanil (Burge, 1972) and its chloroaniline metabolites (Bartha and Bordeleau, 1969).

The influence of metal pollution on soil enzyme activities has also begun to attract the attention of researchers (Ruhling and Tyler, 1973; Tyler, 1974; Gorbanov, 1975; Tyler, 1976; Tabatabai, 1977).

Efforts to correlate soil enzyme activities with fertility have met with limited success (Galstyan and Tatevosyan, 1964; Pauli, 1965; Rawald, 1970; Kunze, 1970) and it is readily apparent that enzymes are just one of many factors affecting soil productivity (see Chapters 1 and 4 for a detailed discussion). The influence of microbial exoenzymes on the success of soil-borne plant pathogens has also been investigated (Kelley and Rodriguez-Kabana, 1976; Morrissey *et al.*, 1976).

Peripheral areas of soil enzyme research have involved forensic science (Thornton and McLaren, 1975) and exobiology (Cameron, 1963; Ponnamperuma and Gabel, 1968).

As evidenced by this volume, soil enzyme research is progressing rapidly. Yet the fact remains that even with the present collection of superficial data concerning fifty or more soil enzymes and detailed information about a dozen, our overall comprehension of the fundamental problems is expanding at a lethargic rate. Such problems as: what are the components of total enzyme activity; what are the different properties of these components; what is their origin; what is the temporal and spatial relationship of enzymes to their source; what mechanisms are involved in the protection of exoenzymes; what is their contribution to mineral cycles; how do they relate to soil fertility and in what ways do natural and synthetic perturbations affect them? These are the sorts of questions which will be puzzling soil enzymologists during the next decade.

References

ACTON C. J., RENNIE D. A. and PAUL E. A. (1963). The relationship of polysaccharides to soil aggregation. *Can. J. Soil Sci.* **43**, 201–209.

ACKERMAN E. (1962). "Biophysical Science." Prentice-Hall, Englewood Cliffs, New Jersey.

ALEXANDER M. (1971). Biochemical ecology of microorganisms. *Ann. Rev. Microbiol.* **25**, 361–392.

ALIEV R. A. and ZVIAGINTSEV P. G. (1974). The effect of proteolytic enzymes on free and adsorbed catalase as well as on the catalase activity of the soil. *Pochvovedenie.* (29), 97–104.

ALLISON F. E. (1968). Soil aggregation—some facts and fallacies as seen by a microbiologist. *Soil Sci.* **106**, 136–144.

ALLISON F. E. (1973). "Soil Organic Matter and its Role in Crop Production." Elsevier, Amsterdam.

AMBROŽ Z. (1970). Factors influencing distribution of some proteolytic enzymes in soils. *Zentralbl. Bakteriol. Abt. Bd. II* **125**, 433–437.

ANDERSON J. R. (1962). Urease activity, ammonia volatisation and related microbiological aspects in some South African soils. *Proc. Ann. Congr. S. Afr. Sugar Technol. Ass.* **36**, 97–105.

ANDERSON J. R. and SLINGER J. M. (1975). Europium chelate and fluorescent brightener staining of soil propagules and their photomicrographic counting—I. Methods. *Soil Biol. Biochem.* **7**, 205–209.

ARISTOVSKAYA T. V. (1973). The use of capillary techniques in ecological studies of microorganisms. *Bull. Ecol. Res. Comm.*, *Stockholm* **17**, 47–52.

ASPIRAS R. B., ALLEN O. N., HARRIS R. F. and CHESTERS G. (1971a). Aggregate stabilization by filamentous microorganisms. *Soil Sci.* **112**, 282–284.

ASPIRAS R. B., ALLEN O. N., HARRIS R. F. and CHESTERS G. (1971b). The role of microorganisms in the stabilization of soil aggregates. *Soil Biol. Biochem.* **3**, 347–353.

AUSMUS B. S. (1973). The use of the ATP assay in terrestrial decomposition studies. *Bull. Ecol. Res. Comm.*, *Stockholm* **17**, 223–234.

BAILEY G. W. and WHITE J. L. (1970). Factors influencing the adsorption, desorption, and movement of pesticides in soil. *Residue Rev.* **32**, 29–92.

BAILEY M., ENARI T.-M. and LINKO M. (1975). "Symposium on Enzymic Hydrolysis of Cellulose." Sitva, Helsinki, Finland.

BARTHA R. and BORDELEAU L. (1969). Cell-free peroxidases in soil. *Soil Biol. Biochem.* **1**, 139–143.

BASCOMB C. L. and BULLOCK P. (1974). Sample preparation and stone content. *In* "Soil Survey Laboratory Methods" (B. W. Avery and C. L. Bascomb, Eds), Soil Surv. Tech. Monog. Vol. 6, Rothamsted Exp. Stn., Herts., England.

BAVER L. D. (1968). The effect of organic matter on soil structure. *Pont. Acad. Scient. Scripta Varia* **32**, 1–31.

BEACHAM I. R., TAYLOR N. S. and YOVELL M. (1976). Enzyme secretion in *Escherichia coli*: Synthesis of alkaline phosphatase and acid hexose phosphatase in the absence of phospholipid synthesis. *J. Bacteriol.* **128**, 522–527.

BECK T. and POCHENRIEDER H. (1963). Experiments concerning the action of toluene on the microflora in soils. *Pl. Soil* **18**, 346–357.

BERGER-LANDEFELDT U. (1965). Activity of some enzymes in soils under different plant associations. *Flora, Jena* **115**, 452–458.

BERRY S. A., JOHNSON K. G. and CAMBELL J. N. (1970). The extracellular nuclease activity of *Micrococcus sodonensis*. IV, Physical studies, characterization as a glycoprotein and involvement with the cell wall. *Biochim. Biophys. Acta* **220**, 269–283.

BHAVANANDAN V. P. and FERNANDO V. (1970). Studies on the use of urea as a fertilizer for tea in Ceylon. 2. Urease activity in tea soils. *Tea Quart.* **41**, 94–106.

BRAMS W. H. and MCLAREN A. D. (1974). Phosphatase reactions in columns of soil. *Soil Biol. Biochem.* **6**, 183–189.

BROCK T. D. (1971). Microbial growth rates in nature. *Bact. Rev.* **35**, 39–58.

BROWN M. E. (1975). Rhizosphere microorganisms-opportunists, bandits or bene-factors. *In* "Soil Microbiology" (N. Walker, Ed.), pp. 21–37, Butterworths, London.

BULL A. T. (1972). Environmental factors influencing the synthesis and excretion of exocellular macromolecules. *J. Appl. Chem. Biotechnol.* **22**, 261–292.

BURGE W. D. (1972). Microbial populations hydrolysing propanil and accumulation of 3, 4-dichloroaniline and 3, 3′, 4, 4′-tetrachloroazobenzene in soils. *Soil Biol. Biochem.* **4**, 379–386.

BURNS R. G. (1975). Factors affecting pesticide loss from soil. *In* "Soil Biochemistry" Vol. 4 (E. A. Paul and A. D. McLaren, Eds), pp. 103–141, Marcel Dekker, New York.

BURNS R. G. (1976). Microbial control of pesticide persistence in soil. *Brit. Crop. Prot. Symp.* **17**, 229–239.

BURNS R. G. (1977). Soil enzymology. *Sci. Prog. Oxf.* **64**, 281–291.

BURNS R. G. 1978a). The soil microenvironment: aggregates, enzymes and pes-ticides. *C.N.R., Laboratorio per la Chimica del Terreno Conferenze* **5**, 1–23.

BURNS R. G. (1978b). The interaction of microorganisms, their substrates and their products with soil surfaces. *In* "Adhesion of Microorganisms to Surfaces" (D. C. Ellwood, J. Melling and P. Rutter, Eds), S.C.I. Monograph Series (in press).

BURNS R. G. and GIBSON W. P. (1978). The disappearance of 2, 4-D, diallate and malathion from soil and soil components. *In* " Agrochemicals in Soils" (A. Banin, Ed.), Springer-Verlag, New York (in press).

BURNS R. G., EL-SAYED M. H. and MCLAREN A. D. (1972a). Extraction of urease-active organo-complex from soil. *Soil Biol. Biochem.* **4**, 107–108.

BURNS R. G., PUKITE A. H. and MCLAREN A. D. (1972b). Concerning the location and persistence of soil urease. *Soil Sci. Soc. Am. Proc.* **36**, 308–311.

BURNS R. G., GREGORY L. J., LETHBRIDGE G. and PETTIT N. M. (1978). The effect of γ-irradiation on soil enzyme stability. *Experientia* (in press).

BUTLER J. H. A. and LADD J. N. (1971). Importance of the molecular weight of humic and fulvic acids in determining their effects on protease activity. *Soil Biol. Biochem.* **3**, 249–257.

CACCO G. and MAGGIONI A. (1976). Multiple forms of acetyl-naphthyl-esterase activity in soil organic matter. *Soil Biol. Biochem.* **8**, 321–325.

CALDWELL D. E. and HIRSCH P. (1973). Growth of microorganisms in two-dimensional steady-state diffusion gradients. *Can. J. Microbiol.* **19**, 53–58.

CAMERON R. E. (1963). The role of soil science in space exploration. *Space Sci. Rev.* **2**, 297–312.

CARLILE M. J. (1970). Nutrition and chemotaxis in the myxomycete *Physarum poly-cephalum;* the effect of carbohydrates on the plasmodium. *J. Gen. Microbiol.* **63**, 221–226.

CASIDA L. E. (1968). Infrared color photography: selective demonstration of bacteria. *Science* **159**, 199–200.

CAWSE P. A. (1975). Microbiology and biochemistry of irradiated soils. *In* "Soil Biochemistry" Vol. 3. (E. A. Paul and A. D. McLaren, Eds) pp. 213–267, Marcel Dekker, New York.

CAWSE P. A. and MABLESON K. M. (1971). The effect of gamma irradiation on the release of carbon dioxide from fresh soil. *Commum. Soil Sci. Pl. Anal.* **2,** 421–431.

CERNÁ S. (1966). Enzymatic activity of soil in relation to its structure. *Acta Univ. Carol. Biol.* **1,** 83–89.

CERNÁ S. (1970). The influence of crushing of soil aggregates on the activity of hydrolytic enzymes in soil. *Acta Univ. Carol. Biol.* **6,** 461–466.

CHALVIGNAC M-A. and MAYAUDON J. (1971). Extraction and study of soil enzymes metabolizing tryptophan. *Pl. Soil.* **34,** 25–31.

CHESHIRE M. V. (1977). Origins and stability of soil polysaccharide. *J. Soil Sci.* **28,** 1–10.

CHET I. and MITCHELL R. (1976). Ecological aspects of microbial chemotactic behavior. *Ann. Rev. Microbiol.* **30,** 221–239.

CHRISPEELS M. J. (1976). Biosynthesis, intracellular transport, and secretion of extracellular macromolecules. *Ann. Rev. Pl. Physiol.* **27,** 19–38.

CLARK F. E. (1967). Bacteria in soil. *In* "Soil Biology" (N. A. Burges and F. A. Raw, Eds), pp. 15–49, Academic Press, London and N.Y.

CONRAD J. P. (1940). Catalytic activity causing the hydrolysis of urea in soil as influenced by several agronomic factors. *Soil Sci. Soc. Am. Proc.* **5,** 238–241.

COOPER P. J. M. (1972). Arylsulphatase activity in northern Nigerian soils. *Soil Biol. Biochem.* **4,** 333–337.

CORTEZ J., LOSSAINT P. and BILLES G. (1972). Biological activity of soils in Mediterranean ecosystems. III. Enzymatic activities. *Rev. Ecol. Biol. Sol.* **9,** 1–19.

COSTERTON J. W., INGRAM J. M. and CHENG K. J. (1974). Structure and function of the cell envelope of Gram-negative bacteria. *Bacteriol. Rev.* **38,** 87–110.

DALAL R. C. (1975). Urease activity in some Trinidad soils. *Soil Biol. Biochem.* **7,** 5–8.

DANCER B. N. and LAMPEN J. O. (1975). Importance of phospholipids in excretion of extracellular penicillinase. *Biochem. Biophys. Res. Comm.* **66,** 1357–1264.

DANIELLI J. F. (1937). Relations between surface pH, ion concentrations and interfacial tension. *Proc. Roy. Soc., London* **B122,** 155–174.

DECAD G. M. and NIKAIDO H. (1976). Outer membrane of Gram-negative bacteria. XII. Molecular-sieving function of cell wall. *J. Bacteriol.* **128,** 325–336.

DOUGLAS L. A. and BREMNER J. M. (1971). A rapid method of evaluating different compounds as inhibitors of urease activity in soils. *Soil Biol. Biochem.* **3,** 309–315.

DOUGLAS L. A., RIAZI-HAMADANI A. and FIELD J. F. B. (1976). Assay of pyrophosphatase activity in soil. *Soil Biol. Biochem.* **8,** 391–393.

DRĂGAN-BULARDA M. and KISS S. (1972). Dextranase activity in soil. *Soil Biol. Biochem.* **4,** 413–416.

DROBNIK J. (1961). On the role of toluene in the measurement of the activity of soil enzymes. *Pl. Soil* **14,** 94–95.

DROZDOWICZ A. (1971). The behaviour of cellulase in soil. *Rev. Microbiol.* **2,** 17–23.

DUDMAN W. F. (1977). The role of surface polysaccharides in natural environments. *In* "Surface Carbohydrates of the Prokaryotic Cell" (I. W. Sutherland, Ed.), pp. 357–414. Academic Press, London.

ELBEIN A. D. (1974). Interactions of polynucleotides and other polyeletrolytes with enzymes and other proteins. *Adv. Enzymol.* **40**, 29–64.

EPSTEIN W. (1975). Membrane transport. *In* "M.T.P. Inter. Rev. Sci., Biochem." *Ser. I, Vol.* 2 (C. F. Fox, Ed.), pp. 249–278. Butterworths, London.

EREN J. and PRAMER D. (1966). Application of immunofluorescent staining to studies of the ecology of soil microorganisms. *Soil Sci.* **101**, 39–45.

EYLAR E. H. (1965). On the biological role of glycoproteins. *J. Theoret. Biol.* **10**, 89–113.

FELBECK G. T. (1971). Chemical and biological characterization of humic matter. *In* "Soil Biochemistry", Vol. 2 (A. D. McLaren and J. Skujiņš, Eds), pp. 36–59. Marcel Dekker, New York.

FERNANDO V. and ROBERTS G. R. (1976). The partial inhibition of soil urease by naturally occurring polyphenols. *Pl. Soil* **44**, 81–86.

FILIP Z. (1973). Clay minerals as a factor influencing the biochemical activity of soil microorganisms. *Folia Microbiol.* **18**, 56–74.

FLAIG W. (1975). An introductory review on humic substances: aspects of research on their genesis, their physical and chemical properties, and their effect on organisms. *Proc. Int. Meet. Humic Substances, Nieuwersluis* **1972**, Pudoc, Wageningen.

FLAIG W., BEUTELSPACHER H. and RIETZ E. (1975). Chemical composition and physical properties of humic substances. *In* "Soil Components" Vol. 1 (J. E. Gieseking, Ed.), pp. 1–211, Springer-Verlag, Berlin.

FLEISCHER S., BRIERLEY G., KLOUWEN H. and SLAUTTERBACK D. B. (1962). Electron transport system. The role of phospholipids in electron transport. *J. Biol. Chem.* **237**, 3264–3272.

FRANZ G. (1973). Comparative investigations on the enzyme activity of some soils in Nordrhein-Wesfalen and Rheinland-Pfalz. *Pedobiologia* **13**, 423–436.

GADAL P. and BOUDET A. (1965). Inhibition of enzymes by tannins from the leaves of *Quercus sessilis*. Inhibition of β-amylase. *C.R. Acad. Sci. Paris,* **260**, 4252–4255.

GALSTYAN A. S. (1965). A method of determining the activity of hydrolytic enzymes in soil. *Soviet Soil Sci.* **2**, 170–175.

GALSTYAN, A. S. and TATEVOSYAN G. S. (1964). Enzyme activity as characteristic of soil type. *In Proc. 8th Int. Congr. Soil Sci.* **3**. Publishing House of the Academy of the Socialist Republic of Romania, Bucharest.

GETZIN L. W. and ROSEFIELD I. (1968). Organophosphorus insecticide degradation by heat labile substances in soil. *J. Agric. Fd Chem.* **16**, 598–601.

GIARDINA M. C., TOMATI U. and PIETROSANTI W. (1970). Hydrolytic activities of the soil treated with paraquat. *Meded. Rijksfac. Landbouwwet., Gent* **35**, 615–626.

GIBSON W. P. and BURNS R. G. (1977). Breakdown of malathion in soil and soil components. *Microbial Ecol.* **3**, 219–230.

GLENN A. R. (1976). Production of extracellular proteins by bacteria. *Ann. Rev. Microbiol.* **30**, 41–62.

GLEW R. H. and HEATH E. C. (1971). Studies on the extracellular alkaline phosphatase of *Micrococcus sodensis*. *J. Biol. Chem.* **246**, 1556–1565.

GNITTKE J. and KUNZE C. (1975). The influence of tannic acid on catalase activity of soil samples. *Zéntralbl. Bakteriol. Abt. II* **130**, 37–40.

GOLDMAN R., GOLDSTEIN L. and KATCHALSKI E. (1971). *In* "Biochemical Aspects of Reactions on Solid Supports". (G. R. Starts, Ed.), pp. 291–424. Academic Press, N.Y.

GOLDSTEIN J. L. and SWAIN T. (1965). Inhibition of enzymes by tannins. *Phytochemistry* **4**, 185–192.

GOLDSTEIN L., PECHT M., BLUMBERG S., ATLAS D. and LEVIN Y. (1970). Water-insoluble enzymes. Synthesis of a new carrier and its utilization for preparation of insoluble derivatives of papain, trypsin, and subtilo-peptidase A. *Biochem.* **9**, 2322–2334.

GORBANOV S. P. (1975). The effect of molybdenum on soil phosphatase activity. *Soil Sci. Agrochem.* **10**, 97–102.

GOULD A. R., MAY B. K. and ELLIOTT W. H. (1975). Release of extracellular enzymes from *Bacillus amyloliquefaciens*. *J. Bact.* **122**, 34–40.

GRAY T. R. G. (1967). Stereoscan electron microscopy of soil microorganisms. *Science* **155**, 1668–1670.

GRAY T. R. G. (1976). Survival of vegetative microbes in soil. *Symp. Soc. Gen. Microbiol.* **26**, 327–364.

GREAVES M. P. and DARBYSHIRE J. F. (1972). The ultrastructure of the mucilaginous layer on plant roots. *Soil Biol. Biochem.* **4**, 443–449.

GREENAWALT J. W. and WHITESIDE T. L. (1975). Mesosomes: membranous bacterial organelles. *Bact. Rev.* **39**, 405–463.

GREENWOOD D. J. (1968). Measurement of microbial metabolism in soil. *In* "Ecology of Soil Bacteria" (T. R. G. Gray and D. Parkinson, Eds), pp. 138–157. Liverpool Univ. Press.

GRIFFITHS E. (1965). Micro-organisms and soil structure. *Biol. Rev.* **40**, 129–142.

GRIFFITHS E. and BURNS R. G. (1968). Effects of gamma irradiation on soil aggregate stability. *Pl. Soil* **28**, 169–172.

GRIFFITHS E. and BURNS R. G. (1972). Interaction between phenolic substances and microbial polysaccharides in soil aggregation. *Pl. Soil* **36**, 599–612.

GUENZI W. D. (1974). "Pesticides in Soil and Water". Soil Sci. Soc. Am. Inc., Madison, Wisconsin.

HAGEN C. A., HAWRYLEWICZ E. J., ANDERSON B. T., TOLKACZ V. K. and CEPHUS M. L. (1968). Use of the scanning electron microscope for viewing bacteria in soil. *Appl. Microbiol.* **16**, 932–934.

HAIDER K., MARTIN J. P. and FILIP Z. (1975). Humus biochemistry. *In* "Soil Biochemistry", Vol. 4 (E. A. Paul and A. D. McLaren, Eds), pp. 195–244. Marcel Dekker, New York.

HANNIJ G. J. and REESE E. T. (1969). "Cellulases and Their Applications". Adv. Chem. Ser. **95**, American Chem. Soc., Washington, D.C.

HARRIS R. F., CHESTERS G. and ALLEN O. N. (1966). Dynamics of soil aggregation. *Adv. Agron.* **18**, 107–169.

HARTER R. D. and STOTZKY G. (1971). Formation of clay-protein complexes. *Soil Sci. Soc. Am. Proc.* **35**, 383–389.

HARTER R. D. and STOTZKY G. (1973). X-ray diffraction, electron microscopy, electrophoretic mobility, and pH of some stable smectite-protein complexes. *Soil Sci. Soc. Am. Proc.* **37**, 116–123.

HARTLEY G. S. and ROE J. W. (1940). Ionic concentrations at interfaces. *Trans. Faraday Soc.* **36**, 101–109.

HATTORI T. and HATTORI R. (1976). The physical environment in soil microbiology: an attempt to extend principles of microbiology to soil microorganisms. *CRC Crit. Rev. Microbiol.* **4**, 423–461.

HEINEN W. and LAUWERS A. M. (1975). Variability of the molecular size of extra-cellular amylase produced by intact cells and protoplasts of *Bacillus caldolyticus*. *Arch. Microbiol.* **106**, 201–207.

HEPPEL L. A. (1971). The concept of periplasmic enzymes. *In* "Structure and Function of Biological Membranes" (L. I. Rothfield, Ed.), pp. 223–247. Academic Press, New York.

HEPPER C. M. (1975). Extracellular polysaccharides of soil bacteria. *In* "Soil Microbiology" (N. Walker, Ed.), pp. 93–110. Butterworths, London.

HIGGINS I. J. and BURNS R. G. (1975). "The Chemistry and Microbiology of Pollution." Academic Press, London and New York.

HISSETT R. and GRAY T. R. G. (1976). Microsites and time changes in soil microbe ecology. *In* "The Role of Terrestrial and Aquatic Organisms in Decomposition Processes." (J. M. Anderson and A. Macfadyen, Eds), pp. 23–29. Blackwell Scientific Publications, Oxford.

HOFSTEN B. V. (1975). Topological effects in enzymatic and microbial degradation of highly ordered polysaccharides. *Symp. Enzymatic Hydrolysis of Cellulose, Finland*, 281–295.

HUGHES J. D. and HASHIMOTO I. (1971). Ammonium pyrophosphate as a source of phosphorus for plants. *Soil Sci. Soc. Am. Proc.* **35**, 643–647.

INGRAM J. M., CHENG K. J. and COSTERTON J. W. (1973). Alkaline phosphatase of *Pseudomonas aeruginosa*: the mechanism of secretion and release of the enzyme from whole cells. *Can. J. Microbiol.* **19**, 1407–1415.

IRVING G. C. J. and COSGROVE D. J. (1976). The kinetics of soil acid phosphatase. *Soil Biol. Biochem.* **8**, 335–340.

JAMES L. K. and AUGENSTEIN L. G. (1966). Adsorption of enzymes at interfaces: film formation and the effect on activity. *Adv. Enzymol.* **28**, 1–40.

JENNY H. and GROSSENBACHER K. (1963). Root-soil boundary zones as seen in the electron microscope. *Soil Sci. Soc. Am. Proc.* **27**, 273–277.

JONES D. and GRIFFITHS E. (1964). The use of soil sections for the study of soil microorganisms. *Pl. Soil* **20**, 232–239.

KABACK H. R. (1970). Transport. *Ann. Rev. Biochem.* **39**, 561–598.

KATCHALSKI E., SILMAN I. and GOLDMAN R. (1971). Effect of the microenvironment on the mode of action of immobilized enzymes. *Adv. Enzymol.* **34**, 445–536.

KELLEY P. M., NEUMANN P. A., SHRIEFER K., CANCEDDA F., SCHLESINGER M. J. and BRADSHAW R. A. (1973). Amino acid sequence of *Escherichia coli* alkaline phospha-tase. Amino- and carboxyl-terminal sequences and variations between two isoenzymes. *Biochemistry* **2**, 3499–3503.

KELLEY W. D. and RODRIGUEZ-KABANA R. (1975). Effects of potassium azide on soil microbial populations and soil enzymatic activities. *Can. J. Microbiol.* **21**, 565–570.

KELLEY W. D. and RODRIGUEZ-KABANA (1976). Competition between *Phytophthora cinnamomi* and *Trichoderma* spp. in autoclaved soil. *Can. J. Microbiol.* **22**, 1120–1127.

KILPATRICK D. C. and STIRLING J. L. (1977). Glycosidases from the culture medium of *Physarum polycephalum. Biochem. J.* **161**, 149–157.

KISHK F. M., EL-ESSAWI T., ABDEL-GHAFAR S. and ABOU-DONIA M. B. (1976). Hydrolysis of methylparathion in soils. *J. Agric. Fd Chem.* **24**, 305–307.

KISS S. and BOARU M. (1965). Some methodological problems of soil enzymology. *Symp. Methods in Soil Biology. Bucharest*, 115–127.

KISS S., DRĂGAN-BULARDA M. and KHAZIEV F. H. (1972). Influence of chloromycetin on the activities of some oligases of soils. *Lucr. Conf. Nat. Stiinta Solului, Iasi 1970*, 451–462.

KISS S., STEFANIC G. and DRĂGAN-BULARDA M. (1974). Soil enzymology in Romania I. *Contrib. Bot., Cluj-Napoca*, pp. 207–219.

KISS S., DRĂGAN-BULARDA M. and RĂDULESCU D. (1975). Biological significance of enzymes accumulated in soil. *Adv. Agron.* **27**, 25–87.

KONONOVA M. M. (1966). "Soil Organic Matter, its Nature, its Role in Soil Formation and in Soil Fertility." 2nd edition, Pergamon Press, Oxford.

KUNZE C. (1970). The effect of streptomycin and aromatic carboxylic acids on the catalase activity in soil samples. *Zentralbl. Bakteriol. Parasitenkd. Abt.* II **124**, 658–661.

KUWABARA S. and LLOYD P. H. (1971). Protein and carbohydrate moieties of a preparation of B-lactamase II. *Biochem. J.* **124**, 215–220.

LADD J. N. and BUTLER J. H. A. (1969). Inhibition and stimulation of proteolytic enzyme activities by soil humic acids. *Aust. J. Soil Res.* **7**, 253–262.

LADD J. N. and BUTLER J. H. A. (1970). The effect of inorganic cations on the inhibition and stimulation of protease activity by soil humic acids. *Soil Biol. Biochem.* **2**, 33–40.

LADD J. N. and BUTLER J. H. A. (1975). Humus–enzyme systems and synthetic, organic polymer-enzymes analogs. *In* "Soil Biochemistry", Vol. 4 (E. A. Paul and A. D. McLaren, Eds), pp. 143–194. Marcel Dekker, New York.

LADD J. N., BRISBANE P. G., BUTLER J. H. A. and AMATO M. (1976). Studies in soil fumigation—III. Effects on enzyme activities, bacterial numbers and extractable ninhydrin reactive compounds. *Soil Biol. Biochem.* **8**, 255–260.

LAMPEN J. O. (1965). Secretion of enzymes by microorganisms. *Symp. Soc. Gen. Microbiol.* **15**, 115–133.

LAMPEN J. O. (1974). Movement of extracellular enzymes across cell membranes. *In* "SEB Symp. (28), Transport at the Cellular Level" (M. A. Sleigh and D. H. Jennings, Eds), pp. 351–374.

LETHBRIDGE G. and BURNS R. G. (1976). Inhibition of soil urease by organophosphorus insecticides. *Soil Biol. Biochem.* **8**, 99–102.

LETHBRIDGE G., BULL A. T. and BURNS R. G. (1978). Assay and properties of 1, 3-β-glucanase in soil. *Soil Biol. Biochem.* **10** (in press).

LEWIS J. A. and STARKEY R. L. (1968). Vegetable tannins, their decomposition and effects on decomposition of some organic compounds. *Soil Sci.* **106**, 241–247.

LINDSAY S. S., WHEELER B., SANDERSON K. E., COSTERTON J. W. and CHENG K. J. (1973). The release of alkaline phosphatase and of lipopolysaccharide during the growth of rough and smooth strains of *Salmonella typhimurium*. *Can. J. Microbiol.* **19**, 335–343.

LOPPES R. (1976). Release of enzymes by normal and wall-free cells of *Chlamydomonas*. *J. Bact.* **128**, 114–116.

LOW A. J. (1954). The study of soil structure in the field and the laboratory. *J. Soil Sci.* **5**, 57–74.

LYR H. (1961). Analytical studies with inhibitors of various enzymes of wood-rotting fungi. *Enzymologia* **23**, 231–248.

MANDELS M., ANDREOTTI R. and ROCHE C. (1976). Measurement of saccharifying cellulase. *In* "Enzymatic Conversion of Cellulosic Materials. Technology and Applications" (E. Gaden, M. H. Mandels, E. T. Reese and L. A. Spano, Eds), pp. 21–33. Interscience, Wiley, New York.

MARSHALL K. C. (1971). Sorptive interactions between soil particles and microorganisms. *In* "Soil Biochemistry", Vol. 2 (A. D. McLaren and J. Skujiņš, Eds) pp. 409–445. Marcel Dekker, New York.

MARTIN J. P. and HAIDER K. (1971). Microbial activity in relation to soil humus formation. *Soil Sci.* **111**, 54–63.

MARTIN J. P., FILIP Z. and HAIDER K. (1976). Effect of montmorillonite and humate on growth and metabolic activity of some actinomycetes. *Soil Biol. Biochem.* **8**, 409–413.

MATO M. C., FABREGAS R. and MENDEZ J. (1971). Inhibitory effect of soil humic acids on indoleacetic acid-oxidase. *Soil Biol. Biochem.* **3**, 285–288.

MAY P. B. and DOUGLAS L. A. (1976). Assay for soil urease activity. *Pl. Soil* **45**, 301–305.

MAYAUDON J. and SARKAR J. M. (1974). Study of diphenol oxidases extracted from a forest litter. *Soil Biol. Biochem.* **6**, 269–274.

MAYAUDON J., BATISTIC L. and SARKAR J. M. (1975). Properties of proteolytically active extracts from fresh soils. *Soil Biol. Biochem.* **7**, 281–286.

MCGARITY J. W. and MYERS M. G. (1967). A survey of urease activity in soils of northern New South Wales. *Pl. Soil* **27**, 217–238.

MCLAREN A. D. (1960). Enzyme action in structurally restricted systems. *Enzymologia* **21**, 356–364.

MCLAREN A. D. (1969). Radiation as a technique in soil biology and biochemistry. *Soil Biol. Biochem.* **1**, 63–73.

MCLAREN A. D. (1972). Consecutive biochemical reactions in soil with particular reference to the nitrogen cycle. *C.N.R., Laboratorio per la Chimica del Terreno, Conferenze* **1**, 1–18.

MCLAREN A. D. and PACKER L. (1970). Some aspects of enzyme reactions in heterogenous systems. *Adv. Enzymol.* **33**, 245–308.

MCLAREN A. D. and PETERSON G. H. (1961). Montmorillonite as a caliper for the size of protein molecules. *Nature* **192**, 960–961.

MCLAREN A. D., PUKITE A. H. and BARSHAD I. (1975). Isolation of humus with enzymatic activity from soil. *Soil Sci.* **119**, 178–180.

MCLAREN A. D., RESHETKO, L. and HUBER, W. (1957). Sterilization of soil by irradiation with an electron beam and some observations on soil enzyme activity. *Soil Sci.* **83**, 497–502.

MEHTA N. C., STREULI H., MULLER, M. and DEUEL H. (1960). Role of polysaccharides in soil aggregation. *J. Sci. Fd Agr.* **11**, 40–47.

MITCHELL P. (1966). Chemiosmotic coupling in oxidative and photosynthetic phosphorylation. *Biol. Rev.* **41**, 445–502.

MORGAN H. W. and CORKE C. T. (1976). Adsorption, desorption and activity of glucose oxidase on selected clay species. *Can. J. Microbiol.* **22**, 684–693.

MURPHY C. P. BULLOCK P., and TURNER R. H. (1977). The measurement and characterisation of voids in soil thin sections by image analysis. Part I Principles and techniques. *J. Soil Sci.* **28**, 498–508.

MORRISSEY R. F., DUGAN E. P. and KOTHS J. S. (1976). Chitinase production by an *Arthrobacter* sp. using cells of *Fusarium roseum*. *Soil Biol. Biochem.* **8**, 23–28.

MORTLAND M. M. (1968). Protonation of compounds at clay mineral surfaces. *Trans. 9th Int. Congr. Soil Sci., Adelaide* **1**, 691–699.

MOSBACH K. (1977). Immobilized Enzymes. "Methods in Enzymology" Vol. 44. Academic Press, London, New York.

MUNNECKE D. M. (1978). Detoxification of pesticides using soluble or immobilized enzymes. *Process Biochem.* **13**, 14–16, 31.

MYERS M. G. and MCGARITY J. W. (1968). The urease activity in profiles of five great soil groups from northern New South Wales. *Pl. Soil* **28**, 25–37.

NAIDER F., BECKER J. M. and KATZIR-KATCHALSKI E. (1974). Utilization of methionine —containing peptides and their derivatives by a methionine—requiring auxotroph of *Saccharomyces cerevisiae*. *J. Biol. Chem.* **249**, 9–20.

NAKAE T. and NIKAIDO H. (1975). Outer membrane as a diffusion barrier in *Salmonella typhimurium*. Penetration of oligo- and polysaccharides into isolated outer membrane and cells with degraded peptidoglycan layer. *J. Biol. Chem.* **250**, 7359–7365.

NANNIPIERI P., CECCANTI B., CERVELLI S. and SEQUI P. (1974). Use of 0·1M pyrophosphate to extract urease from a podzol. *Soil Biol. Biochem.* **6**, 359–362.

NEAL J. L. (1973). Influence of selected grasses and herbs on soil phosphatase activity. *Can. J. Soil, Sci.* **53**, 119–121.

NISHIO M. (1970). The distribution of nitrifying bacteria in soil aggregates. *Soil Sci. Pl. Nutr.* **16**, 24–27.

OPARIN A. I. (1937). Direction regulation of the action of invertase in the living plant cell. *Enzymologia* **4**, 13–23.

OU L-T and MARQUIS R. E. (1970). Electromechanical interactions in cell walls of Gram-positive cocci. *J. Bact.* **101**, 92–101.

PALADE G. (1975). Intracellular aspects of the process of protein synthesis. *Science* **189**, 347–358.

PANCHOLY S. K. and RICE E. L. (1973a). Carbohydrases in soil as affected by successional stages of revegetation. *Soil Sci. Soc. Am. Proc.* **37**, 227–229.

PANCHOLY S. K. and RICE E. L. (1973b). Soil enzymes in relation to old field succession: amylase, cellulase, invertase, dehydrogenase and urease. *Soil Sci. Soc. Am. Proc.* **37**, 47–50.

PAULI F. W. (1965). The biological assessment of soil fertility. *Pl. Soil* **22**, 337–351.

PAULSON K. N. and KURTZ L. T. (1969). Locus of urease activity in soil. *Soil Sci. Soc. Am. Proc.* **33**, 897–901.

PAYNE J. W. (1976). Peptides and microorganisms. *Adv. Microb. Physiol.* **13**, 55–113.

PAYNE J. W. and GILVARG C. (1968). Size restriction in peptide utilization in *Escherichia coli. J. Biol. Chem.* **243**, 6291–6299.

PAYNE J. W. and MATTHEWS D. M. (1974). *In* "Peptide Transport in Protein Nutrition" (D. M. Matthews and J. W. Payne, Eds) Ch. 1. North Holland, Amsterdam.

PAZUR J. H. and ARONSON N. N. (1972). Glycoenzymes: Enzymes of glycoprotein structure. *Adv. Carbohydrate Chem. Biochem.* **27**, 301–341.

PETERSON G. H. (1962a). Respiration of soil sterilized by ionizing radiations. *Soil Sci.* **94**, 71–74.

PETERSON G. H. (1962b). Microbial activity in heat and electron sterilized soil seeded with microorganisms. *Can. J. Microbiol.* **8**, 519–524.

PETTIT N. M., SMITH A. R. J., FREEDMAN R. B. and BURNS R. G. (1976). Soil urease: activity, stability and kinetic properties. *Soil Biol. Biochem.* **8**, 479–484.

PETTIT N. M., GREGORY, L. J., FREEDMAN R. B. and BURNS R. G. (1977). Differential stabilities of soil enzymes: assay and properties of phosphatase and arylsulphatase. *Biochim. Biophys. Acta* **485**, 357–366.

POCHON J., TARDIEUX P. and D'AGUILAR J. (1969). Methodological problems in soil biology. *In* "Soil Biology Reviews of Research", pp. 13–63. UNESCO, Paris.

POLLOCK M. R. (1962). Exoenzymes: *In* "The Bacteria", Vol. IV (I. C. Gunsalus and R. Y. Stanier, Eds) pp. 121–178. Academic Press, New York.

POLLOCK M. R. and RICHMOND M. H. (1962). Low cyst(e)ine content of bacterial extracellular proteins: its possible physiological significance. *Nature* **194**, 446–449.

PONEC V., KNOR Z. and ČERNÝ S. (1974). *Adsorption on Solids.* CRC Press, Cleveland, Ohio.

PONNAMPERUMA C. and GABEL N. W. (1968). Current status of chemical studies on the origin of life. *Space Life Sci.* **1**, 64–96.

POWLSON D. S. and JENKINSON D. S. (1975). The effects of biocidal treatments on metabolism in soil—II. Gamma irradiation, autoclaving, air-drying and fumigation. *Soil Biol. Biochem.* **8**, 179–188.

PRASAD R., RAJALE G. B. and LAKHDIVE B. A. (1971). Nitrification retarders and slow-release nitrogen fertilizers. *Adv. Agron.* **23**, 337–383.

RADULESCU D. and KISS S. (1971). Enzyme activities in the sediments of the Zanoga and Zanoguta lakes (Retezat Mountain Massif). *In* "Progress in Palinologia Romaneasca", pp. 243–248. Acad. R. S. Romania, Bucharest.

RAMÍREZ-MARTÍNEZ J. R. and MCLAREN A. D. (1966). Some factors influencing the determination of phosphatase activity in native soils and in soils sterilized by irradiation. *Enzymologia* **31**, 23–38.

RAWALD W. (1970). Enzyme activity in soil as a component of the biological activity in soil with particular reference to the evaluation of soil fertility status, and aspects of the tasks of soil enzymological research. *Zentralbl. Bakteriol. Parasitenkd. Abt. II* **125**, 363–384.

REESE E. T. (1976). History of the cellulase program at the US Army National Development Center. *In* "Enzymatic Conversion of Cellulosic Materials" (E. Gaden, M. H. Mandels, E. T. Reese and L. A. Spano, Eds). Interscience, Wiley, New York.

RIEBELING P., THAUER R. K. and JUNGERMANN K. (1975). The internal—alkaline pH gradient, sensitive to uncoupler and ATPase inhibitor, in growing *Clostridium pasteurianum*. *Eur. J. Biochem.* **55**, 445–453.

ROBERGE M. R. (1970). Behaviour of urease added to unsterilized, steam-sterilized, and gamma radiation-sterilized black spruce humus. *Can. J. Microbiol.* **16**, 865–870.

ROBERGE M. R. (1971). Respiration of a black spruce humus sterilized by heat or irradiation. *Soil Sci.* **111**, 124–128.

ROSS D. J. (1966). A survey of activities of enzymes hydrolysing sucrose and starch in soils under pastures. *J. Soil Sci.* **17**, 1–15.

ROSS D. J. (1968). Some observations on the oxidation of glucose by enzymes in soil in the presence of toluene. *Pl. Soil* **28**, 1–11.

ROSS D. J. (1974). Glucose oxidase activity in soil and its possible interference in assays of cellulase activity. *Soil Biol. Biochem.* **6**, 303–306.

ROSS D. J. (1976). Invertase and amylase activities in ryegrass and white clover plants and their relationships with activities in soils under pasture. *Soil Biol. Biochem.* **8**, 351–356.

ROSS D. J. and MCNEILLY B. A. (1972). Some influences of different soils and clay minerals on the activity of glucose oxidase. *Soil Biol. Biochem.* **4**, 9–18.

ROSWELL T. (1973). "Modern Methods in the Study of Microbial Ecology." *Bull. Ecol. Res. Comm.* **17**, NFR, Stockholm.

ROWELL M. J., LADD J. N. and PAUL E. A. (1973). Enzymatically active complexes of proteases and humic acid analogues. *Soil Biol. Biochem.* **5**, 699–703.

RUHLING A. and TYLER G. (1973). Heavy metal pollution and decomposition of spruce needle litter. *Oikos* **24**, 402–416.

RYŠAVÝ P. and MACURA J. (1972). The assay of β-galactosidase in soil. *Folia Microbiol.* **17**, 370–374.

SACHS H. (1969). Neurosecretion. *Adv. Enzymol.* **32**, 327–372.

SALTON M. R. J. (1971). The bacterial membrane. *In* "Biomembranes", Vol. 1 (L. A. Manson, Ed.), pp. 1–65. Plenum Press, New York and London.

SALTON M. R. J. (1974). Membrane associated enzymes in bacteria. *Adv. Microb. Physiol.* **11**, 213–283.

SALTON M. R. J. and OWEN P. (1976). Bacterial membrane structure. *Ann. Rev. Microbiol.* **30**, 451–482.

SANTORO T. and STOTZKY G. (1967). Influence of cations on flocculation of clay minerals by microbial metabolites as determined by the electrical sensing zone particle analyzer. *Soil. Sci. Soc. Am. Proc.* **31**, 761–765.

SATYANARAYANA T. and GETZIN L. W. (1973). Properties of a stable cell-free esterase from soil. *Biochemistry* **12**, 1566–1572.

SAUNDERS W. M. H. and METSON A. J. (1971). Seasonal variation of phosphorus in soil and pasture. *N. Z. J. Agric. Res.* **14**, 307–328.

SCARBOROUGH G. A., RUMLEY M. K. and KENNEDY E. P. (1968). The function of adenosine 5'-triphosphate in the lactose transport system of *Escherichia coli*. *Proc. Nat. Acad. Sci. US* **60**, 951–958.

SCHERRER R. and GERHARDT P. (1971). Molecular sieving by the *Bacillus megaterium* cell wall and protoplast. *J. Bacteriol.* **107**, 718–735.

SCHLESINGER M. J. (1968). Secretion of alkaline phosphatase subunits by spheroplasts of *Escherichia coli*. *J. Bacteriol.* **96**, 727–733.

SCHMIDT E. L. (1973). Fluorescent antibody techniques for the study of microbial ecology. *Bull. Ecol. Res. Comm., Stockholm* **17**, 67–76.

SCHNITZER M. and KHAN S. U. (1972). "Humic Substances in the Environment." Marcel Dekker, New York.

SENECA H., PEER P. and NALLY R. (1962). Microbial urease. *Nature* **193**, 1106–1107.

SINGH A. P. (1975). Cell envelope alterations associated with sensitivity to anti-bacterial agents of gamma-irradiation resistant mutants of *Escherichia coli*. *J. Gen. Appl. Microbiol.* **21**, 195–210.

SINGH A. P. and REITHMEIER R. A. F. (1975). Leakage of periplasmic enzymes from cells of heptose—deficient mutants of *Escherichia coli*, associated with alterations in the protein component of the outer membrane. *J. Gen. Appl. Microbiol.* **21**, 109–118.

SIZER I. W. (1972). Medical applications of microbial enzymes. *Adv. Appl. Microbiol.* **15**, 13–38.

SKUJIŅŠ J. (1967). Enzymes in soil. *In* "Soil [Biochemistry" (A. D. McLaren and G. H. Peterson, Eds) 371–414. Marcel Dekker, New York.

SKUJIŅŠ J. (1973). Dehydrogenase: an indicator of biological activities in arid soils. *Bull. Ecol. Res. Commun., Stockholm* **17**, 235–241.

SKUJIŅŠ J. (1976). Extracellular enzymes in soil. *CRC Crit. Rev. Microbiol.* **4**, 383–421.

SKUJIŅŠ J. J. and MCLAREN A. D. (1969). Assay of urease activity using ^{14}C-urea in stored, geologically preserved, and in irradiated soils. *Soil Biol. Biochem* **1**, 89–99.

SKUJIŅŠ J. J. and MCLAREN A. D. (1968). Persistance of enzymatic activities in stored and geologically preserved soils. *Enzymologia* **34**, 213–225.

SKUJIŅŠ J. J., BRAAL L. and MCLAREN A. D. (1962). Characterization of phosphatase in a terrestrial soil sterilized with an electron beam. *Enzymologia* **25**, 125–133.

SKUJIŅŠ J. J., PUKITE A. and MCLAREN A. D. (1973). Adsorption and reactions of chitinase and lysozyme on chitin. *Molec. Cell Biochem.* **2**, 221–228.

SMILEY K. L. and STRANDBERG G. W. (1972). Immobilized enzymes. *Adv. Appl. Microbiol.* **15**, 13–38.

SMITH D. W., FLIERMANS C. B. and BROCK T. D. (1973). An isotopic technique for measuring the autotrophic activity of soil microorganisms *in situ*. *Bull. Ecol. Res. Comm., Stockholm* **17**, 243–246.

SPEIR T. W. and ROSS D. J. (1975). Effects of storage on the activities of protease, urease, phosphatase and sulphatase in three soils under pasture. *N. Z. Jl. Sci.* **18**, 231–237.

STERNBERG D. (1976). β-Glucosidase of *Trichoderma*: Its biosynthesis and role in saccharification of cellulose. *Appl. Environ. Microbiol.* **31**, 648–654.

STEWART B. J. and LEATHERWOOD J. M. (1976). Depressed synthesis of cellulase by *Cellulomonas*. *J. Bacteriol.* **128**, 609–615.

STOTZKY G. (1974). Activity, ecology, and population dynamics of microorganisms in soil. *In* "Microbial Ecology" (A. Laskin and H. Lechevalier, Eds), pp. 57–135. CRC Press, Cleveland, Ohio.

SUTHERLAND I. W. (1972). Bacterial exopolysaccharides. *Adv. Microb. Physiol.* **8**, 143–213.

SWABY R. J. (1949). The relationship between microorganisms and soil aggregation. *J. Gen. Microbiol.* **3**, 236–254.

TABATABAI M. A. (1977). Effects of trace elements on urease activity in soils. *Soil Biol. Biochem.* **9**, 9–13.

TABATABAI M. A. and BREMNER J. M. (1972). Assay of urease activity in soils. *Soil Biol. Biochem.* **4**, 479–487.

TABATABAI M. A. and SINGH B. B. (1976). Rhodanese activity of soils. *Soil Sci. Soc. Am. J.* **40**, 381–385.

THENTE B. (1970). Effects of toluene and high-energy radiation on urease activity in soil. *Lantbrukshögsk. Annlr* **36**, 401–418.

THORNTON J. I. and MCLAREN A. D. (1975). Enzymic characterization of soil evidence. *J. Forensic Sci.* **20**, 674–692.

TODD R. L., CROMACK K. and STORMER J. C. (1973). Chemical exploration of the micro-habitat by electron probe microanalysis of decomposer organisms. *Nature, London* **243**, 544–546.

TORRIANI A. (1960). Influence of inorganic phosphatase in the formation of phosphatases by *Escherichia coli*. *Biochim. Biophys. Acta.* **38**, 460–469.

TYAGNY-RYADNO M. G. (1968). Distribution of microorganisms and nutrients in soil aggregates. *Visn. Sil'-s'kogospod. Nauki* **11**, 46–51.

TYLER G. (1974). Heavy-metal pollution and soil enzymatic activity. *Pl. Soil* **41**, 303–311.

TYLER G. (1976). Influence of vanadium on soil phosphatase activity. *J. Environ. Qual.* **5**, 216–217.

VERMA L., MARTIN J. P. and HAIDER K. (1975). Decomposition of carbon-14-labeled proteins, peptides, and amino acids; free and complexed with humic polymers. *Soil Sci. Soc. Am. Proc.* **39**, 279–284.

WADSO I. (1973). Microcalorimetry and its application in biological sciences. *In* "New Techniques in Biophysics and Cell Biology", Vol. 2 (R. Pain and B. Smith, Eds) Wiley, New York.

WAGNER G. H. (1974). Observations of fungal growth in soil using a capillary pedoscope. *Soil Biol. Biochem.* **6**, 327–333.

WAGNER G. H. (1975). Microbial growth and carbon turnover. *In* "Soil Biochemistry", Vol. 3 (E. A. Paul and A. D. McLaren, Eds) 269–305. Marcel Dekker, New York.

WEBER J. B. (1972). Interaction of organic pesticides with particular matter in aquatic and soil systems. *In* "Fate of Organic Pesticides in the Aquatic Environment", Adv. Chem. Ser. **111**, 55–120.

WEETALL H. H. (1975). Immobilised enzymes and their application in the food and beverage industry. *Proc. Biochem.* **10**, July/Aug., 3–24.

WILLIAMS A. H. (1963). *In* "Enzyme Inhibition by Phenolic Compounds" (J. B. Pridham, Ed.) pp. 87–95. Pergamon Press, Oxford.

WILLIAMS S. T. and MAYFIELD C. I. (1971). Studies on the ecology of actinomycetes in soil. III. The behaviour of neutrophilic streptomycetes in acid soil. *Soil Biol. Biochem.* **3**, 197–208.

ZABORSKY O. (1973). "Immobilized Enzymes." CRC Press, Cleveland, Ohio.

ZANTUA M. I. and BREMNER J. M. (1975a). Comparison of methods of assaying urease activity in soils. *Soil Biol. Biochem.* **7**, 291–295.

ZANTUA M. I. and BREMNER J. M. (1975b). Preservation of soil samples for assay of urease activity. *Soil Biol. Biochem.* **7,** 297–299.

ZANTUA M. I. and BREMNER J. M. (1976). Production and persistence of urease activity in soils. *Soil Biol. Biochem.* **8,** 369–374.

APPENDIX

Methodology of Soil Enzyme Measurement and Extraction

M. R. ROBERGE

*Laurentian Forest Research Centre, Department of Fisheries and the Environment
Sainte-Foy, Québec, Canada*

I. Introduction

Enzymes derived from the microflora, the root system of plants and plant residues play a major role in soils. The accuracy of any assessment of their activity is considerably influenced by conditions during storage of soil after its collection, methods of preparing samples prior to enzyme measurement and extraction, conditions during the actual enzyme determination and procedures used for analysis of the substrates and the products. It is also difficult to ascertain from this in-vitro approach the function of enzymes in soil under natural conditions. The gradual accumulation of knowledge on the methodology of soil enzyme measurement and extraction has been pointed out in reviews by Kiss (1958), Hofmann (1963), Durand (1965), Skujiņš (1967) and Kiss et al. (1975). The purpose of this appendix chapter is to detail some of the methods currently used for the measurement and, where relevant, the extraction of soil enzymes. In addition, the effect of sterilants, storage of soil, pH, temperature and substrate concentration on each enzyme assay are outlined and the estimation of soil respiratory activity is discussed.

II. Methodology

1. Amylase

1.1. Assay (Hofmann, 1963; Kelley and Rodriguez-Kabana, 1975).

The soil was air-dried and stored over a desiccant in a refrigerator prior to assay. To 5 g of air-dried soil, contained in a 50 ml Erlenmeyer flask, 1·5 ml of toluene was added; the mixture was shaken and allowed to stand for 15 min. To this was added 10 ml of distilled water and 5 ml of 2% (w/v) solution of soluble starch. The flask was then stoppered and placed in an incubator. After 5 h at 37°C, the flask was opened, 15 ml of distilled water added, the contents mixed and 10 ml of the suspension centrifuged to produce a clear supernatant; 1 ml of the supernatant was then analysed for reducing sugars by a modified Nelson-Somogyi method.

1.2. Discussion

Amylase activity was reduced whether soil was stored air-dried (Ross, 1965b; Ambrož, 1970) or moist at room temperature (Ross, 1965b; Pancholy and Rice, 1972), at 4°C (Ross, 1965b; Pancholy and Rice, 1972), or at −20°C (Ross, 1965b). This reduction tended to increase with length of storage (Ross, 1965b; Pancholy and Rice, 1972). Minimum loss of activity occurred in moist soil stored at 4°C (Ross, 1965b; Pancholy and Rice, 1972). According to Galstyan (1965, 1974), amylase activity was slightly decreased by toluene, and continued to decrease with an increase of toluene and with longer contact times. Radulescu (1972) reported that amylase activity was lower in reaction mixtures buffered with acetic acid–sodium acetate solution at pH 5·6 than in mixtures in which distilled water was used instead of buffer. With the appropriate substrate concentration, amylase activity increased linearly with incubation time (Galstyan, 1965).

1.3. Extraction

According to Kiss et al. (1975), Ukhtomskaya in 1952 and Shcherbakova et al. in 1971 extracted from soil a protein exhibiting amylase activity.

2. Arylsulphatase

2.1. Assay (Tabatabai and Bremner, 1970a, b).

The soil was sieved through a 2 mm screen and stored in a moist condition at −10°C.

The equivalent of 1 g of air-dried soil was placed in a 50 ml Erlenmeyer flask and 4 ml of 0·5 M sodium acetate buffer at pH 5·8, 0·25 ml of toluene and 1 ml of 0·005 M p-nitrophenyl sulphate solution were added. The flask was swirled for a few seconds to mix the contents, stoppered, and placed in an incubator at 37°C. After 1 h the stopper was removed, 1 ml of 0·5 M $CaCl_2$ and 4 ml of 0·5 M NaOH were added and the flask swirled for a few more seconds. Finally, the soil suspension was filtered and the intensity of the yellow colour in the filtrate measured.

2.2 Discussion

When soil was stored in an air-dried state a marked increase in arylsulphatase activity was observed by Tabatabai and Bremner (1970b), although a slight decrease was reported by Speir and Ross (1975). Activity declined when soil was stored moist at 23°C, 5°C and −10°C; reduction tending to increase with length of storage (Tabatabai and Bremner, 1970b).

Toluene stimulated arylsulphatase activity which was not subsequently affected when the amount of toluene was increased from 0·1 to 1·0 ml g^{-1} soil (Tabatabai and Bremner, 1970a). Activity was reduced by gamma irradiation, with a dosage which was at least twice as large as that normally required for sterilizing soils (Tabatabai and Bremner, 1970b). This finding agreed with that of Popenoe and Eno (1962) who showed that sulphatase activity was progressively reduced by increasing doses of radiation up to 2·048 Mrad.

According to Tabatabai and Bremner (1970a), arylsulphatase activity was decreased when water was used instead of sodium acetate buffer at pH 5·8. The decrease was related to the difference between the soil pH and the buffer pH and was minimal for a soil with a pH close to 6·2.

Arylsulphatase activity was higher with sodium acetate buffer than with phosphate or modified universal buffer, and was maximal with 0·5 M sodium acetate buffer having a pH of 6·2 (Tabatabai and Bremner, 1970a).

Arylsulphatase activity increased linearly with time when the concentration of p-nitrophenyl sulphate was varied from 0·25 ml to 2·0 mlg^{-1} soil (Tabatabai and Bremner, 1970a). Activity also increased from 27°C to 67°C and decreased from 67°C to 77°C (Tabatabai and Bremner, 1970a). Cooper (1972) found that arylsulphatase activity increased when soils were continually moist and decreased when soils dried out.

A detailed discussion of the factors influencing arylsulphatase activity can be found in Chapter 6.

3. Catalase

3.1. Assay (Kuprevich, 1951; Johnson and Temple, 1964).

A 2 g (oven-dry) soil sample was placed in a 125 ml Erlenmeyer flask with 40 ml of distilled water and put on a rotary shaker. To this was added 5 ml of 30% H_2O_2 and the slurry was shaken for 20 min. The remaining peroxide was then stabilized by adding 5 ml of 3 N H_2SO_4, the contents of the flask filtered, and a 25 ml aliquot titrated with 0·05 M K_2MnO_4.

3.2. Discussion

Catalase activity declined when soil was air-dried (Baranovskaya, 1954; Latypova and Kurbatov, 1961; Johnson and Temple, 1964; Latypova, 1965). According to Skujiņš (1967), Scharrer found that catalase activity was diminished by ultraviolet irradiation.

Phosphate buffer and tris buffer at pH 7·1 reduced catalase activity and the magnitude of the effect varied with buffer but was not dependent upon any obvious relation to concentration or pH (Johnson and Temple, 1964).

3.3. Extraction

According to Kiss et al. (1975), Ukhtomskaya reported in 1952 the extraction of catalase from soil.

4. Cellulase

4.1. Assay (Pancholy and Rice, 1973).

The soil was screened through a 2 mm sieve, 5 g were placed in a 50 ml flask and 0·5 ml of toluene added. The contents were mixed thoroughly and after 15 min 10 ml of acetate buffer at pH 5·9 were added followed by 10 ml of 1% carboxymethylcellulose. The flask was then incubated for 24 h at 30°C. At the end of this period approximately 50 ml of distilled water were added, the suspension filtered and distilled water added to increase the volume of filtrate to 100 ml. The reducing sugar content was then determined by the Nelson-Somogyi method.

4.2. Discussion

Cellulase activity was reduced in the presence of toluene (Markus, 1955; Kiss et al., 1962a; Kozlov and Kislitsina, 1967). Gamma irradiation at minimum sterilizing dosage did not bring about any significant changes in cellulase activity although at heavier dosages some reductions occurred (Kiss et al., 1974).

According to Tomescu (1970), cellulase activity of a soil at a pH of 6·8 was slightly enhanced when the pH of phosphate buffer was increased from 4·0 to 7·0, but was then always higher than at pH 7·6. Apparently the concentration of carboxymethylcellulose in the reaction mixture should be at least 0·4% for there to be a linear increase of cellulase activity with time (Tomescu, 1970).

5. Dehydrogenase

5.1. Assay (Casida et al., 1964; Kiss and Boaru, 1965; Kiss et al., 1969)

A 3 g air-dried soil sample, including 0·03 g of added calcium carbonate, was saturated with 0·5 ml of a 3% (w/v) solution of triphenyl tetrazolium chloride and 1·25 to 1·75 ml of distilled water. The contents were mixed thoroughly and incubated at 37°C for 24 h. Following incubation, the triphenyl formazan formed was extracted with ethanol.

5.2. Discussion.

Dehydrogenase activity was reduced when soil was stored air-dried at room temperature (Ross, 1970; Ross and McNeilly, 1972), a reduction which tended to increase with storage (Casida et al., 1964; Peterson, 1967; Ross, 1970). Nevertheless, dehydrogenase activity was still detectable in some air-dried soils that had been stored for a few years (Skujiņš and McLaren, 1968). Activity was also reduced when soil was stored moist at room temperature (Casida et al., 1964; Peterson, 1967; Ross, 1970; Pancholy and Rice, 1972), at 4°C (Peterson, 1967; Ross, 1970; Pancholy and Rice, 1972); it was generally reduced at −20°C (Ross, 1970; Ross, 1972; Ross and McNeilly, 1972). Reduction tended to increase with length of storage (Pancholy and Rice, 1972) and minimum loss of activity occurred at 4°C (Peterson, 1967; Ross, 1970; Pancholy and Rice, 1972). According to Ross (1971), dehydrogenase activity was decreased by toluene.

Dehydrogenase activity was higher with 0·5 M hepes, tes, or tris buffers at pH 7·6 than in mixtures containing calcium carbonate or water; it was not affected by actual volume of buffer up to 2·5 ml (Ross, 1971). The optimum pH for measurement was between 7·4 and 8·5 (Galstyan and Markosyan, 1965; Ross, 1971).

When the concentration of triphenyl tetrazolium chloride was varied between 0·25 and 2·0 ml, dehydrogenase activity increased up to about 1·0 ml. Excess concentrations may, in fact, be inhibitory (Lenhard, 1956; Thalmann, 1968). The rates of triphenyl formazan formation declined during lengthy incubation periods (Peterson, 1967; Thalmann, 1968; Ross, 1971).

Incubation at 37°C increased dehydrogenase activity over that occurring lower temperatures (Casida et al., 1964; Cerna, 1972) and activity was greater in systems incubated anaerobically than in those incubated aerobically (Ross, 1971; Cerna, 1972).

6. Glucosidase

6.1. Assay (Tomescu, 1966).

Soil was sieved through a 1 mm screen and a 20 g sample placed in a stoppered flask. Toluene and cellobiose were then added together with 0·067 M phosphate buffer. After ten days at 28°C, 5 ml of 10% trichloroacetic acid was added and the glucose produced measured using Schoorl's method (Schoorl, 1920).

6.2. Discussion

Glucosidase activity was slightly reduced whether soil was stored air-dried (Hoffmann, 1959) or moist at 4°C (Tyler, 1974). According to Galstyan (1965, 1974) and to Kiss et al. (1972), glucosidase activity was slightly lower with than without toluene.

Glucosidase activity of a soil of pH 6·8 increased with increasing phosphate buffer pH from 5·5 to 7·2, and decreased at higher pH (Tomescu, 1966).

With appropriate substrate concentration, activity increased linearly with time (Galstyan, 1965).

7. Invertase

7.1. Assay (Hofmann and Seegerer, 1951; Ross, 1965a).

Ten grams of fresh soil sieved through a 2 mm sieve were mixed in a 250 ml stoppered conical flask. The soil was then allowed to stand with 2·5 ml of toluene for 15 min at 20°C before shaking for 24 h and incubating at 37°C with 20 ml of 0·5 M acetate–phosphate buffer at pH 5·5 and 20 ml of 5% sucrose solution. After incubation, the contents were filtered and 0·15 ml of the filtrate added within 10 min to 5·0 ml of Shaffer-Hartmann reagent (Shaffer and Hartmann, 1921) and 0·05 ml 2 M NaOH. The volume was made up to 10 ml with water and the solution heated for 15 min in a boiling water bath. The reagent was calibrated against glucose.

7.2. Discussion

Invertase activity was reduced when soil was stored air-dried (Hoffmann, 1959; Drobnik, 1960 and 1961; Kiss et al., 1962a; Kleinert, 1962; Peterson and Astafieva, 1962; Ross, 1965b, 1968; Nizova, 1969; Ambrož, 1970). Drying in vacuo at 40°C also lowered activity (Scheffer and Twachtmann, 1953) as did moist storage at room temperature (Pancholy and Rice, 1972), at 4°C (Ross, 1965b; Pancholy and Rice, 1972), or at −20°C (Ross, 1965b). The reduction tended to increase with longer storage (Ross, 1965b; Nizova, 1969; Pancholy and Rice, 1972) and was minimum when soil was stored moist at 4°C (Ross, 1965b; Pancholy and Rice, 1972).

Invertase activity was slightly reduced by toluene (Kiss, 1958; Drobnik, 1960, 1961; Kiss et al., 1962a; Peterson and Astafieva, 1962; Kiss, 1964; Galstyan, 1965; Voets and Dedeken, 1965; Voets et al., 1965; Ross, 1968; Kiss et al., 1971, 1972; Galstyan, 1974). Soil samples pretreated with toluene, either at room temperature for 15 min or at 50°C for 3 h, showed identical invertase activity (Kiss, 1964). Soil samples pre-sterilized by radiation and treated with toluene showed a lower activity than those which had only been radiation-sterilized (Voets and Dedeken, 1965). The same authors also demonstrated that invertase activity was higher in toluene-treated than in irradiated soil. Radiation also slightly reduced activity (Dommergues, 1960; Voets and Dedeken, 1965; Voets et al., 1965; Kiss et al., 1974). However, at dosages heavier than the minimum sterilizing one, marked reductions occurred (Kiss et al., 1974).

According to Domocos et al. (1966), Soreanu et al. (1970), Radulescu (1972), and Rogobete et al. (1972), invertase activity was lower in reaction mixtures buffered with acetic acid–sodium acetate solution at pH 5·6 than in mixtures in which distilled water was used instead of buffer.

With appropriate substrate concentration, invertase activity increased linearly with incubation time (Kiss, 1958; Galstyan, 1965); invertase activity was curvilinear at low substrate concentrations and linear at high substrate concentrations (Kiss, 1958; Stefanic, 1972). Activity increased with increasing temperature from 5°C to 35°C (Kiss, 1958).

7.3. Extraction

According to Kiss et al. (1975), Ukhtomskaya in 1952 and Shcherbakova et al. in 1971 extracted from soil a protein exhibiting invertase activity.

8. Levansucrase

8.1. Assay (Kiss, 1961).

Three grams of air-dried soil were incubated with 2 ml of toluene for 15 min, after which, 10 ml of 3 M acetate buffer at pH 5·6 containing 10% sucrose was added. Levansucrase activity was determined photocolorimetrically after four months incubation at 37°C.

8.2. Discussion

Levansucrase activity was more intense in the absence than in the presence of toluene (Kiss et al., 1962b; Kiss, 1964; Kiss et al., 1972). Soil samples pretreated with toluene at 50°C for 3 h showed a higher levansucrase activity than when the toluene-pre-treatment was at room temperature for 15 min (Kiss, 1964). Gamma radiation at minimum sterilizing dosage did not bring about any significant changes in levansucrase activity; at heavier dosages some reductions occur (Kiss et al., 1974).

9. Lipase and carboxylesterase

9.1. Assay (Pancholy and Lynd, 1973).

A 10 g soil sample in a 50 ml Erlenmeyer flask was amended with 4% olive oil and incubated in a saturated atmosphere at 30°C for fifteen days. Following incubation, the sample was extracted with 20 ml of 0·1 M sodium phosphate buffer solution at pH 7·5, with rotary shaking at 120 rpm at 30°C for 30 min. To 3·1 ml of the filtered soil extract, 0·1 ml of 0·1 M 4-methyl umbelliferone butyrate was added and vortex mixed. The rates and magnitude of fluorescence increase were then determined.

9.2. Discussion

According to Pokorná (1964), lipase activity was lower in the presence of toluene than in its absence. Carboxylesterase activity was lower in a soil irradiated with at least twice the sterilizing dose, decreased during storage after irradiation but much was still present three months after irradiation (Getzin and Rosefield, 1968).

Lipase activity was higher with buffer than without buffer and when measured at temperatures varying between 25°C and 60°C, reached its maximum at 37°C (Pokorna, 1964). Activity decreased when the water content of soil was lowered from 90% to 12% of its water-holding capacity (Pokorná, 1964).

9.3. Method of extraction–Carboxylesterase (Getzin and Rosefield, 1968, 1971).

One hundred and twenty grams air-dried equivalents of moist soil were placed in each of two 1000 ml Erlenmeyer flasks with 600 ml of 0·2 M NaOH and shaken for 30 min. The soil suspension was centrifuged at 16 000 g for 10 min and the supernatant containing the enzyme (in this instance responsible for the breakdown of the organophosphorus insecticide, malathion) was filtered and immediately adjusted to pH 7·9 with 3 M HCl. To assay the esterase 1·5 μmol of malathion and desired amounts of filtrate were added in 5 or 10 ml of 0·075 M tris–HCl buffer at pH 7·0. The mixture was incubated for 4–6 h at 37°C in a water bath shaker and subsequently extracted with 10 ml of hexane by shaking for 30 min. Residual malathion concentrations were determined by gas–liquid chromatography of the extract.

9.4. Discussion

Carboxylesterase was not extracted with 1 M sodium pyrophosphate, oxalate, chloride, bicarbonate, and citrate, 0·1% EDTA, 0·1 N aluminium sulphate, ion exchange resins, charcoal, or buffered solutions within a pH range of 2–10. In addition, these reagents and adsorbents did not inactivate the enzyme because extracted soil residues retained the capacity to degrade malathion (Getzin and Rosefield, 1968).

Getzin and Rosefield (1968, 1971) showed that Carboxylesterase extracted from a soil, which had been irradiated with at least twice the sterilizing dose, was four to five times less active than that extracted from non-irradiated soil. Enzyme extracts also lost activity when they were exposed to radiation.

The extract lost little or no activity upon prolonged storage at 4°C or at −10°C, but was largely destroyed by lyophilization (Getzin and Rosefield, 1971).

Carboxylesterase activity was maximum above pH 6·8 and was equal in 0·05 M citrate buffer, 0·05 M phosphate buffer, and 0·075 M tris–HCl buffer (Getzin and Rosefield, 1971). According to the same authors activity was linear up to 11 h when the amount of extract was adjusted to have 0·5 unit of enzyme and was then incubated at 37°C with 1·5 μmol of malathion in 5 ml of 0·075 M tris–HCl buffer at pH 7·0.

10. Nitrate reductase

10.1. Assay (Cawse and Cornfield, 1971)

The fresh soil was passed through a 2 mm sieve. Water was added two days before irradiation to moisten the soil to 60% water-holding capacity and a 10–50 g sample was then treated with 0·8 Mrad at a temperature of 22°C. Ammonium, nitrite, and nitrate were determined after incubation for 64 h at 25°C in either polystyrene or polypropylene tubes which were not air-tight. For analysis of ammonium, nitrite and nitrate, the soil was shaken for 1 h with 100 ml of 1 M KCl, but water was used as the extractant when nitrite or nitrate only were determined. A distillation method was employed for separate analysis of ammonium and nitrate after nitrite had been removed by addition of 1 ml of 10% (w/v) sulphamic acid.

10.2. Discussion

A decrease in nitrate reductase activity was observed in air-dried stored soil. An increase occurred in soil stored at 30% water-holding capacity at room temperature moistened to 60% water-holding capacity and irradiated (but not sterilized) before enzyme measurement (Cawse and Cornfield, 1971).

Nitrate reductase activity was measurable in the absence of toluene (Galstyan and Markosyan, 1966), and in gamma-irradiated soil (Cawse and Cornfield, 1969, 1971). There was a higher nitrate reductase activity in irradiated but non-sterile soil, than in non-irradiated soil (Cawse and Cornfield, 1971).

According to Nömmik (1956), Bremner and Shaw (1958), and Cawse and Cornfield (1971), nitrate reductase activity increased with pH from 3·6 to 8·2. The nitrate concentration positively influenced nitrate reductase activity (Nömmik, 1956; Cawse and Cornfield, 1971). Nitrate reductase activity increased with temperatures between 5°C and 40°C (Nömmik, 1956; Bremner and Shaw, 1958; Cawse and Cornfield, 1971).

Higher moisture in excess of 50% water-holding capacity, larger soil particles and poor aeration all increased nitrate reductase activity (Nömmik, 1956; Bremner and Shaw, 1958; Cady and Bartholomew, 1960;

Greenwood, 1962; Cawse and Crawford, 1967; Mahendrappa and Smith, 1967; Cawse and Cornfield, 1971).

11. Peroxidase

11.1. Assay and Method of Extraction (Bartha and Bordeleau, 1969; Bordeleau and Bartha, 1972a, b).

The soil sample was passed through a 2 mm sieve and used fresh, without drying. Fifty grams of soil were suspended in 50–200 ml of 0·2 M phosphate buffer at pH 6·0. The suspension was then agitated for 5 min on a rotary shaker and filtered. The resulting extract was filter-sterilized. The peroxidase assay was as follows: 0·3 ml 0·06% hydrogen peroxide in 0·05 M phosphate buffer at pH 6·0, 0·05 ml 0·5% o-dianisidine in methanol, and 2·7 ml soil extract were combined in a 1 cm spectrophotometric cuvette. The ingredients were mixed and the increase in optical density continuously recorded at 460 nm and 25°C.

11.2 Discussion

Peroxidase activity was much reduced when air-dried soils were used instead of fresh or moist-stored soils (Ross, 1968; Bartha, 1971). Toluene generally decreased peroxidase activity (Drobnik, 1961; Hofmann and Hoffmann, 1961; Ross, 1968; Bartha and Bordeleau, 1969) and a much greater amount was extractable by sonification than by shaking (Burge, 1973). The filter-sterilized extract showed a 5–15% lower peroxidase activity than the non-sterile soil extract (Bartha and Bordeleau, 1969).

Phosphate buffer was generally more efficient at extracting peroxidase than was acetate buffer, but the optimum activity occurred when using 0·2 M acetate buffer at pH 5·0. Extraction efficiency increased rapidly with phosphate buffer concentration from 0·05 M up to 0·2 M. Above this concentration up to 1·0 M, the increase was not consistent and only slight. The pH optimum of 0·2 M phosphate buffer was 6·0 and not 7·0 or 8·0 (Bordeleau and Bartha, 1972a, b).

According to Kiss et al. (1975), Ukhtomskaya also reported in 1952 the extraction of peroxidase from soil.

12. Phosphatase

12.1. Assay (Ramírez-Martínez and McLaren, 1966a).

Fresh soil samples were used, all being screened through a 2 mm sieve. A 1 g sample was treated with 2 ml of 0·005 M Na-β-naphthylphosphate solution and 2 ml of modified universal buffer at pH 7·0. The amount of

β-naphthol released into the soil extract was determined by fluorimetric measurements after incubation for 1 h at pH 7·0 and 25°C.

12.2. Discussion

Phosphatase activity was reduced by air-drying (Jackman and Black, 1952b; Geller and Dobrotvorska, 1960; Ambrož, 1970; Goian, 1971; Speir and Ross, 1975). It was also slightly reduced during storage in a moist state at 4°C (Tyler, 1974; Speir and Ross, 1975) and at −10°C (Ponomareva *et al.*, 1972).

Toluene slightly decreased phosphatase activity (Thomson and Black, 1948; Geller and Dobrotvorska, 1960; Halstead *et al.*, 1963; Voets and Dedeken, 1965; Voets *et al.*, 1965; Suciu, 1970; Thomson and Black, 1970) but not in radiation-sterilized soil (Voets and Dedeken, 1965). The same authors also demonstrated that phosphatase activity was identical in toluene-treated and in radiation-sterilized soil. Irradiation at minimum sterilizing dosage brought about a slight reduction (Voets and Dedeken, 1965; Voets *et al.*, 1965; Kiss *et al.*, 1974); at heavier dosages greater reduction generally occurred (McLaren *et al.*, 1962; Skujiņš *et al.*, 1962; Ramírez-Martínez and McLaren, 1966a; Bowman *et al.*, 1967; Kiss *et al.*, 1974); with an increase in soil moisture content during irradiation even greater reduction occurred (Ramírez-Martínez and McLaren, 1966b).

Most often the highest phosphatase activity was observed around neutral pH (Rogers, 1942; Jackman and Black, 1951, 1952a; Kroll *et al.*, 1955; Drobnikova, 1961; Halstead, 1964; Galstyan and Markosyan, 1965; Ramírez-Martínez and McLaren, 1966b; Gilliam and Sample, 1968; Hossner and Melton, 1970); it tended to occur at lower pH when the original pH of the soil was acid, and at higher pH when the original pH of the soil was alkaline.

Phosphatase activity increased with the amount of substrate to a maximum which depended on the soil, the type of substrate, and the incubation conditions (Jackman and Black, 1952b; Kroll *et al.*, 1955; Ramírez-Martínez and McLaren, 1966b).

Optimum temperatures for activity varied from 37°C to 60°C according to soil, substrate, and other incubation conditions (Rogers, 1942; Thomson and Black, 1948; Jackman and Black, 1952a; Drobnikova, 1961; Ramírez-Martínez and McLaren, 1966b; Hossner and Phillips, 1971).

According to Jackman and Black (1952a) and Ramírez-Martínez and McLaren (1966b), phosphatase activity decreased when soil moisture content dropped below 10–20% water-holding capacity; it was greater under flooded conditions than at field capacity moisture content (Racz and Savant, 1972) or at 1/3 atmospheric moisture content (Hossner and Phillips, 1971).

Speir and Ross have discussed this enzyme at length in Chapter 6 of this book.

12.3. Extraction

According to Kiss et al. (1975), Rotini reported in 1955 that he obtained phosphatase-active extracts from different soils.

13. Polyphenol oxidase

13.1. Method of extraction (Mayaudon et al., 1973a, b).

A sample of 450 g of fresh soil screened through a 1 mm sieve was placed in portions of 75 g in six 1 L pyrex bottles. To each bottle, 100 ml of 0·3 M K_2HPO_4 and 50 ml of 0·3 M EDTA at pH 7·0 was added and the pH brought to 8·0. The bottles were then agitated on a horizontal shaker for 1 h and allowed to stand for 16 h at 0°C before centrifugation at 2500 g. The supernatant was filtered through a Buchner funnel containing two Whatman no. 3 filter papers. For each 100 ml of filtrate, 60 g of $(NH_4)_2SO_4$ were added. This extract, kept at 0°C, was brought to pH 7·0 by addition of 1 M KH_2PO_4 and allowed to stand for 2 h at 0°C. The precipitate was centrifuged at 0°C, separated from the supernatant by siphonage, and weighed moist. To each gram of precipitate, 10 ml of 0·2 M K_2HPO_4 was added and the pH brought to 7·0. The extract was dialysed for 24 h at 0°C against 0·02 M phosphate buffer, centrifuged at 20 000 g for 20 min in sterile conditions, and filtered through a 0·2 μm sterile Gelman membrane. The extract was stored at 0°C and used within 24 h. It contained the enzyme system responsible for the transformation of tryptophan to indole-3-acetic acid (Pilet and Chalvignac, 1970; Chalvignac, 1971; Chalvignac and Mayaudon, 1971), for the decarboxylation of aromatic amino acids (Mayaudon et al., 1973a) as well as polyphenol oxidases responsible for the oxidation of d-catechin, p-cresol, catechol, DL-3 (3, 4-dihydroxyphenyl) alanine, p-phenylene diamine and p-quinol (Mayaudon et al., 1973b; Mayaudon and Sarkar, 1974a, b, 1975).

13.2 Discussion

When stored moist at room temperature, soils of low organic-matter content lost less polyphenol oxidase activity than soils of high organic-matter content; when soils were stored under refrigeration, polyphenol oxidase activity was slightly reduced (Chalvignac, 1971).

Much polyphenol oxidase activity persisted in the presence of toluene (Pilet and Chalvignac, 1970; Chalvignac and Mayaudon, 1971). Activity increased with pH from 3·1 to 7·0 (and decreased with a pH increase from 7·0 to 7·8) with the substrate concentration, and with temperature from 4°C to 37°C (Mayaudon et al., 1973b).

14. Protease

14.1. Assay (Ladd and Butler, 1972).

Soils were air-dried, ground, screened through a 2 mm sieve, and stored at room temperature in sealed metal containers. Prior to assay soils were re-moistened with distilled water to a moist crumb consistency and stored at 25°C in the dark in glass jars sealed with polythene covers for not less than three weeks and not more than three months. A 0·5 g sample of the moist soil was then suspended in 1·8 ml of 0·1 M tris–sodium borate buffer at pH 8·1 in a 12·5 ml tissue culture tube with a screw cap. Two millilitres of 0·002 M benzyloxycarbonyl phenylalanyl leucine were added to the tube which was tightly capped, submerged at 40°C in a water bath and shaken lengthwise (100–200 oscillations min^{-1}). Incubation times ranged from 10–60 min and were selected to ensure that less than 30% of the substrate was hydrolysed. At the appropriate time, the tube was cooled quickly to about 20°C and enzyme activity stopped by adding 0·2 ml of 5 M HCl. The reaction mixture, still in the culture tube, was centrifuged at 2000 g for 20 min; 2·0 ml samples of the supernatant were treated with a ninhydrin reagent and absorbances at 570 nm, determined with an EEL Automatic Colorimeter, were related to those of similarly-treated leucine standards.

14.2 Discussion

Air-drying resulted in considerable loss of protease activity (Ambrož, 1966, 1970; Speir and Ross, 1975) as did storing moist at room temperature (Ladd, 1972), at 4°C (Speir and Ross, 1975), or at −8°C (Ambrož, 1970).

Toluene decreased protease activity (Drobnik, 1960; Beck and Poschenrieder, 1963; Voets and Dedeken, 1964, 1965; Voets et al., 1965; Ambrož, 1966), and inhibited soil deaminase activity (Voets and Dedeken, 1964; Dedeken and Voets, 1965; Ambrož, 1966). Variation in the amount of toluene between 0·05 to 1 ml. g^{-1} of soil had no effect on activity (Ambrož, 1966). Protease activity observed in radiation-sterilized soil was not changed by addition of toluene (Voets and Dedeken, 1965). The same authors also demonstrated that activity was identical in both toluene-treated and radiation-sterilized soil. Gamma irradiation, at about the minimum sterilizing dosage, slightly reduced protease activity (Voets et al., 1965; Ladd, 1972); it hindered deaminase activity (McLaren et al., 1957; Voets et al., 1965) but not all of the enzymes participating in ammonification and nitrification (Popenoe and Eno, 1962; Eno and Popenoe, 1963; Cawse, 1968; Roberge and Knowles, 1968a).

According to Ladd (1972) and Ladd and Butler (1972), protease was less active in distilled water extracts than in tris–sodium borate buffer; it was optimally active at a pH around 8·0.

Hoffmann and Teicher (1957), Ladd (1972), and Ladd and Butler (1972) all found that maximal protease activity occurred at 55–60°C. Increasing the moisture content from 30 to 90% of the water-holding capacity of the soil increased activity of protease (Latypova, 1965) which was maximal when the reaction mixture was agitated (Ladd and Butler, 1972).

14.3. Extraction

It is possible to extract from soil proteins with a gelatinase activity (Fermi, 1910), with glycinase activity (Subrahmanyan, 1927), and with cathepsin-like and pepsin-like activity (Antoniani et al., 1954).

15. Respiratory activity

15.1. Assay (Roberge, 1971).

Fresh soil was ground to pass through a 4 mm sieve, placed in polythene bags, and stored at 5°C in moist conditions in darkness. Soil samples, equivalent to 5 g oven-dried at 105°C, were weighed into 125 ml Warburg flasks, brought to 60% water-holding capacity, and sterilized by 1·1 Mrad of gamma radiation. The samples were stored at 5°C immediately after sterilization and the respiratory assays started within 12 h. To measure the respiratory activity, 2 ml of 0·36 M KOH and a folded filter paper was placed in the centre well of the Warburg flask. After 1 h incubation at 20°C respiratory activity was measured during 6 h.

15.2. Discussion

Air-dried soil, subsequently remoistened, generally had a greater respiratory activity than the same soil which had not been air-dried (Vandecaveye and Katznelson, 1938; Gray and Taylor, 1939; Bodily, 1944; Koepf, 1954; Birch, 1958, 1959; Bernier, 1960; Soulides and Allison, 1961; Peterson, 1962; Ross and McNeilly, 1972). The longer the drying period the greater the increase (Birch, 1959; Soulides and Allison, 1961). Respiratory activity increased with time during the first few hours following re-moistening, decreased thereafter, and eventually became generally lower than that of a fresh soil (Bernier, 1960). Freezing also generally increased respiratory activity (Gooding and McCalla, 1945; Soulides and Allison, 1961; Mack, 1963; Ross, 1964; Ivarson and Sowden, 1970; Ross and McNeilly, 1972). The increase was greater as the temperature decreased to −100°C and as the moisture content approached the water-holding capacity (Mack, 1963). The length of storage in a freezing state had little effect (Ross and McNeilly, 1972). Respiratory activity was reduced when moist soil was stored at 30°C and 10°C (Bernier, 1960; Weetman, 1962; Ross and McNeilly, 1972; Tyler, 1974). This reduction was greater at 30°C than at 10°C, and at 10°C than at 5°C (Bernier, 1960).

Toluene reduced respiratory activity (Rogers, 1942; Drobnik, 1961; Halstead *et al.*, 1963; Ross, 1968) as did irradiation (Peterson, 1962; Popenoe and Eno, 1962; Gregers-Hansen, 1964; Roberge, 1971). The reduction was of short duration and respiratory activity became higher than that of irradiated soil when doses lower than the sterilizing dose were used (Dommergues, 1960; Popenoe and Eno, 1962). The reduction increased with an increase in the irradiation dose (Popenoe and Eno, 1962; Roberge, 1971), and with a longer storage after irradiation sterilization (Roberge, 1971).

Decreasing acidity from pH 4·0 to 7·0 increased respiratory activity (Weetman, 1962; Ivarson and Sowden, 1970) but there was no further increase at alkaline pHs (Weetman, 1962). Respiratory activity also increased with higher temperatures to a maximum determined by soil type (Bernier, 1960; Weetman, 1962; Wiant, 1967a; Ross, 1968; Ivarson and Sowden, 1970; Roberge, 1976).

According to Webley (1947), Rovira (1953), Gaarder (1957), Bernier (1960), Weetman (1962), Wiant (1967b), Ross and Boyd (1970), and Roberge (1976), respiratory activity was generally maximal at about 60% water-holding capacity. Respiration decreased as the size of the soil particles increased from 1 to 3 mm (Bernier, 1960).

16. Urease

16.1. Assay (*Roberge and Knowles, 1968*a).

Fresh soil was ground to pass through a 4 mm sieve, placed in polythene bags, and stored at 5°C in moist conditions in darkness. Samples, equivalent to 5 g oven-dried at 105°C, were weighed into 125 ml Erlenmeyer flasks, brought to 60% water-holding capacity and sterilized by 1·1 Mrad of gamma radiation. The samples were stored at 5°C immediately after sterilization and assays started within 12 h. The assays were carried out with 3500 ppm of urea-N over an 8 h reaction time at 20°C and at 60% water-holding capacity. Urea hydrolysis products were obtained by suspending the soil of each Erlenmeyer flask in 50 ml of 0·5 M K_2SO_4 acidified to pH 1·5 with 1 M H_2SO_4. Aliquots of the extract were made alkaline with NaOH and distilled steam.

16.2. Discussion

Urease activity was generally decreased by soil desiccation (Stojanovic, 1959; Vasilenko, 1962; McGarity and Myers, 1967; Skujiņš and McLaren, 1969; Zantua and Bremner, 1975b); by storage in moist state at room temperature (McGarity and Myers, 1967; Pancholy and Rice, 1972), at about 3°C (Roberge and Knowles, 1968b; Pancholy and Rice, 1972; Tyler, 1974; Speir and Ross, 1975), and at −10°C (Tagliabue, 1958). The reduction tended to increase with longer storage (Skujiņš and McLaren, 1969; Pancholy and Rice, 1972), and was less when stored at 4°C than at 21°C (Pancholy and Rice, 1972).

Generally urease activity decreased when toluene was added to soil (Rotini, 1935; Conrad, 1940; Anderson, 1962; Galstyan, 1965; Voets *et al.*, 1965; McGarity and Myers, 1967; Roberge, 1968; Bhavanandan and Fernando, 1970; Thente, 1970; Douglas and Bremner, 1971; Tabatabai and Bremner, 1972; Galstyan, 1974; Zantua and Bremner, 1975a). The decrease was greater with more toluene and with longer contact time (Galstyan, 1965; Roberge, 1968). It was also greater with larger soil moisture content, when buffer instead of water was used and with larger amounts of urea (Roberge, 1968). A decrease in urease activity was also noted when soils sterilized by radiation were treated with toluene (Voets and Dedeken, 1965; Roberge, 1968). The same authors demonstrated that urease activity was lower in toluene-treated than in irradiated soil. Irradiation also reduced urease activity (McLaren *et al.*, 1957; Voets and Dedeken, 1965; Roberge and Knowles, 1968b; Skujiņš and McLaren, 1969; Roberge, 1970; Thente, 1970). The reduction was of short duration and urease activity became greater than that of non-irradiated soil when doses lower than the sterilizing dose were used (McLaren *et al.*, 1957; Vela and Wyss, 1962; Roberge and Knowles, 1968b; Thente, 1970). The reduction increased with an increasing dose in excess of the sterilizing dose (McLaren *et al.*, 1957; Roberge and Knowles, 1968b; Thente, 1970). Urease activity decreased the higher the soil moisture content during irradiation (Thente, 1970), the lower the radiation flux (Roberge and Knowles, 1968b), the longer the storage and the higher the temperature during storage after irradiation (Roberge and Knowles, 1968a, b). Urease added to soil was partly inactivated by subsequent irradiation (Roberge, 1970).

In the presence of buffer, there was an increase in urease activity with rising pH up to a maximum generally around neutrality (Hofmann and Schmidt, 1953; McLaren *et al.*, 1957; Vasilenko, 1962; Galstyan and

Markosyan, 1965; Roberge and Knowles, 1968b; Delaune and Patrick, 1970; Tabatabai and Bremner, 1972). In the absence of phosphate and phthalate buffers, however, urease activity was higher than when the soil was buffered at either the initial or the final pH value of the unbuffered system (Roberge, 1968; Roberge and Knowles, 1968b); it was however lower than when buffered at around neutral pH (Roberge and Knowles, 1968b; Zantua and Bremner, 1975a). Varying the phosphate buffer concentration between 0·25 and 1·00 M had no effect on urease activity (Roberge and Knowles, 1968b). Tham (tris–H_2SO_4) buffer gave a higher urease activity than phosphate, citrate, and universal buffers (Tabatabai and Bremner, 1972).

An increase in the amount of urea added resulted in an increased urease activity (Fisher and Parks, 1958; Overrein and Moe, 1967; Roberge and Knowles, 1968a, b; Douglas and Bremner, 1971; Tabatabai and Bremner, 1972), until a maximum was reached, which was followed by a decrease (Overrein and Moe, 1967; Roberge and Knowles, 1968a, b). The increase, the maximum, and the decrease were different with different soils (Roberge and Knowles, 1968a, b). With appropriate substrate concentration urease activity increased linearly with incubation time (Galstyan, 1965; Roberge and Knowles, 1968b; Skujiņš and McLaren, 1969).

Urease activity increased from 5°C to 37°C. Above 37°C no further increase occurred (Fisher and Parks, 1958; Overrein and Moe, 1967; Roberge and Knowles, 1968b; Tabatabai and Bremner, 1972).

When the water content of the soil was increased to over 60% of its water-holding capacity, there was a decrease in urease activity (Roberge and Knowles, 1968b; Delaune and Patrick, 1970; Douglas and Bremner, 1971) High or low soil aeration had a minor depressing effect on urease activity (Overrein and Moe, 1967; Delaune and Patrick, 1970).

The reader is referred to Chapter 5 for greater detail.

16.3 Method of extraction (Burns et al., 1972a, b).

Twenty-five grams of soil were suspended in 250 ml of distilled water and sonicated in a water bath for 20 min. A mixture of salts, 0·95 M citrate, 0·05 M sodium dihydrogen phosphate, 0·05 M glycine, and 2·0 M sodium chloride, were slowly added to the soil suspension with constant stirring. After dissolution of the salts had occurred, the pH is adjusted to 6·3 and 1 ml of toluene added. The mixture was next agitated for 2 h at 10°C and then centrifuged at 18 000 *g* for 30 min. The supernatant soil was passed through a bacteriological filter and then through a Coors porcelain filter. The sedimented soil was extracted four more times, at pH 6·5, once buffered at 0·25 M in Na phosphate (0·01 M in glycine), and three times buffered at 0·05 M. A shaking time of 30 min used on each occasion. Each additional extract was also filtered but the five filtrates were dialysed separately, initially against running tap water at 15–16°C for three days, and then against distilled water at 4°C for one day. The resulting precipitates were concentrated by centrifugation at 18 000 g for 60 min. The enzyme-active residues were suspended in 0·001 M Na phosphate at pH 7·0 to a final volume of between 5 and 20 ml depending on their viscosity. Urease activity was determined using a modified Conway diffusion dish method (McLaren et al., 1957).

16.4. Discussion

Urease was not extractable with soil solutions except when fairly large amounts of urease were added (Roberge, 1970). It was however extractable with phosphate buffer (Briggs and Segal, 1963) and with tetra-sodium pyrophosphate (Lloyd, 1975).

17. Uricase

17.1. Method of extraction (Martin-Smith, 1963).

The fresh soil was sieved through a 2 mm sieve, and stored at −15°C. Portions of 10 g were placed in 250 ml stoppered conical flasks, followed by 15 ml of 0·1 M phosphate buffer solution at pH 7·0 and ten drops of toluene. The flasks were left, without shaking, for about 16 h at 24°C, and then filtered, 2 ml aliquots of the clear soil filtrate were incubated stationary in 250 ml conical flasks for approximately 24–48 h with sodium urate solution containing 1 mg uric acid ml^{-1} of water and toluene.

17.2. Discussion

When soil was stored at −15°C, uricase activity remained fairly constant over a period of one month but extracts stored under the same condition lost 50% of their activity (Martin-Smith, 1963). Toluene slightly decreased uricase activity, while inhibiting completely allantoinase and allantoicase activity (Durand, 1961).

Efficiency of extraction with 0·1 M buffers at pH 7·2 varied in the ascending order: borate < tris < carbonate < phosphate; there was no difference when the buffer concentration was varied from 0·1 M to 2 M, but at lower concentrations efficiency of extraction decreased (Martin-Smith, 1963). Both an increase in temperature (to a maximum of 30°C) and in the duration of the extraction were found to enhance the removal of uricase from soil (Martin-Smith, 1963).

III. Conclusions

In my opinion the best criterion for adopting a particular method of enzyme measurement and extraction is the degree of correlation obtained between enzymatic activities as they occur under laboratory conditions and under natural conditions (but see Chapter 8, for an alternative viewpoint). Further expansion of research on methods for assessing the relationships mentioned above is needed.

Through field experimentation much has been learned from different soil, climatic and management conditions concerning the factors that should be taken into account in enzyme measurement and extraction in the laboratory, especially in regard to substrate concentration, temperature, moisture and aeration. Through laboratory experimentation, on the other hand, much has been learned concerning the way these factors influence enzyme activity and much has been discovered about the factors associated with the handling and treatment of soil before performing enzymatic tests: such as sieving, storing and inhibiting microbial growth. Although still more needs to be known before enzyme assays are both reliable and agriculturally relevant some broad conclusions can be drawn:

(1) Larger size soil particles may increase certain enzymatic activities and decrease others but the generally-used 2 mm sieve for mineral soil and 4 mm sieve for organic appear adequate.

(2) Enzymatic activities are decreased by storage, more if the soil is in an air-dried state then in a moist state, and more at room temperature or frozen than at 5°C. Fresh soil should be used if at all possible, if not soil should be stored moist at 5°C.

(3) Toluene generally reduces enzyme activities. However, regardless of its concentration and its contact time with soil, toluene does not sterilize the soil totally and only limits the growth of microbes in an inconsistent way. The surviving microbes can both synthesize enzymes and consume the products of enzymatic reactions. Some of the decrease observed in enzyme activities in the presence of toluene is possibly due to a lower synthesis of enzyme and (or) to an uptake of products of enzymatic reactions by the surviving microorganisms; some may even be attributable to the inhibitory effect of toluene on the enzyme. These shortcomings show that toluene does not meet the stringent requirements for enzymatic studies of soil and should no longer be used in this context.

(4) Radiation usually reduces enzyme activities. It renders soil aseptic without totally destroying or inhibiting enzymes whilst inducing negligible soil chemical and physical changes. Radiation thus permits subsequent study of soil enzymes. Some of the decrease observed in enzyme activities

of radiation-sterilized soil is possibly due to a lack of synthesis of new enzymes, to the inactivation of the enzymes which occurs during the period between the irradiation and the enzymatic assays and to some damaging effect of radiation on soil enzymes. Soil sample size, radiation flux, soil temperature and moisture during and after irradiation, time lapse between the sterilization and enzymatic tests, and the use of the exact dose of radiation necessary to sterilize the soil must be carefully chosen and examined before results are interpreted. Even if the radiation sterilization of soil is efficiently effected it is not certain that the results obtained will correlate any better with enzymatic activities occurring under natural conditions. Thus the use of radiation sterilization in enzymatic tests of soil is questionable.

(5) Are buffers necessary in soil enzymatic assays and if so, at what pH should the soil be buffered, with what buffer and at what concentration? This is a problem which has also been discussed elsewhere in this volume. Maximum enzymatic activities often occur in buffered soil at a pH around 7·0. They tend to occur at a lower pH when the original pH of the soil is acid and at a higher pH when the original pH of the soil is alkaline. The activities then measured are often superior to those observed without buffer. When the soils are buffered at another pH, the enzymatic activities are, on the contrary, often inferior to those observed without buffer. The inferiority is more or less related to the difference between the soil pH and the buffer pH and appears minimum for a soil with a pH close to neutrality. No buffer and no concentration seem to give higher activities for all enzymes and all soils. It is evident from soil enzyme assays carried out with buffer that for an adequate use of buffer and for a precise comparison of enzymatic activities of various soils, the maximum enzymatic activities for each soil need to be determined by using different buffers, different pHs, and different concentrations. If this is not done, it may be preferable to dispense with the use of buffers in enzyme assays altogether.

References

AMBROŽ Z. (1966). Some of the reasons for differences in occurrence of proteases in soils. *Rostl. Výroba* **39**, 1203–1210.

AMBROŽ Z. (1970). Effect of drying and wetting on the enzymatic activity of soils. *Rostl. Vyroba* **16**, 869–876.

ANDERSON J. R. (1962). Urease activity, ammonia volatilization and related microbiological aspects in some South African soils. *Proc. 36th Congr. S. Afr. Sug. Technol. Ass.* 97–105.

ANTONIANI C., MONTANARI T. and CAMORIANO A. (1954). Soil enzymology. I. Cathepsin-like activity. A preliminary note. *Ann. Fac. Agr. Univ. Milano* **3**, 99–101.

BARANOVSKAYA A. V. (1954). Catalase activity in some soils of the forest and steppe zone. *Pochvovedenie* (11), 41–48.

BARTHA R. (1971). Altered propanil biodegradation in temporarily air-dried soil. *J. Agric. Fd Chem.* **19**, 394–395.

BARTHA R. and BORDELEAU L. (1969). Cell-free peroxidases in soil. *Soil Biol. Biochem.* **1**, 139–143.

BECK T. and POSCHENRIEDER H. (1963). Experiments on the effect of toluene on the soil microflora. *Pl. Soil* **18**, 346–357.

BERNIER B. (1960). Observations sur le métabolisme respiratoire de quelques humus forestiers. *Fonds Rech. For. Univ. Laval Contr.* **5**.

BHAVANANDAN V. P. and FERNANDO V. (1970). Studies on the use of urea as a fertilizer for tea in Ceylon. 2. Urease. *Tea Quart,* **41**, 94–106.

BIRCH H. F. (1958). The effect of soil drying on humus decomposition and nitrogen availability. *Pl. Soil* **10**, 9–31.

BIRCH H. F. (1959). Further observations on humus decomposition and nitrification. *Pl. Soil* **11**, 262–286.

BODILY H. L. (1944). The activity of microorganisms in the transformation of plant materials in soil under various conditions. *Soil Sci.* **57**, 341–349.

BORDELEAU L. M. and BARTHA R. (1972a). Biochemical transformations of herbicide-derived anilines in culture medium and in soil. *Can. J. Microbiol.* **18**, 1857–1864.

BORDELEAU L. M. and BARTHA R. (1972b). Biochemical transformations of herbicide-derived anilines: purification and characterization of causative enzymes. *Can. J. Microbiol.* **18**, 1865–1871.

BOWMAN B. T., THOMAS R. L. and ELRICK D. E. (1967). The movement of phytic acid in soil cores. *Soil Sci. Soc. Am. Proc.* **31**, 477–481.

BREMNER J. M. and SHAW K. (1958). Denitrification in soil. *J. Agric. Sci., Camb.* **51**, 40–52.

BRIGGS M. H. and SEGAL L. (1963). Preparation and properties of a free soil enzyme. *Life Sci.* **2**, 69–72.

BURGE W. D. (1973). Transformation of propanil-derived 3, 4-dichloroaniline in soil to 3, 3′, 4, 4′-tetrachloroazobenzene as related to soil peroxidase activity. *Soil Sci. Soc. Am. Proc.* **37**, 392–395.

BURNS R. G., EL-SAYED M. H. and MCLAREN A. D. (1972a). Extraction of an urease-active organo-complex from soil. *Soil Biol. Biochem.* **4**, 107–108.

BURNS R. G., PUKITE A. H. and MCLAREN A. D. (1972b). Concerning the location and persistence of soil urease. *Soil Sci. Soc. Soc. Am. Proc.* **36**, 308–311.

CADY F. B. and BARTHOLOMEW W. V. (1960). Sequential products of anaerobic denitrification in Norfolk soil material. *Soil Sci. Soc. Am. Proc.* **24**, 477–482.

CASIDA L. E. Jr., KLEIN D. A. and SANTORO T. (1964). Soil dehydrogenase activity. *Soil Sci.* **98**, 371–376.

CAWSE P. A. (1968). Effects of gamma radiation on accumulation of mineral nitrogen in fresh soils. *J. Sci. Fd Agric.* **19**, 395–398.

CAWSE P. A. and CORNFIELD A. H. (1969). The reduction of ^{15}N-labelled nitrate to nitrite by fresh soils following treatment with gamma radiation. *Soil Biol. Biochem.* **1**, 267–274.

CAWSE P. A. and CORNFIELD A. H. (1971). Factors affecting the formation of nitrite in γ-irradiated soils and its relationship with denitrifying potential. *Soil Biol. Biochem.* **3**, 111–120.

CAWSE P. A. and CRAWFORD D. V. (1967). Accumulation of nitrite in fresh soils after gamma irradiation. *Nature* **216**, 1142–1143.

CERNÁ S. (1972). Effect of some ecological factors on dehydrogenase activity in soil. *Rostl. Výroba* **18**, 101–106.

CHALVIGNAC M-A. (1971). Stabilité et activité d'un système enzymatique dégradant le tryptophane dans divers types de sols. *Soil Biol. Biochem.* **3**, 1–7.

CHALVIGNAC M-A. and MAYAUDON J. (1971). Extraction and study of soil enzymes metabolizing tryptophan. *Pl. Soil* **34**, 25–31.

CONRAD J. P. (1940). Hydrolysis of urea in soils by thermolabile catalysis. *Soil Sci.* **49**, 253–263.

COOPER P. J. M. (1972). Arylsulphatase activity in Northern Nigerian soils. *Soil Biol. Biochem.* **4**, 333–337.

DEDEKEN M. and VOETS J. P. (1965). Recherches sur le métabolisme des acides aminés dans le sol. I. Le métabolisme de glycine, alanine, acide aspartique et acide glutaminique. *Ann. Inst. Pasteur* **109**, 103–111.

DELAUNE R. D. and PATRICK W. H. (1970). Urea conversion to ammonia in water-logged soils. *Soil Sci. Soc. Am. Proc.* **34**, 603–607.

DOMMERGUES Y. (1960). Influence du rayonnement infra-rouge et du rayonnement solaire sur la teneur en azote minéral et sur quelques caractéristiques biologiques des sols. *Agron. Trop.* (Nogent-sur-Marne) **15**, 381–389.

DOMOCOS T., MURGASEANU P. and ROSA V. (1966). Saccharase activity of several soils in Banat. *Symp. Soil Biol., Cluj* **1966**, 113–124.

DOUGLAS L. A. and BREMNER J. M. (1971). A rapid method of evaluating different compounds as inhibitors of urease activity in soils. *Soil Biol. Biochem.* **3**, 309–315.

DROBNIK J. (1960). Primary oxidation of organic matter in the soil. I. The form of respiration curves with glucose as the substrate. *Pl. Soil* **12**, 199–211.

DROBNIK J. (1961). On the role of toluene in the measurements of the activity of soil enzymes. *Pl. Soil* **14**, 94–95.

DROBNIKOVA V. (1961). Factors influencing in determination of phosphates in soil. *Folia Microbiol., Prague* **6**, 260–267.

DURAND G. (1961). Sur la dégradation des bases puriques et pyrimidiques dans le sol: dégradation aérobie de l'acide urique. *C. R. Acad. Sci. Paris* **252**, 1687–1689.

DURAND G. (1965). Les enzymes dans le sol. *Rev. Écol. Biol. Sol.* **2**, 141–205.

ENO C. F. and POPENOE H. (1963). The effect of gamma radiation on the availability of nitrogen and phosphorus in soil. *Soil Sci. Soc. Am. Proc.* **27**, 299–301.

FERMI C. (1910). Sur la présence des enzymes dans le sol, les eaux, et dans les poussières. *Zentrabl. Bakteriol. Parasitenkd.*, Abt. II **26**, 330–334.

FISHER W. B. Jr. and PARKS W. L. (1958). Influence of soil temperature on urea hydrolysis and subsequent nitrification. *Soil Sci. Soc. Am. Proc.* **22**, 247–248.

GAARDER T. (1957). Studies in soil respiration in Western Norway, the Bergen district. *Univ. i Bergen Skrifter* (3).

GALSTYAN A. S. (1965). A method of determining the activity of hydrolytic enzymes in soil. *Soviet Soil Sci.* **2**, 170–175.

GALSTYAN A. S. (1974). "Fermentativnaya Aktivnost' Pochv Armenii." Ayastan, Erevan.

GALSTYAN A. S. and MARKOSYAN L. V. (1965). Study of the optimum pHs of soil enzymes. *Izv. Akad. Nauk. Armyan. SSR. Biol. Nauki* **18**, (7) (1), 21—27.

364   M. R. ROBERGE

GALSTYAN A. S. and MARKOSYAN L. V. (1966). Determination of nitrate-reductase activity of soil. *Dokl. Akad. Nauki. Arm. SSR* **43**, 147–150.

GELLER I. A. and DOBROTVORSKA O. M. (1960). Phosphatase activity of soils. *Ukrain. Akad. Skog. Nauki* **1**, 38–42.

GETZIN L. W. and ROSEFIELD I. (1968). Organophosphorus insecticide degradation by heat-labile substances in soil. *J. Agric. Fd Chem.* **16**, 598–601.

GETZIN L. W. and ROSEFIELD I. (1971). Partial purification and properties of a soil enzyme that degrades the insecticide malathion. *Biochim. Biophys. Acta* **235**, 442–453.

GILLIAM J. W. and SAMPLE E. G. (1968). Hydrolysis of pyrophosphate in soils: pH and biological effects. *Soil Sci.* **106**, 352–357.

GOIAN M. (1971). Activitatea fosfatazică si microorganismele solului. *Cerc. Biol. în partea de vest a României* **1**, 423–426.

GOODING T. H. and MCCALLA T. M. (1945). Loss of carbon dioxide and ammonia from crop residues during decomposition. *Soil Sci. Soc. Am. Proc.* **10**, 185–190.

GRAY P. H. H. and TAYLOR C. B. (1939). The aerobic decomposition of glucose in podzol soils. *Can. J. Res.* **17**, 109–124.

GREENWOOD D. J. (1962). Nitrification and nitrate dissimilation in soil. II. Effect of oxygen concentration. *Pl. Soil* **17**, 378–391.

GREENWOOD D. J. and GOODMAN D. (1964). Oxygen diffusion and aerobic respiration in soil spheres. *J. Sci. Fd Agric.* **15**, 579–588.

GREGERS-HANSEN B. (1964). Decomposition of diethylstilboestrol in soil. *Pl. Soil* **20,** 50–58.

HALSTEAD R. L. (1964). Phosphatase activity of soils as influenced by lime and other treatments. *Can. J. Soil Sci.* **44**, 137–144.

HALSTEAD R. L., LAPENSEE J. M. and IVARSON K. C. (1963). Mineralization of soil organic phosphorus with particular reference to the effect of lime. *Can. J. Soil Sci.* **43**, 97–106.

HOFFMANN G. (1959). Distribution and origin of some enzymes in soil. *Z. PflanzenErnähr. Düng. Bodenkd.* **85**, 97–104.

HOFFMANN G. and TEICHER K. (1957). Enzyme systems of our arable soils. VII. Proteases. *Z. PflanzenErnähr. Düng. Bodenkd.* **77**, 243–251.

HOFMANN E. (1963). Die Analyse von Enzymen in Boden. *In* "Moderne Methoden des Pflanzenanalyse", Vol. 6 (K. Peach, M. V. O. Tracy and H. F. Linskens, Eds), pp. 416–423, Springer, Berlin.

HOFMANN E. and HOFFMANN G. (1961). The reliability of E. Hofmann's methods for determining enzyme activity in soil. *Pl. Soil* **14**, 96–99.

HOFMANN E. and SCHMIDT W. (1953). Uber das Enzymsystem unserer Kulturböden. II. Urease. *Biochem. Z.* **324**, 125–127.

HOFMANN E. and SEEGERER A. (1951). Uber das Enzymsystem unserer Kulturböden. I. Saccharase. *Biochem. Z.* **322**, 174–179.

HOSSNER L. R. and MELTON J. R. (1970). Pyrophosphate hydrolysis of ammonium, calcium, and calcium ammonium pyrophosphates in selected Texas soils. *Soil Sci. Soc. Am. Proc.* **34**, 801–805.

HOSSNER L. R. and PHILLIPS D. P. (1971). Pyrophosphate hydrolysis in flooded soil. *Soil Sci. Soc. Am. Proc.* **35**, 379–383.

IVARSON K. C. and SOWDEN F. J. (1970). Effect of frost action and storage of soil at freezing temperatures on the free amino acids, free sugars and respiratory activity of soil. *Can. J. Soil Sci.* **50**, 191–198.

JACKMAN R. H. and BLACK C. A. (1951). Hydrolysis of iron, aluminium calcium and magnesium inositol phosphates by phytase at different pH values. *Soil Sci.* **72**, 261–265.

JACKMAN R. H. and BLACK C. A. (1952a). Phytase activity in soils. *Soil Sci.* **73**, 117–125.

JACKMAN R. H. and BLACK C. A. (1952b). Hydrolysis of phytate phosphorus in soils. *Soil Sci.* **73**, 167–171.

JOHNSON J. L. and TEMPLE K. L. (1964). Some variables affecting the measurement of catalase activity in soil. *Soil Sci. Soc. Am. Proc.* **28**, 207–209.

KELLEY W. D. and RODRIGUEZ-KABANA R. (1975). Effects of potassium azide on soil microbial populations and soil enzymatic activities. *Can. J. Microbiol.* **21**, 565–570.

KISS S. (1958). Talajenzimek. *In* "Talajtan" (M. J. Csapo, Ed.), pp. 491–622, Agro-Silvica, Bucharest.

KISS S. (1961). Über die Anwesenheit von Lävansucrase in Boden. *Naturwissenshaften.* **48**, 700.

KISS S. (1964). Contributions to the study of specific substrate action on the production of soil enzymes by microorganisms. *Trans. 8th Int. Congr. of Soil Sci., Bucharest* **3**, 705–710.

KISS S. and BOARU M. (1965). Methods for the determination of dehydrogenase activity in soil. *Symp. on Methods in Soil Biology, Bucharest,* 137–143.

KISS S., BOSICA I. and MELIUSZ P. (1962a). Eficacitatea toluenului ca agent antiseptic in determinarea activitatii enzimelor din sol. *Stud. Univ. Babes-Bolyia, Ser. Biol.* **2**, 65–70.

KISS S., BOSICA I. and POP M. (1962b). Despre degradarea enzimatica a licheninei in sol. *Contrib. Bot. Univ. Babes-Bolyai Cluj,* **1962**, 335–340.

KISS S., BOARU M. and STOIATA M. (1969). Enzyme activity in vineyard soils. *Revue Roumaine de Biologie, Série de Botanique* **14**, 127–132.

KISS S., DRĂGAN-BULARDA M. and RĂDULESCU D. (1971). Semnificatia biologica a enzimelor acumulate in sol. *Contrib. Bot. Univ. Babes-Bolyai, Cluj* **1971**, 377-397.

KISS S., DRĂGAN-BULARDA M. and KHAZIEV M. (1972). Influenta cloromicetinei asupra activitatii unor oligaze din sol. *Lucr. Conf. Nat. Stiinta Solului, Iasi* **1970**, 451–462.

KISS S., STEFANIC G. and DRĂGAN-BULARDA M. (1974). Soil enzymology in Romania (Part I). *Contrib. Bot. Univ. Babes-Bolyai Cluj* **1974**, 207–209.

KISS S., DRĂGAN-BULARDA M. and RĂDULESCU D. (1975). Biological significance of enzymes accumulated in soils. *Adv. Agron.* **27**, 25–87.

KLEINERT H. (1962). Determination of the invertase activity of soil, and the decrease of activity under constant conditions of storage. *Albrecht-Thaer-Orch.* **6**, 477–484.

KOEPF H. (1954). Die biologische Aktivität des Bodens und ihre experimentelle Kennzeichnung. *Z. PflanzenErnähr. Düng. Bodenkd.* **64**, 138–146.

KOZLOV K. A. and KISLITSINA V. P. (1967). Effect of ecological conditions on enzymatic activity of soils and micro-organisms. *Stiinta Solului* **5**, 217–226.

KROLL L., KRAMER M. and LÖRINCZ E. (1955). Enzyme analysis with phenylphosphate as a substrate applied in soil and fertilizer examinations. *Agrokem. Talajt.* **4**, 173–182.

KUPREVICH V. F. (1951). The biological activity of soils and methods for its determination. *Dokl. Akad. Nauki SSSR* **79**, 863–866.

LADD J. N. (1972). Properties of proteolytic enzymes extracted from soil. *Soil Biol. Biochem.* **4**, 227–237.

LADD J. N. and BUTLER J. H. A. (1972). Short-term assays of soil proteolytic enzyme activities using proteins and dipeptide derivatives as substrates. *Soil Biol. Biochem.* **4**, 19–30.

LATYPOVA R. M. (1965). Effect of environmental conditions on activity of soil enzymes. *Trudy Beloruss. Sel'skokhoz Akad.* **37**, 60–65.

LATYPOVA R. M. and KURBATOV I. M. (1961). The fermentative activity of peat-treated soils. *Trudy Beloruss. Sel'skokhoz. Akad.* **34**, 95–101.

LENHARD G. (1956). Dehydrogenaseaktivität des Bodens als Mass fur die Mikroorganismentatigkeit im Bodens. *Z. Pflanzenernähr. Düng. Bodenkd.* **73**, 1–11.

LLOYD A. B. (1975). Extraction of urease from soil. *Soil Biol. Biochem.* **7**, 357–358.

MACK A. R. (1963). Biological activity and mineralization of nitrogen in three soils as induced by freezing and drying. *Can. J. Soil Sci.* **43**, 316–324.

MAHENDRAPPA M. K. and SMITH R. L. (1967). Some effects of moisture on denitrification in acid and alkaline soils. *Soil Sci. Soc. Am. Proc.* **31**, 212–215.

MARKUS L. (1955). The determination of carbohydrates in vegetable matter with anthrone reagent. II. Measurement of cellulase activity in soils and manures. *Agrokem. Talajt.* **4**, 207–216.

MARTIN-SMITH M. (1963). Uricolytic enzymes in soil. *Nature* **197**, 361–362.

MAYAUDON J. and SARKAR J. M. (1974a). Etude des diphénol oxydases extraites d'une litière de forêt. *Soil Biol. Biochem.* **6**, 269–274.

MAYAUDON J. and SARKAR J. M. (1974b). Chromatographie et purification des diphénol oxydases du sol. *Soil Biol. Biochem.* **6**, 275–285.

MAYAUDON J. and SARKAR J. M. (1975). Laccases de *Polyporus versicolor* dans le sol et la litière. *Soil Biol. Biochem.* **7**, 31–34.

MAYAUDON J., EL HALFAWI M. and BELLINCK C. (1973a). Décarboxylation des acides aminés aromatiques-1-^{14}C par les extraits de sol. *Soil Biol. Biochem.* **5**, 355–367.

MAYAUDON J., EL HALFAWI M. and CHALVIGNAC M. A. (1973b). Propriétés des diphénol oxydases extraites des sols. *Soil Biol. Biochem.* **5**, 369–383.

MCGARITY J. W. and MYERS M. G. (1967). A survey of urease activity in soils of northern New South Wales. *Pl. Soil* **27**, 217–238.

MCLAREN A. D., RESHETKO L. and HUBER W. (1957). Sterilization of soil by irradiation with an electron beam and some observations on soil enzyme activity. *Soil. Sci.* **83**, 497–502.

MCLAREN A. D., LUSE R. A. and SKUJIŅŠ J. J. (1962). Sterilization of soil by irradiation and some further observations on soil enzyme activity. *Soil Sci. Soc. Am. Proc.* **26**, 371–377.

NIZOVA A. A. (1969). Saccharase activity in derno-podzolic heavy loamy soil. *Mikrobiologiya* **38**, 336–339.

NÖMMIK H. (1956). Investigations on denitrification in soil. *Acta Agric. Scand.* 195–228.

OVERREIN L. N. and MOE P. G. (1967). Factors affecting urea hydrolysis and ammonia volatilization in soil. *Soil Sci. Soc. Am. Proc.* **31**, 57–61.

PANCHOLY S. K. and LYND J. Q. (1973). Interactions with soil lipase activation and inhibition. *Soil Sci. Am. Proc.* **37**, 51–52.

PANCHOLY S. K. and RICE E. L. (1972). Effect of storage conditions on activities of urease, invertase, amylase, and dehydrogenase in soil. *Soil Sci. Soc. Am. Proc.* **36**, 536–537.

PANCHOLY S. K. and RICE E. L. (1973). Soil enzymes in relation to old field succession: amylase, cellulase, invertase, dehydrogenase, and urease. *Soil Sci. Soc. Am. Proc.* **37**, 47–50.

PETERSON G. H. (1962). Respiration of soil sterilized by ionizing radiations. *Soil Sci.* **94**, 71–74.

PETERSON N. V. (1967). Dehydrogenase activity in the soil as a mirror of the activity of its microflora. *Mikrobiologiya* **36**, 518–525.

PETERSON N. V. and ASTAF'EVA E. V. (1962). Method of determining saccharase activity of soil. *Mikrobiologiya* **31**, 918–922.

PILET P. E. and CHALVIGNAC M-A. (1970). Effet d'un extrait enzymatique tellurique sur la croissance et le catabolisme auxinique des racines du *Lens culinaris. Ann. Inst. Pasteur* **118**, 349–355.

POKORNÁ V. (1964). Method of determining the lipolytic activity of upland and lowland peats and muds. *Soviet Soil Sci.* **1**, 85–87.

PONOMAREVA N. S., PIROGOVA T. I., NIKULINA V. D. and ANTONOVA V.K. (1972). Phosphatase activity of high solonetzes of the wooded steppe of the Omsk region. *Agrokhimiya* (6), 102–108.

POPENOE H. and ENO C. F. (1962). The effect of gamma radiation on the microbial population of the soil. *Soil Sci. Soc. Am. Proc.* **26**, 164–167.

RACZ G. J. and SAVANT N. K. (1972). Pyrophosphate hydrolysis in soil as influenced by flooding and fixation. *Soil Sci. Soc. Am. Proc.* **36**, 678–682.

RADULESCU D. (1972). Enzymological analysis of the peat in Avrig, Sibiu district. *Third Symp. Soil Biol., Bucharest* **1972**, 113–118.

RAMÍREZ-MARTÍNEZ J. R. and MCLAREN A. D. (1966a). Determination of soil phosphatase activity by a fluorimetric technique. *Enzymologia* **30**, 243–253.

RAMÍREZ-MARTÍNEZ J. R. and MCLAREN A. D. (1966b). Some factors influencing the determination of phosphatase activity in native soils and in soils sterilized by irradiation. *Enzymologia* **31**, 23–28.

ROBERGE M. R. (1968). Effects of toluene on microflora and hydrolysis of urea in a black spruce humus. *Can. J. Microbiol.* **14**, 999–1003.

ROBERGE M. R. (1970). Behavior of urease added to unsterilized, steam-sterilized, and gamma radiation-sterilized black spruce humus. *Can. J. Microbiol.* **16**, 865–870.

ROBERGE M. R. (1971). Respiration of a black spruce humus sterilized by heat or irradiation. *Soil Sci.* **111**, 124–128.

ROBERGE M. R. (1976). Respiration rates for determining the effects of urea on the soil-surface organic horizon of a black spruce stand. *Can. J. Microbiol.* **22**, 1328–1335.

ROBERGE M. R. and KNOWLES R. (1968a). Urease activity in a black spruce humus sterilized by gamma radiation. *Soil Sci. Soc. Am. Proc.* **32**, 518–521.

ROBERGE M. R. and KNOWLES R. (1968b). Factors affecting urease activity in a black spruce humus sterilized by gamma radiation. *Can. J. Soil Sci.* **48**, 355–361.

ROGERS H. T. (1942). Dephosphorylation of organic phosphorus compounds by soil catalysts. *Soil Sci.* **54**, 439–446.

ROGOBETE G., FAUR A. and ROSCA V. (1972). Cercetarea activitatii enzimatice a unei lacovisti recent luata in cultura (Janu-Mare, Timis). *Lucr. Conf. Nat. Stiinta Solului, Iasi* **1970**, 463–473.

ROSS D. J. (1964). Effects of low-temperature storage on the oxygen uptake of soil. *Nature* **204**, 503–504.

ROSS D. J. (1965a). A seasonal study of oxygen uptake of some pasture soils and activities of enzymes hydrolysing sucrose and starch. *J. Soil Sci.* **16**, 73–85.

ROSS D. J. (1965b). Effects of air-dry, refrigerated, and frozen storage on activities of enzymes hydrolysing sucrose and starch. *J. Soil Sci.* **16**, 86–94.

ROSS D. J. (1968). Some observations on the oxidation of glucose by enzymes in soil in the presence of toluene. *Pl. Soil* **28**, 1–11.

ROSS D. J. (1970). Effects of storage on dehydrogenase activities of soils. *Soil Biol. Biochem.* **2**, 55–61.

ROSS D. J. (1971). Some factors influencing the estimation of dehydrogenase activities of some soils under pasture. *Soil Biol. Biochem.* **3**, 97–110.

ROSS D. J. (1972). Effects of freezing and thawing of some grassland topsoils on oxygen uptake and dehydrogenase activities. *Soil Biol. Biochem.* **4**, 115–117.

ROSS D. J. and BOYD I. W. (1970). Influence of moisture and aeration on oxygen uptakes in Warburg respiratory experiments with litter and soil. *Pl. Soil* **33**, 251–256.

ROSS D. J. and MCNEILLY B. A. (1972). Effects of storage on oxygen uptakes and dehydrogenase activities of beech forest litter and soil. *NZ Jl Sci.* **15**, 453–462.

ROTINI O. T. (1935). La transformazione enzimatica dell'-urea nell terreno. *Ann. Labor. Ric. Ferm. Spallanzani Milano* **3**, 143–154.

ROVIRA A. D. (1953). Use of the Warburg apparatus in soil metabolism studies. *Nature* **172**, 29–30.

SCHEFFER VON F. and TWACHTMANN R. (1953). Erfahrungen mit der Enzymmethod nach Hofmann. *Z. PflanzenErnähr. Düng. Bodenkd.* **62**, 158–171.

SCHOORL N. (1920). La signification de la réfraction spécifique pour la chimie analytique. *Rec. Trav. Chim. Pays-Bas* **39**, 594–600.

SHAFFER P. A. and HARTMANN A. F. (1921). The iodometric determination of copper and its use in sugar analysis. II. Methods for the determination of reducing sugars in blood, urine, milk, and other solutions. *J. Biol. Chem.* **45**, 365–390.

SKUJIŅŠ J. (1967). Enzymes in soil. *In* "Soil Biochemistry" (A. D. McLaren and G. H. Peterson, Eds), pp. 371–414, Marcel Dekker, New York.

SKUJIŅŠ J. J. and MCLAREN A. D. (1968). Persistence of enzymatic activities in stored and geologically preserved soils. *Enzymologia* **34**, 213–225.

SKUJIŅŠ J. J. and MCLAREN A. D. (1969). Assay of urease activity using ^{14}C-urea in stored, geologically preserved, and in irradiated soils. *Soil Biol. Biochem.* **1**, 89–99.

SKUJIŅŠ J. J., BRAAL L. and MCLAREN A. D. (1962). Characterization of phosphatase in a terrestrial soil sterilized with an electron beam. *Enzymologia* **25**, 125–133.

SOREANU I., DOMOCOS T. and FAUR A. (1970). Contributii la studiul enzimatic al saraturilor din Banat. *Microbiologia* **1**, 477–482.

SOULIDES D. A. and ALLISON F. E. (1961). Effect of drying and freezing soils on carbon dioxide productions, available mineral nutrients, aggregation, and bacterial population. *Soil Sci.* **91**, 291–298.

SPEIR T. W. and ROSS D. J. (1975). Effects of storage on the activities of protease, urease, phosphatase, and sulphatase in three soils under pasture. *NZ Jl Sci.* **18**, 231–237.

STEFANIC G. (1972). The spectrophotometrical determination of the soil saccharase activity by the dinitrosalicylic acid reagent. *Third Symp. Soil Biol., Bucharest* **1972**, 146–148.

STOJANOVIC B. J. (1959). Hydrolysis of urea in soil as affected by season and by added urease. *Soil Sci.* **88**, 251–255.

SUBRAHMANYAN V. (1927). Biochemistry of water-logged soils. Part II. The presence of deaminase in water-logged soils and its role in the production of ammonia. *J. Agric. Sci.* **17**, 449–467.

SUCIU M. (1970). Studii asupra cineticii fosfomono-esterazci din sol. *Lucr. de Diploma. Univ. Babes-Bolyai, Cluj.*

TABATABAI M. A. and BREMNER J. M. (1970a). Arylsulfatase activity of soils. *Soil Sci. Soc. Am. Proc.* **34**, 225–229.

TABATABAI M. A. and BREMNER J. M. (1970b). Factors affecting soil arylsulfatase activity. *Soil Sci. Soc. Am. Proc.* **34**, 427–429.

TABATABAI M. A. and BREMNER J. M. (1972). Assay of urease activity in soils. *Soil Biol. Biochem.* **4**, 479–487.

TAGLIABUE L. (1958). Cryoenzymological research on urease in soils. *Chimica, Milano* **34**, 488–491.

THALMANN A. (1968). Zur Methodik des Bestimmung der Dehydrogenase aktivität im Boden mittels Triphenyltetrazoliumchlorid (TTC). *Landw. Forsch.* **21**, 249–258.

THENTE B. (1970). Effects of toluene and high energy radiation on urease activity in soil. *Lantbr. Högsk. Annlr* **36**, 401–418.

THOMSON L. M. and BLACK C. A. (1948). The effect of temperature on the mineralization of soil organic phosphorus. *Soil Sci. Soc. Am. Proc.* **12**, 323–326.

THOMSON E. J. and BLACK C. A. (1970). Changes in extractable organic phosphorus in soil in the presence and absence of plants. III. Phosphatase effects. *Pl. Soil* **32**, 335–348.

TOMESCU E. (1966). L'influence du pH sur l'activité de la cellobiase du sol. *Symp. Soil Biol., Cluj,* **1966**, 101–104.

TOMESCU E. (1970). Contributii la metoda de determinare a celulazei din sol. *Microbiol. Lucr. Conf. Nat. Microbiol. Gen. Appl., Bucharest,* **1968**, 509–513.

TYLER G. (1974). Heavy metal pollution and soil enzymatic activity. *Pl. Soil* **41**, 303–311.

VANDECAVEYE S. C. and KATZNELSON H. (1938). Microbial activity in soil. V. Microbial activity and organic matter transformation in Palouse and Helmer soils. *Soil Sci.* **46**, 139–167.

VASILENKO Y. S. (1962). Urease activity in the soil. *Soviet Soil Sci.* **11**, 1267–1272.

VELA G. R. and WYSS O. (1962). The effect of gamma radiation on nitrogen transformation in soil. *Bact. Proc.* **62**, 24.

VOETS J. P. and DEDEKEN M. (1964). Recherches sur les phénomènes biologiques de la protéolyse dans le sol. *Ann. Inst. Pasteur* **107**, suppl. 3, 320–329.

VOETS J. P. and DEDEKEN M. (1965). Influence of high-frequency and gamma radiation on the microflora and enzymes of soil. *Meded. Landbouhogesch, Opzoekingsta Staat Gent* **30,** 2037–2049.

VOETS J. P., DEKEDEN M. and BESSEMS E. (1965). The behaviour of some amino acids in gamma irradiated soils. *Naturwissenshaften.* **52,** 476.

WEBLEY D. M. (1947). A technique for the study of oxygen availability to microorganisms in soil and its possible use as an index of soil aeration. *J. Agric. Sci.* **37,** 249–256.

WEETMAN G. F. (1962). Nitrogen relations in a black spruce (*Picea mariana* Mill.) stand subject to various fertilizers and soil treatments. *Woodl. Res. Index Pulp. Pap. Res. Inst. Can.* **129.**

WIANT H. V. Jr. (1967a). Influence of temperature on the rate of soil respiration. *J. For.* **65,** 489–490.

WIANT H. V. Jr. (1967b). Influence of moisture content on soil respiration. *J. For.* **65,** 902–903.

ZANTUA M. I. and BREMNER J. M. (1975a). Comparison of methods of assaying urease activity in soils. *Soil Biol. Biochem.* **7,** 291–295.

ZANTUA M. I. and BREMNER J. M. (1975b). Preservation of soil samples for assay of urease activity. *Soil Biol. Biochem.* **7,** 297–299.

Enzyme Nomenclature

EC Number	Recommended Name	Other Names used in this Volume	Systematic Name
Oxidoreductases			
1.1.1.47	glucose dehydrogenase		β-D-glucose: NAD(P)$^+$ 1-oxidoreductase
1.1.3.4.	glucose oxidase		β-D-glucose: oxygen 1-oxidoreductase
1.2.2.2.	pyruvate dehydrogenase (cytochrome)		pyruvate: ferricytochrome b_1 oxidoreductase
1.2.3.1.	aldehyde oxidase		aldehyde: oxygen oxidoreductase
1.7.3.3.	urate oxidase	uricase	urate: oxygen oxidoreductase
1.7.99.1	hydroxylamine reductase		ammonia: (acceptor) oxidoreductase
1.7.99.3	nitrite reductase		nitric-oxide: (acceptor) oxidoreductase
1.7.99.4	nitrate reductase		nitrite: (acceptor) oxidoreductase
1.10.3.3	ascorbate oxidase	ascorbase	L-ascorbate: oxygen oxidoreductase
1.11.1.6	catalase		hydrogen-peroxide: hydrogen-peroxide oxidoreductase
1.11.1.7	peroxidase		donor: hydrogen-peroxide oxidoreductase
1.13.11.11	tryptophan 2,3-dioxygenase	tryptophanase	L-tryptophan: oxygen 2,3-oxidoreductase (decyclizing)
1.14.14.1	flavoprotein-linked monooxygenase	aniline oxidase	RH, reduced-flavoprotein: oxygen oxidoreductase (RH-hydroxylating)
1.14.18.1	monophenol monooxygenase	tyrosinase[a] polyphenol oxidase[a] catachol oxidase[a] o-diphenol oxidase[a] f-diphenol oxidase[b] laccase[b]	monophenol, dihydroxyphenylalanine: oxygen oxidoreductase
Transferases			
2.4.1.5	dextransucrase		sucrose: 1,6-α-D-glucan 6-α-glucosyltransferase
2 4.1.10	levansucrase		sucrose: 2,6-β-D-fructan 6-β-fructosyltransferase
2.6.1.—	aminotransferases	transaminases	
2.8.1.1	thiosulphate sulphurtransferase	rhodanese	thiosulphate: cyanide sulphurtransferase

EC Number	Recommended Name	Other Names used in this Volume	Systematic Name
Hydrolases			
3.1.1.1	carboxylesterase	malathion esterase	carboxylic-ester hydrolase
3.1.1.2	arylesterase		aryl-ester hydrolase
3.1.1.3	triacylglycerol lipase	lipase	tracylglycerol acyl-hydrolase
3.1.1.6	acetylesterase		acetic-ester hydrolase
3.1.1.7	acetylcholinesterase		acetylcholine hydrolase
3.1.3.1	alkaline phosphatase	phosphomonoesterase	orthophosphoric-monoester phosphohydrolase (alkaline optimum)
3.1.3.2	acid phosphatase	phosphomonoesterase	orthophosphoric-monoester phosphohydrolase (acid optimum)
3.1.3.26	6-phytase	phytase	myo-inositol hexakisphosphate 6-phosphohydrolase
3.1.4.—	phosphoric diester hydrolases	phosphodiesterases	
3.1.6.1	arylsulphatase	sulphatase	aryl-sulphate sulphohydrolase
3.2.1.1	α-amylase		1,4-α-D-glucan glucanohydrolase
3.2.1.2	β-amylase		1,4-α-D-glucan maltohydrolase
3.2.1.4	cellulase	endo-1,4-β-glucanase	1,4-(1,3;1,4)-β-D-glucan 4-glucanohydrolase
3.2.1.6	endo-1,3(4)-β-D-glucanase	1,3-β-glucanase laminarinase lichenase (see also 3.2.1.73)	1,3-(1,3;1,4)-β-D-glucan 3(4)-glucanohydrolase
3.2.1.7	inulinase	inulase	2,1-β-D-fructan fructanohydrolase
3.2.1.8	endo-1,4-β-xylanase	xylanase	1,4-β-D-xylan xylanohydrolase
3.2.1.10	oligo-1,6-glucosidase	dextrinase	dextrin 6-α-glucanohydrolase
3.2.1.11	dextranase		1,6-α-D-glucan 6-glucanohydrolase
3.2.1.14	chitinase		poly(1,4-β-(2-acetamido-2-deoxy-D-glucoside)) glucanohydrolase
3.2.1.15	polygalacturonase	pectinase	poly(1,4,α-D-galacturonide) glycanohydrolase
3.2.1.20	α-glucosidase	maltase	α-D-glucoside glucohydrolase

EC Number	Recommended Name	Other Names used in this Volume	Systematic Name
3.2.1.21	β-glucosidase	gentiobiase cellobiase emulsin	β-D-glucoside glucohydrolase
3.2.1.22	α-galactosidase	melibiase	α-D-galactoside galactohydrolase
3.2.1.23	β-galactosidase	lactase	β-D-galactoside galactohydrolase
3.2.1.26	β-fructofuranosidase	sucrase invertase saccharase β-fructosidase	β-D-fructofuranoside fructohydrolase
3.2.1.65	levanase		2,6,β-D-fructan fructanohydrolase
3.2.1.73	lichenase		1,3-1,4-β-D-glucan-4-glucanohydrolase
3.4.23.2	pepsin B	gelatinase	cleaves acetyl-L-phenylalanyl-L-diiodotyrosine
3.5.1.1	asparaginase		L-asparagine amidohydrolase
3.5.1.2	glutaminase		L-glutamine amidohydrolase
3.5.1.5	urease		urea amidohydrolase
3.5.1.13	aryl acylamidase		aryl-acylamide amidohydrolase
3.5.2.5	allantoinase		allantoin amidohydrolase
3.5.3.4	allantoicase		allantoate amidinohydrolase
3.6.1.1	inorganic pyrophosphatase		pyrophosphate phosphohydrolase
3.6.1.3	adenosinetriphosphatase	ATPase	ATP phosphohydrolase
3.6.1.10	endopolyphosphatase	metaphosphatase polyphosphatase	polyphosphate polyphosphohydrolase
Lyases			
4.1.1.12	aspartate 4-decarboxylase	aspartate 1-decarboxylase[c]	L-aspartate 4-carboxy-lyase
4.1.1.15	glutamate decarboxylase		L-glutamate 1-carboxy-lyase
4.1.1.25	tyrosine decarboxylase		L-tyrosine carboxy-lyase
4.1.1.28	aromatic-L-amino-acid decarboxylase	DOPA decarboxylase tryptophan decarboxylase	aromatic-L-amino-acid carboxy-lyase

a previously 1.10.3.1 o-diphenol : oxygen oxidoreductase
b previously 1.10.3.2 p-diphenol : oxygen oxidoreductase
c previously 4.1.1.11 aspartate 1-decarboxylase

Subject Index

A

Acetylcholinesterase, 268

Acetylesterase, 16

Activation energies of soil enzymes, 5, 59, 65, 154–155

Adsorptive mechanisms in soil, 300–301

Aggregates in soil
dextranase attack on, 136, 299
enzyme activities associated with, 64, 71, 72, 74, 77, 136, 137, 205, 299
formation of, 118, 298
irradiation effects on, 10
microbial distribution within, 298–299

Aldehyde oxidase, 62

Amylases
assay of, 69, 342
air-drying soil prior to, 69, 342
toluene use during, 342
components of, 13
correlation with other soil properties, 69, 126
effect on activity of:
cultivation, 125–126, 127–128
fertilizers, 127, 129–133
fungicides, 139
herbicides, 137–139, 262–263
insecticides, 139
liming, 129–131
season, 69, 120–121, 297
soil depth, 69, 126
vegetation, 27, 57, 69, 120 *et seq.*, 297
extraction from soil of, 343
origins of, 6, 13, 69, 129

in rhizosphere, 119, 120, 131, 133
soil fertility and, 27, 124–125

Aniline oxidase, *see* flavoprotein-linked monooxygenase

Aromatic amino acid decarboxylases, 21, 30, 79

Aryl acylamidase, 255–256

Arylesterase, 66

Arylsulphatase
assay of, 68, 222, 343
air-drying soil prior to, 69, 223, 343, 344
γ-irradiation prior to, 223, 321
reproducibility of, 325
toluene use during, 223, 343
components of, 16
correlation with other soil properties, 68, 224–225
effect on activity of:
season, 69, 225–226
soil depth, 68, 225
extraction of, 226–227
kinetics of, 28, 68, 103, 228–229
origins of, 16, 221, 229–232
properties of, 223–224
in rhizosphere, 233
soil colloids and, 68, 227–228
terminology of, 222

Ascorbate oxidase, 63

Asparaginase, 6, 17, 76, 261

Aspartate decarboxylase, 21 *see also* glutamate decarboxylase

B

Buffers, use of in soil enzyme assays (*see* methodology)